MECHANICS
OF COMPOSITE
MATERIALS

MECHANICS OF COMPOSITE MATERIALS

ROBERT M. JONES
*Professor of Engineering Science
and Mechanics
Virginia Polytechnic Institute
and State University
Blacksburg, Virginia*

⬤ HEMISPHERE PUBLISHING CORPORATION
A member of the Taylor & Francis Group

New York   Washington   Philadelphia   London

MECHANICS OF COMPOSITE MATERIALS

Library of Congress Cataloging in Publication Data

Jones, Robert M., date.
    Mechanics of composite materials.

    Bibliography: p.
    1. Composite materials.   2. Laminated materials.
1. Title.
TA418.9.C6J59   62.1'1   74-14576
ISBN 0-89116-490-1

To my neglected family:
Donna, Mark, Karen, and Christopher

CONTENTS

PREFACE

Composite materials are ideal for structural applications where high strength-to-weight and stiffness-to-weight ratios are required. Aircraft and spacecraft are typical weight-sensitive structures in which composite materials are cost-effective. When the full advantages of composite materials are utilized, both aircraft and spacecraft will be designed in a manner much different from the present.

The study of composite materials actually involves many topics, such as, for example, manufacturing processes, anisotropic elasticity, strength of anisotropic materials, and micromechanics. Truly, no one individual can claim a complete understanding of all these areas. Any practitioner will be likely to limit his attention to one or two subareas of the broad possibilities of analysis versus design, micromechanics versus macromechanics, etc.

The objective of this book is to introduce the student to the basic concepts of the mechanical behavior of composite materials. Actually, only an overview of this vast set of topics is offered. The balance of subject areas is intended to give a fundamental knowledge of the broad scope of composite materials. The mechanics of laminated fiber-reinforced composite materials are developed as a continuing example. Many important topics are ignored in order to restrict the coverage to a one-semester graduate course. However, the areas covered do provide a firm foundation for further study and research and are carefully selected to provide continuity and balance. Moreover, the subject matter is chosen to exhibit a high degree of comparison between theory and experiment in order to establish confidence in the derived theories.

The whole gamut of topics from micromechanics and macromechanics through lamination theory and examples of plate bending, buckling, and vibration problems is treated so that the physical significance of the concepts is made clear. A comprehensive introduction to composite materials and motivation for their use in current structural applications is given in Chapter 1. Stress-strain relations for a lamina are displayed with engineering material constants in Chapter 2. Strength theories are also compared with experimental results. In Chapter 3, micromechanics is introduced by both the mechanics of materials approach and the elasticity approach. Predicted moduli are compared with measured values. Lamination theory is presented in Chapter 4 with the aid of a new laminate classification scheme. Laminate stiffness predictions are compared with experimental results. Laminate strength concepts, as well as interlaminar stresses and design are also discussed. In Chapter 5, bending, buckling, and vibration of a simply supported plate with various lamination characteristics is examined to display the effects of coupling stiffnesses in a physically meaningful problem. Miscellaneous topics such as fatigue, fracture mechanics, and transverse shear effects are addressed in Chapter 6. Appendices on matrices and tensors, maxima and minima of functions of a single variable, and typical stress-strain curves are provided.

This book was written primarily as a graduate-level textbook, but is well suited as a guide for self-study of composite materials. Accordingly, the theories presented are simple and illustrate the basic concepts, although they may not be the most accurate. Emphasis is placed on analyses compared with experimental results, rather than on the most recent analysis for the material currently "in vogue." Accuracy may suffer, but educational objectives are better met. Many references are included to facilitate further study. The background of the reader should include an advanced mechanics of materials course or separate courses in which three-dimensional stress-strain relations and plate theory are introduced. In addition, knowledge of anisotropic elasticity is desirable, although not essential.

Many people have been most generous in their support of this writing effort. I would like to especially thank Dr. Stephen W. Tsai, of the Air Force Materials Laboratory, for his inspiration by example over the past ten years and for his guidance throughout the past several years. I deeply appreciate Steve's efforts and those of Dr. R. Byron Pipes of the University of Delaware and Dr. Thomas Cruse of Pratt and Whitney Aircraft, who reviewed the manuscript and made many helpful comments. Still others contributed material for the book. My thanks to Marvin Howeth of General Dynamics, Fort Worth, Texas, for many photographs; to John Pimm of LTV Aerospace Corporation for the photograph in Sec. 4.7; to Dr. Nicholas Pagano of the Air Force Materials Laboratory for many figures; to Dr. R. Byron Pipes of the University of Delaware for many photographs and figures in Sec. 4.6; and to Dr. B. Walter Rosen of Materials Sciences Corporation, Blue Bell, Pennsylvania, for the photo in Sec. 3.5. I also appreciate the permission of Technomic Publishing Company, Inc., of Westport, Connecticut, to reprint throughout the text many figures which have appeared in the various Technomic books and in the *Journal of Composite Materials* over the past several years. I am very grateful for support by the Air Force Office of Scientific Research (Directorate of Aerospace Sciences) and the Office of Naval Research (Structural Mechanics Program) of my research on laminated plates and shells discussed in Chapters 5 and 6. I am also indebted to several classes at the Southern Methodist University Institute of Technology and the Naval Air Development Center, Warminster, Pennsylvania, for their patience and help during the development of the class notes that led to this book. Finally, I must single out Harold S. Morgan for his numerous contributions and Marty Kunkle for her manuscript typing (although I did some of the typing myself!).

R.M.J.

MECHANICS
OF COMPOSITE
MATERIALS

Chapter 1
INTRODUCTION TO COMPOSITE MATERIALS

1.1 INTRODUCTION

The word "composite" in composite material signifies that two or more materials are combined on a macroscopic scale to form a useful material. The key is the macroscopic examination of a material. Different materials can be combined on a microscopic scale, such as in alloying, but the resulting material is macroscopically homogeneous. The advantage of composites is that they usually exhibit the best qualities of their constituents and often some qualities that neither constituent possesses. The properties that can be improved by forming a composite material include:

- strength
- stiffness
- corrosion resistance
- wear resistance
- attractiveness
- weight

- fatigue life
- temperature-dependent behavior
- thermal insulation
- thermal conductivity
- acoustical insulation

Naturally, not all of the above properties are improved at the same time nor is there usually any requirement to do so.

Composite materials have a long history of usage. Their beginnings are unknown, but all recorded history contains references to some form of composite material. For example, straw was used by the Israelites to strengthen mud bricks. Plywood was used by the ancient Egyptians when they realized that wood could be rearranged to achieve superior strength and resistance to thermal expansion as well as to swelling owing to the presence of moisture. Medieval swords and armor were constructed with layers of different materials. More recently, fiber-reinforced resin composites that have high strength-to-weight and stiffness-to-weight ratios have become important in weight-sensitive applications such as aircraft and space vehicles.

Composite materials are classified and characterized in Sec. 1.2. The mechanical behavior of composite materials is described in Sec. 1.3. The scope of this book is then limited in Sec. 1.4 to laminated fiber-reinforced composite materials, and the basic terminology is defined. Manufacturing processes for laminated fiber-reinforced composites are described briefly in Sec. 1.5. Finally, the current advantages and the potential advantages of laminated fiber-reinforced composite materials are discussed in Sec. 1.6.

1.2 CLASSIFICATION AND CHARACTERISTICS OF COMPOSITE MATERIALS

There are three commonly accepted types of composite materials:

1. *Fibrous composites* which consist of fibers in a matrix
2. *Laminated composites* which consist of layers of various materials
3. *Particulate composites* which are composed of particles in a matrix

These types of composite materials are described and discussed in the following sections. (I am indebted to Professor A. G. H. Dietz [Ref. 1-1] for the character of the presentation in this section.)

1.2.1 Fibrous Composites

Long fibers in various forms are inherently much stiffer and stronger than the same material in bulk form. For example, ordinary plate glass fractures at stresses of only a few thousand pounds per square inch, yet glass fibers have strengths of 400,000 to 700,000 psi in commercially available forms and about 1,000,000 psi in laboratory prepared forms. Obviously, then, the geometry of a fiber is somehow crucial to the evaluation of its strength and must be considered in structural applications. More properly, the paradox of a fiber having different properties from the bulk form is due to the more perfect structure of a fiber. The crystals are aligned in the fiber along the fiber axis. Moreover, there are fewer internal defects in fibers than in bulk material. For example, in materials that have dislocations, the fiber form has fewer dislocations than the bulk form.

Properties of fibers

A fiber is characterized geometrically not only by its very high length-to-diameter ratio but by its near crystal-sized diameter. Strengths and stiffnesses of a few selected fiber materials are shown in Table 1-1. Many common materials are listed for the purpose of comparison. Note that the density of each material is listed since the strength-to-density and stiffness-to-density ratios are commonly used as indicators of the effectiveness of a fiber, especially in weight-sensitive applications such as aircraft and space vehicles. Entries in Table 1-1 are arranged in increasing average S/ρ and E/ρ.

Properties of whiskers

A whisker has essentially the same near crystal-sized diameter as a fiber, but generally is very short and stubby although the length-to-diameter ratio can be

TABLE 1-1. Fiber and wire properties

Fiber or wire	Density, ρ lb/in.^{3} (kN/m^3)	Tensile strength, S 10^3 lb/in.^{2} (GN/m^2)	S/ρ 10^5 in. (km)	Tensile stiffness, E 10^6 lb/in.^{2} (GN/m^2)	E/ρ 10^7 in. (Mm)
Aluminum	.097 (26.3)	90 (.62)	9 (24)	10.6 (73)	11 (2.8)
Titanium	.170 (46.1)	280 (1.9)	16 (41)	16.7 (115)	10 (2.5)
Steel	.282 (76.6)	600 (4.1)	21 (54)	30 (207)	11 (2.7)
E-glass	.092 (25.0)	500 (3.4)	54 (136)	10.5 (72)	11 (2.9)
S-glass	.090 (24.4)	700 (4.8)	78 (197)	12.5 (86)	14 (3.5)
Carbon	.051 (13.8)	250 (1.7)	49 (123)	27 (190)	53 (14)
Beryllium	.067 (18.2)	250 (1.7)	37 (93)	44 (300)	66 (16)
Boron	.093 (25.2)	500 (3.4)	54 (137)	60 (400)	65 (16)
Graphite	.051 (13.8)	250 (1.7)	49 (123)	37 (250)	72 (18)

Source: Adapted from Dietz (Ref. 1-1). *By permission of the American Society for Testing and Materials,* 1965.

in the hundreds. Thus, a whisker is an even more obvious example of the crystal-bulk material property difference paradox. That is, a whisker is more perfect than a fiber and exhibits even higher properties. Whiskers are obtained by crystallization on a very small scale resulting in a nearly perfect alignment of crystals. Materials such as iron have crystalline structures with a theoretical strength of 2,900,000 psi, yet structural steels, which are mainly iron, have strengths ranging from 75,000 psi to upwards of 100,000 psi. The discrepancy is due to imperfections in the crystalline structure of steel. Those imperfections are called dislocations and are easily moved for ductile materials. The movement of dislocations changes the relation of the crystals and hence the strength and stiffness of the material. For a nearly perfect whisker, there are few dislocations. Thus, whiskers of iron have significantly higher strengths than steel in bulk form. A representative set of whisker properties is given in Table 1-2.

TABLE 1-2. Whisker properties

Whisker	Density, ρ lb/in.^{3} (kN/m^3)	Theoretical strength, S_T 10^6 lb/in.^{2} (GN/m^2)	Experimental strength, S_E 10^6 lb/in.^{2} (GN/m^2)	S_E/ρ 10^5 in. (km)	Tensile stiffness, E 10^6 lb/in.^{2} (GN/m^2)	E/ρ 10^7 in. (Mm)
Copper	.322 (87.4)	1.8 (12)	.43 (3.0)	13 (34)	18 (124)	5.6 (1.4)
Nickel	.324 (87.9)	3.1 (21)	.56 (3.9)	17 (44)	31 (215)	9.6 (2.4)
Iron	.283 (76.8)	2.9 (20)	1.9 (13)	67 (170)	29 (200)	10.2 (2.6)
B$_4$C	.091 (24.7)	6.5 (45)	.97 (6.7)	106 (270)	65 (450)	71 (18)
SiC	.115 (31.2)	12 (83)	1.6 (11)	139 (350)	122 (840)	106 (27)
Al$_2$O$_3$	.143 (38.8)	6 (41)	2.8 (19)	196 (490)	60 (410)	42 (11)
C	.060 (16.3)	14.2 (98)	3 (21)	500 (1300)	142 (980)	237 (60)

Source: Adapted trom Sutton, Kosen, and Flom (Ref. 1-2). *By permission of Society of Plastic Engineers Journal.*

Several mechanisms can be used to grow whiskers. These mechanisms generally relate to the liquid phase, the vapor phase, or solid diffusion of the materials involved. One of the most common and promising processes results in direct formation of whiskers in a metal matrix. Consider a mixture of two metals, say A and B in Fig. 1-1, that are liquid at elevated temperatures. There is a composition, known as the eutectic composition, at which the mixture instantly freezes into a combination of two solid phases, α and A_3B. Phase α is a solid solution of B in A and has X percent of B. Since A_3B has 25 percent B, obviously α has less B than the eutectic composition. That is, some areas of the composite are poor in B whereas others are rich in B. At the eutectic composition, the physical structure of the two zones of material is a near perfect array of fibers of α in a matrix of A_3B if the freezing takes place in a special manner. For example, in Fig. 1-2 the mixture of A and B is placed in a container and a heat source is moved up the container. The solidification that takes place after the heat source passes has fibers perpendicular to the solid-liquid interface. Thus, the fibers are vertical in Fig. 1-2. If the composition is on either side of the eutectic composition, then other growths called dendrites occur in addition to the fibers of α. The resulting structure is both less perfect and weaker. Thus, the use of eutectic composites of two materials is limited to a fixed composition for those two materials. Examples of eutectic composites are Al—Al_3Ni, Cu—Cr, and Ta—Ta_2C.

Properties of matrices
Naturally, fibers and whiskers are of little use unless they are bound together to take the form of a structural element which can take loads. The binder material

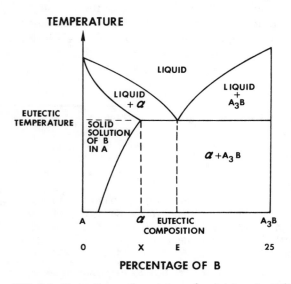

FIG. 1-1. Phase diagram for mixture of metal A and metal B.

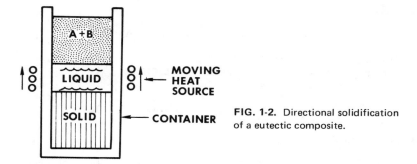

FIG. 1-2. Directional solidification of a eutectic composite.

is usually called a matrix. The purpose of the matrix is manifold: support, protection, stress transfer, etc. Typically, the matrix is of considerably lower density, stiffness, and strength than the fibers or whiskers. However, the combination of fibers or whiskers and a matrix can have very high strength and stiffness, yet still have low density.

A typical organic epoxy matrix material such as Narmco 2387 (Ref. 1-3) has a density of .044 lb/in.^{3} (11.9 kN/m^3), compressive strength of 23,000 lb/in.^{2} (.158 GN/m^2), compressive modulus of 560,000 lb/in.^{2} (3.86 GN/m^2), tensile strength of 4,200 lb/in.^{2} (.029 GN/m^2), and tensile modulus of 490,000 lb/in.^{2} (3.38 GN/m^2). Metal matrices such as aluminum have their usual bulk-form properties.

1.2.2 Laminated Composites

Laminated composites consist of layers of at least two different materials that are bonded together. Lamination is used to combine the best aspects of the constituent layers in order to achieve a more useful material. The properties that can be emphasized by lamination are strength, stiffness, low weight, corrosion resistance, wear resistance, beauty or attractiveness, thermal insulation, acoustical insulation, etc. Such claims are best represented by the examples of the following paragraphs in which bimetals, clad metals, laminated glass, plastic-based laminates, and laminated fibrous composites are described.

Bimetals

Bimetals are laminates of two different metals with significantly different coefficients of thermal expansion. Under change in temperature, bimetals warp or deflect a predictable amount and are therefore well suited for use in temperature measuring devices. For example, a simple thermostat can be made from a cantilever strip of two metals bonded together as shown in Fig. 1-3. There, metal A with coefficient of thermal expansion α_A is bonded to metal B with α_B less than α_A. Thus, on increase of temperature over the temperature at which bonding was performed, metal A tends to expand more than metal B; hence, the strip bends down. If the temperature were decreased from the reference temperature, metal A would tend to shrink more than metal B, and consequently,

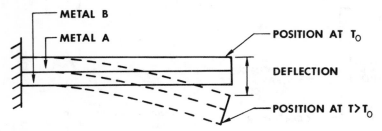

FIG. 1-3. Cantilevered bimetallic strip (thermostat).

the strip would bend up. This is a simple example of coupling between bending and extension. Obviously, application of a uniform temperature to a single homogeneous metal causes only extension. However, application of uniform temperature to two bonded dissimilar metals was just shown to result in bending as well as extension. Coupling between bending and extension is a typical result when materials are laminated, as will be seen in Chap. 4.

Clad metals

The cladding or sheathing of one metal with another is done to obtain the best properties of both. For example, high-strength aluminum alloys do not resist corrosion; however, pure aluminum and some aluminum alloys are very corrosion resistant. Thus, a high-strength aluminum alloy covered with a corrosion-resistant aluminum alloy is a composite material with unique and attractive advantages over its constituents.

Recently, aluminum wire, clad with about 10 percent copper, was introduced as a replacement for copper wire in the electrical wiring market. Aluminum wire by itself is economical and lightweight, but overheats and is difficult to connect. On the other hand, copper wire is expensive and relatively heavy, but stays cool and can be connected easily. The copper-clad aluminum wire is lightweight and connectable, stays cool, and is cheaper than copper wire. Moreover, copper-clad aluminum wire is much less susceptible to the ever-present, construction site problem of theft because of lower salvage value than copper wire. Copper-clad aluminum wire comes 3/16-inch in diameter and can be drawn as fine as .005-inch in diameter without affecting the percentage of copper cladding. An initial disadvantage of debonding during drawing has been substantially overcome, although careful control is still necessary.

Laminated glass

The concept of protection of one layer of material by another as described under "Clad metals" can be extended in a rather unique way to safety glass. Ordinary window glass is durable enough to retain its transparency under the extremes of weather. However, glass is quite brittle and is dangerous because it can break into many sharp-edged pieces. On the other hand, a plastic called polyvinyl butryal is very tough (deforms to high strains without fracture), but is very flexible and susceptible to scratching. Safety glass is a layer of polyvinyl butyral sandwiched between two layers of glass. The glass in the composite protects the plastic from scratching and gives it stiffness. The plastic provides the toughness of the composite. Thus, together, the glass and plastic protect each other in different ways and lead to a composite with properties that are vastly improved over those of its constituents.

Plastic-based laminates

Many materials can be saturated with various plastics and subsequently treated for many purposes. The common product Formica is merely layers of heavy kraft paper impregnated with a phenolic resin overlaid by a plastic-saturated decorative sheet which, in turn, is overlaid with a plastic-saturated cellulose mat. Heat and pressure are used to bind the layers together. A useful variation on the theme is obtained when an aluminum layer is placed between the decorative layer and the kraft paper layer to quickly dissipate the heat of, for example, a burning cigarette on a kitchen counter.

Layers of glass or asbestos fabrics can be impregnated with silicones to yield a composite with significant high-temperature properties. Glass or nylon fabrics can be laminated with various resins to yield an impact- and penetration-resistant composite that is uniquely suitable as lightweight personnel armor. The list of examples is seemingly endless, but the purpose of illustration is served by the preceding examples.

Laminated fibrous composites

Laminated fibrous composites are a hybrid class of composites involving both fibrous composites and lamination techniques. A more common name is laminated fiber-reinforced composites. Here, layers of fiber-reinforced material are built up with the fiber directions of each layer typically oriented in different directions to give different strengths and stiffnesses in the various directions. Thus, the strengths and stiffnesses of the laminated fiber-reinforced composite can be tailored to the specific design requirements of the structural element being built. Examples of laminated fiber-reinforced composites include Polaris missile cases, fiberglass boat hulls, aircraft wing panels and body sections, tennis rackets, golf club shafts, etc.

1.2.3 Particulate Composites

Particulate composites consist of particles of one or more materials suspended in a matrix of another material. The particles can be either metallic or nonmetallic as can the matrix. Common combinations of these possibilities are described in the following paragraphs.

Nonmetallic in nonmetallic composites

The most common example of a nonmetallic particle system in a nonmetallic matrix, indeed the most common composite material, is concrete. Concrete is particles of sand and rock that are bound together by a mixture of cement and water that has chemically reacted and hardened. The strength of the concrete is normally ascribable to the rock. The rate of accumulation of strength up to that of the rock is varied by changing the type of cement in order to slow or speed the chemical reaction. Many books have been written on concrete and on a variation, reinforced concrete, that could be considered a fibrous composite as well as a particulate composite.

Flakes of nonmetallic materials such as mica or glass can form an effective composite material when suspended in a glass or plastic, respectively. Flakes have a primarily two-dimensional geometry with accompanying strength and stiffness in two directions, as opposed to only one for fibrous composites. Ordinarily, flakes are packed parallel to one another with a resulting higher density than fiber packing concepts. Accordingly, less matrix material is required to bind the flakes. Flakes overlap so much that a flake composite is much more impervious to liquids than an ordinary composite of the same constituent materials. Mica in glass composites are extensively used in electrical applications because of good insulating and machining qualities. Glass flakes in plastic resin matrices have a potential similar to, if not higher than that of glass-fiber composites. Even higher stiffnesses and strengths should be attainable with glass-flake composites than with glass-fiber composites because of the higher packing density. However, surface flaws have reduced the strength of glass-flake composites from that currently obtained with glass-fiber composites.

Metallic in nonmetallic composites

Rocket propellants consist of inorganic particles such as aluminum powder and perchlorate oxidizers in a flexible organic binder such as polyurethane or polysulfide rubber. The particles comprise as much as 75 percent of the propellant leaving only 25 percent for the binder. The objective is a steadily burning reaction. Thus, the composite must be uniform in character and must not crack; otherwise, burning would take place in unsteady bursts that could actually develop into explosions but would at the very least adversely affect the trajectory of the rocket. The thrust of a rocket is proportional to the burning surface area; thus, solid propellants are cast with, for example, a star-shaped hole as in Fig. 1-4 instead of a circular hole. Many stress-analysis problems arise in connection with support of the solid propellant in a rocket-motor casing and with internal

FIG. 1-4. Solid rocket propellant shapes.

stresses due to dissimilar particle and binder stiffnesses. The internal stresses can be reduced by attempting to optimize the shape of the burning cross section; again, a reason for a noncircular hole.

Metal flakes in a suspension are common. For example, aluminum paint is actually aluminum flakes suspended in paint. Upon application, the flakes orient themselves parallel to the surface giving very good coverage. Similarly, silver flakes can be applied to give good electrical conductivity.

Cold solder is metal powder suspended in a thermosetting resin. The composite is strong and hard and conducts heat and electricity. Inclusion of copper in an epoxy resin increases the conductivity immensely. Many metallic additives to plastic are being developed to increase the thermal conductivity, lower the coefficient of thermal expansion, and decrease wear.

Metallic in metallic composites

Unlike an alloy, a metallic particle in a metallic matrix does not dissolve. Lead particles are commonly used in copper alloys and steel to improve the machineability (metal comes off in shaving rather than chip form). In addition, lead is a natural lubricant in bearings made from copper alloys.

Many metals are naturally brittle at room temperature so must be machined when hot. However, particles of these metals, such as tungsten, chromium, molybdenum, etc., can be suspended in a ductile matrix. The resulting composite is ductile, yet has the elevated temperature properties of the brittle constituents. The actual process used to suspend the brittle particles is called liquid sintering and involves infiltration of the matrix material around the brittle particles. Fortunately, in the liquid sintering process, the brittle particles become rounded and therefore naturally more ductile.

Nonmetallic in metallic composites

Nonmetallic particles such as ceramics can be suspended in a metal matrix. The resulting composite is called a cermet. Two common classes of cermets are oxide-based and carbide-based composites.

As a departure from the present classification scheme, oxide-based cermets can be either oxide particles in a metal matrix or metal particles in an oxide matrix. Such cermets are used in tool making and in high-temperature applications where erosion resistance is required.

Carbide-based cermets have particles of carbides of tungsten, chromium, and titanium. Tungsten carbide in a cobalt matrix is used in machine parts requiring very high hardness such as wire-drawing dies, valves, etc. Chromium carbide in a cobalt matrix has high corrosion and abrasion resistance; it also has a coefficient of thermal expansion close to that of steel so is well-suited for use in valves. Titanium carbide in either a nickel or a cobalt matrix is often used in high-temperature applications such as turbine parts. Cermets are also used as nuclear reactor fuel elements and control rods. Fuel elements can be uranium oxide particles in stainless steel ceramic whereas boron carbide in stainless steel is used for control rods.

Dispersion-hardened alloys differ from cermets in that the particles are smaller and of a lesser percentage of the total volume (usually about 3 percent by volume). However, the dispersed particles govern the strength of the composite. The particles are mechanically dispersed as opposed to precipitation-hardened alloys. Usually, the particles are metal oxides or intermetallic compounds (and therefore sometimes fall in another classification under particulate composites). If the composite is cold-worked after dispersion, strength is increased and dislocations are immobilized.

1.2.4 Summary

Numerous multiphase composites exhibit more than one characteristic of the classes, fibrous, laminated, or particulate composites, just discussed. For example, reinforced concrete is both particulate (because the concrete is composed of aggregate in a cement paste binder) and fibrous (because of the reinforcement). Also, laminated fiber-reinforced composites are obviously both laminated and fibrous composites. Thus, any classification scheme is arbitrary and imperfect. Nevertheless, the system should serve to acquaint the reader with the broad possibilities of composite materials.

1.3 MECHANICAL BEHAVIOR OF
COMPOSITE MATERIALS

Composite materials have many characteristics that are different from more conventional engineering materials. Some characteristics are merely modifications of conventional behavior; others are totally new and require new analytical and experimental procedures. Most common engineering materials are *homogeneous* and *isotropic*:

A *homogeneous* body has uniform properties throughout, i.e., the properties are not a function of *position* in the body.

An *isotropic* body has material properties that are the same in every direction at a point in the body, i.e., the properties are not a function of *orientation* at a point in the body.

Bodies with temperature-dependent isotropic material properties are not homogeneous when subjected to a temperature gradient, but still are isotropic.

In contrast, composite materials are often both *inhomogeneous* (or heterogeneous — the two terms will be used interchangeably) and *nonisotropic* (orthotropic or, more generally, anisotropic):

An *inhomogeneous* body has nonuniform properties over the body, i.e., the properties are a function of *position* in the body.

An *orthotropic* body has material properties that are different in three mutually perpendicular directions at a point in the body and, further, have three mutually perpendicular planes of material symmetry. Thus, the properties are a function of *orientation* at a point in the body.

An *anisotropic* body has material properties that are different in all directions at a point in the body. There are no planes of material property symmetry. Again, the properties are a function of *orientation* at a point in the body.

Some composite materials have very simple forms of inhomogeneity. For example, laminated safety glass has three layers each of which is homogeneous and isotropic; thus, the inhomogeneity of the composite is a step function in the direction perpendicular to the plane of the glass. Also, some particulate composites are inhomogeneous, yet isotropic although some are anisotropic. Other composite materials are typically more complex.

Because of the inherent heterogeneous nature of composite materials, they are conveniently studied from two points of view: micromechanics and macromechanics:

> *Micromechanics* is the study of composite material behavior wherein the *interaction* of the constituent materials is examined on a microscopic scale.
> *Macromechanics* is the study of composite material behavior wherein the material is presumed *homogeneous* and the effects of the constituent materials are detected only as averaged apparent properties of the composite.

In this book, attention will first be focused on macromechanics as it is the most readily appreciated of the two and the more important topic in design analysis. Subsequently, micromechanics will be investigated in order to gain an appreciation for how the constituents of composites can be proportioned and arranged to achieve certain specified strengths and stiffnesses.

Use of both the concepts of macromechanics and micromechanics allows the *tailoring* of a composite material to meet a particular structural requirement with little waste of material capability. The ability to tailor a composite material to its job is one of the most significant advantages of a composite over an ordinary material. Tailoring of a composite material yields only the stiffness and strength required in a given direction. In contrast, an isotropic material is, by definition, constrained to have excess strength and stiffness in any direction other than that of the largest requirement.

The inherent anisotropy (most often only orthotropy) of composite materials leads to mechanical behavior characteristics that are quite different from those of conventional isotropic materials. The behavior of isotropic, orthotropic, and anisotropic materials under loadings of normal stress and shear stress is shown in Fig. 1-5 and discussed in the following paragraphs.

For isotropic materials, normal stress causes extension in the direction of the applied stress and contraction in the perpendicular direction. Also, shear stress causes only shearing deformation.

For orthotropic materials, like isotropic materials, normal stress in a principal material direction (along one of the intersections of three orthogonal planes of material symmetry) results in extension in the direction of the applied stress and contraction perpendicular to the stress. However, due to different properties in the two principal material directions, the contraction can be either more or less than the contraction of a similarly loaded isotropic material with the same elastic modulus in the direction of the load. Shear stress causes shearing deformation, but the magnitude of the deformation is independent of the various Young's moduli and the Poisson's ratios. That is, the shear modulus of an orthotropic material is, unlike isotropic materials, not dependent on other material properties.

For anisotropic materials, application of a normal stress leads not only to extension in the direction of the stress and contraction perpendicular to it, but to shearing deformation. Conversely, shearing stress causes extension and contraction in addition to the distortion of shearing deformation. This coupling between both loading modes and both deformation modes is also characteristic of orthotropic materials subjected to normal stress in a nonprincipal material direction. For example, cloth is an orthotropic material composed of two sets of

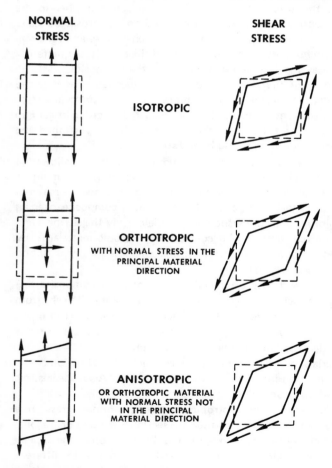

NORMAL STRESS

SHEAR STRESS

ISOTROPIC

ORTHOTROPIC
WITH NORMAL STRESS IN THE PRINCIPAL MATERIAL DIRECTION

ANISOTROPIC
OR ORTHOTROPIC MATERIAL WITH NORMAL STRESS NOT IN THE PRINCIPAL MATERIAL DIRECTION

FIG. 1-5. Mechanical behavior of anisotropic materials.

interwoven fibers at right angles to each other. If cloth is subjected to a stress at 45° to a fiber direction, both stretching and distortion occur.

Coupling between deformation modes and loading modes creates problems that are not easily overcome and, at the very least, cause a reorientation of thinking. For example, the conventional ASTM dog-bone tensile specimen shown in Fig. 1-6 obviously cannot be used to determine the tensile moduli of

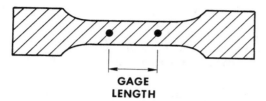

GAGE
LENGTH

FIG. 1-6. ASTM dog-bone tensile specimen.

anisotropic materials. For an isotropic material, loading on a dog bone is actually a prescribed lengthening which is only coincidentally a prescribed stress due to the symmetry of an isotropic material. However, for an anisotropic material, only the prescribed lengthening occurs due to the lack of symmetry of the material and the clamped ends of the specimen. Accordingly, shearing stresses result in addition to normal stresses. Furthermore, the specimen has a tendency to bend. Thus, the strain measured in the specimen gage length in Fig. 1-6 cannot be used with the axial load to determine the axial stiffness or modulus. Techniques more sophisticated than the ASTM dog-bone test must be used to determine the mechanical properties of a composite material.

The foregoing characteristics of the mechanical behavior of composite materials have been presented without proof. In subsequent chapters, these characteristics will be demonstrated to exist, and further observations will be made.

1.4 BASIC TERMINOLOGY OF LAMINATED FIBER-REINFORCED COMPOSITE MATERIALS

For the remainder of this book, emphasis will be placed on fiber-reinforced composite laminates. The fibers are long and continuous as opposed to whiskers; hence, the name filamentary composite is often used. The concepts developed herein are applicable mainly to fiber-reinforced composite laminates, but are also valid for other laminates and whisker composites with some fairly obvious modifications. That is, fiber-reinforced composite laminates are used as a uniform example throughout this book, but concepts used to analyze their behavior are often applicable to other composite materials. In many instances, the applicability will be made clear as an example complementary to the principal example of fiber-reinforced composite laminates.

The basic terminology of fiber-reinforced composite laminates will be introduced in the following paragraphs. For a lamina, the configurations and functions of the constituent materials, fibers and matrix, will be described. The characteristics of the fibers and matrix are then discussed. Finally, a laminate is defined to round out this introduction to the characteristics of fiber-reinforced composite laminates.

1.4.1 Laminae

A *lamina* is a flat (sometimes curved as in a shell) arrangement of unidirectional fibers or woven fibers in a matrix. Two typical laminae are shown in Fig. 1-7 along with their principal material axes which are parallel and perpendicular to the fiber directions. The fibers, or filaments, are the principal reinforcing or load-carrying agent. They are typically strong and stiff. The matrix can be organic, ceramic, or metallic. The function of the matrix is to support and protect the fibers and to provide a means of distributing load among and transmitting load between the fibers. The latter function is especially important if a fiber breaks as in Fig. 1-8. There, load from one side of a broken fiber is transferred to the matrix and subsequently to the other side of the broken fiber as well as to adjacent fibers. The mechanism for the load transfer is the shearing stress developed in the matrix; the shearing stress resists the pulling out of the broken fiber. This load-transfer mechanism is the means by which whisker-reinforced composites carry any load at all above the inherent matrix strength.

The properties of the lamina constituents, the fibers and the matrix, have been only briefly discussed so far. Their stress-strain behavior is typified as one of the classes depicted in Fig. 1-9. Fibers generally exhibit linear elastic behavior, although reinforcing steel bars in concrete are more nearly elastic-perfectly plastic. Aluminum and some composites exhibit elastic-plastic behavior which is really nonlinear elastic behavior if there is no unloading. Commonly, resinous matrix materials are viscoelastic if not viscoplastic. The various stress-strain relations are sometimes referred to as constitutive relations as they describe the mechanical constitution of the material.

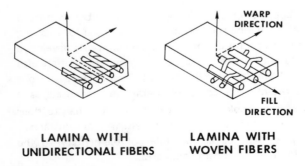

LAMINA WITH
UNIDIRECTIONAL FIBERS

LAMINA WITH
WOVEN FIBERS

FIG. 1-7. Two principal types of laminae.

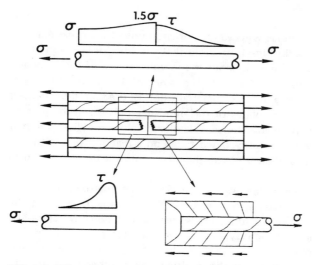

FIG. 1-8. Effect of broken fiber on matrix and fiber stresses.

Fiber-reinforced composites such as boron-epoxy and graphite-epoxy are usually treated as linear elastic materials since the fibers provide the majority of the strength and stiffness. Refinement of that approximation requires consideration of some form of plasticity, viscoelasticity, or both (viscoplasticity). Very little work has been done to implement those idealizations of composite material behavior in structural applications.

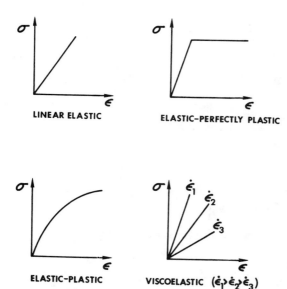

FIG. 1-9. Various stress-strain behaviors.

1.4.2 Laminates

A *laminate* is a stack of laminae with various orientations of principal material directions in the laminae as in Fig. 1-10. Note that the fiber orientation of the layers in Fig. 1-10 is not symmetric about the middle surface of the laminate. This situation will be discussed in Chap. 4. The layers of a laminate are usually bound together by the same matrix material that is used in the laminae. Laminates can be composed of plates of different materials or, in the present context, layers of fiber-reinforced laminae. A laminated circular cylindrical shell can be constructed by winding resin-coated fibers on a mandrel first with one orientation to the shell axis, then another, and so on until the desired thickness is built up.

A major purpose of lamination is to tailor the directional dependence of strength and stiffness of a material to match the loading environment of the structural element. Laminates are uniquely suited to this objective since the principal material directions of each layer can be oriented according to need. For example, six layers of a ten-layer laminate could be oriented in one direction and the other four at $90°$ to that direction; the resulting laminate then has a strength and extensional stiffness roughly 50 percent higher in one direction than the other. The ratio of the extensional stiffnesses in the two directions is approximately 6:4, but the ratio of bending stiffnesses is unclear since the order of lamination is not specified in the example. Moreover, if the laminae are not arranged symmetrically about the middle surface of the laminate, stiffnesses exist that describe coupling between bending and extension. These characteristics are discussed in Chap. 4.

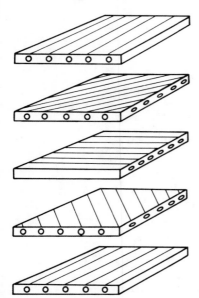

FIG. 1-10. Laminate construction.

A potential problem in the construction of laminates is the introduction of shearing stresses between layers. The shearing stresses arise due to the tendency of each layer to deform independently of its neighbors because all may have different properties (at least from the standpoint of orientation of principal material directions). Such shearing stresses are largest at the edges of a laminate and may cause delamination there. As will be shown in Chap. 4, the transverse normal stress can also cause delamination.

1.5 MANUFACTURE OF LAMINATED FIBER-REINFORCED COMPOSITE MATERIALS

Unlike most conventional materials, there is a very close relation between the manufacture of a composite material and its end usage. The manufacture of the material is often actually part of the fabrication process for the structural element or the complete structure. Thus, complete description of the manufacturing process is not possible nor is it even desirable. The discussion of manufacturing of laminated fiber-reinforced composite materials is restricted in this section to how the fibers and matrix materials are assembled to make a lamina and how, subsequently, laminae are assembled to make a laminate.

1.5.1 Initial Form of Constituent Materials

The fibers and matrix material can be obtained commercially in a variety of forms, both individually and as laminae. Fibers are available individually or as roving which is a continuous, bundled but not twisted group of fibers. The fibers can be unidirectional or interwoven. Fibers are often saturated with resinous material such as polyester resin which is subsequently used as a matrix material. This process is referred to as preimpregnation, and such forms of preimpregnated fibers are called "prepregs." For example, unidirectional fibers in an epoxy matrix are available in tape form (prepreg tape) where the fibers run in the lengthwise direction of the tape (see Fig. 1-11). The fibers are held in position not only by the matrix but by a removable backing that also prevents the tape from sticking together in the roll. Similarly, prepreg cloth or mats are available in which the fibers are interwoven and then preimpregnated with resin. Other variations on these principal forms of fibers and matrix do exist.

1.5.2 Manufacturing Procedures

The two principal steps in the manufacture of laminated fiber-reinforced composite materials are layup and curing. Layup is the arranging of fibers in laminae and laminae in layers or laminates. Various methods of layup are discussed in the following paragraphs. Curing is the drying, or polymerization, of the resinous matrix material to form a permanent bond between fibers and between laminae. Curing will not be discussed except for the general statement

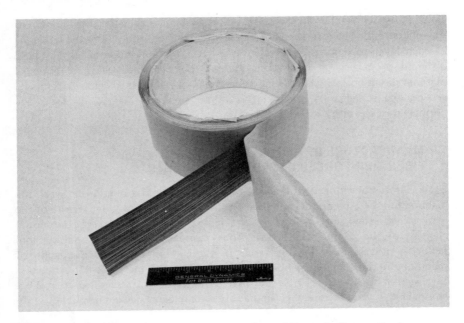

FIG. 1-11. Boron/epoxy prepreg tape. (*Courtesy General Dynamics Corporation.*)

that it can occur unaided or can consist of applying heat and/or pressure to speed the polymerization process. In the latter case, curing is performed by processes such as vacuum bag, autoclave, hydroclave, tool press, etc. Secondary steps in the manufacturing process include bonding after curing, coating, machining, and assembly.

There are three principal layup processes for laminated fiber-reinforced composite materials: winding and laying, molding, and continuous lamination. The choice of a layup process (as well as a curing process) depends on many factors: process effectiveness, part size, cost, schedule, familiarity with particular techniques, etc.

Winding and laying operations include filament winding, tape laying, and cloth winding or wrapping. Filament winding consists of passing a fiber through a liquid resin and then winding it on a mandrel (see Fig. 1-12). The fibers are wrapped at different orientations on the mandrel to yield many layers and hence strength and stiffness in many directions. Subsequently, the entire assembly, including the mandrel, is cured after which the mandrel is removed. Tape laying starts with a tape consisting of fibers in a preimpregnated form held together by a removable backing material. The tape is unwound to form the desired shape in the desired orientations of tape layers. The tape is very similar to the glass-reinforced, heavy-duty box sealing tape that has come into use recently. An example of a tape laying operation is shown in Fig. 1-13. Cloth winding or laying begins with preimpregnated cloth which is unrolled and deposited in the desired form and orientation. Cloth winding or laying is more inflexible and inefficient than filament winding or tape laying in achieving specified goals of strength and stiffness because of the bidirectional character of the fibers in the cloth. Cloth layers are often used as filler layers where strength and stiffness are not critical.

Molding operations are of lesser importance for laminated fiber-reinforced composite materials than the various winding and laying operations, but are discussed because they are used to make other common composite materials. Molding can begin with hand or automated deposition of preimpregnated fibers in layers. Often, the prepreg layers are also precut. Subsequently, the layers are compressed under elevated temperature to form the final laminate. Molding is used, for example, to fabricate radomes to close tolerances in thickness. The

FIG. 1-12. Filament winding a rocket motor case. (*Courtesy Structural Composites Industries Inc.*)

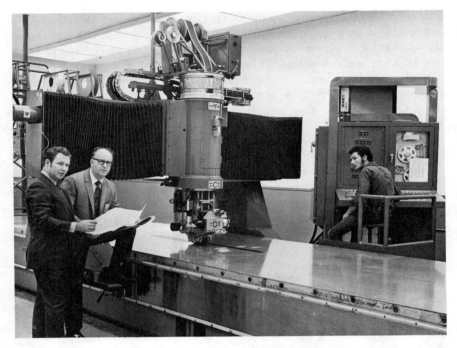

FIG. 1-13. Automatic tape laying operation. (*Courtesy General Dynamics Corporation.*)

F-111 boron/epoxy fuselage frame assembly shown in Fig. 1-14 is another common molded composite part. Actually, the upper one-third of the frame is molded and the lower two-thirds is laid-up tape.

Continuous lamination involves the bonding of sheets of laminae with subsequent application of pressure by rolling. This is the way plywood is made. Prepreg cloth can also be laminated in this manner, as can prepreg mats.

1.5.3 Quality Control

The special nature of composite materials makes it essential for the design analyst to be aware of certain types of material defects that occur during manufacture or fabrication. Such defects are intimately related to the achievement of design strengths and stiffnesses of the finished product.

Some of the common defects that must be controlled and subsequently related to the performance of the structural element are:

- interlaminar voids due to air entrapment, delamination, lack of resin, etc.
- incomplete curing of resin
- excess resin between layers
- excess matrix voids and porosity

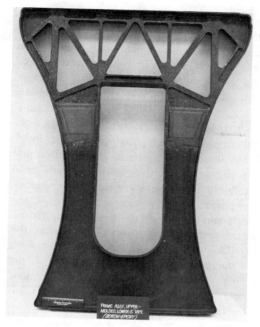

FIG. 1-14. Molded F-111 fuselage frame assembly.
(*Courtesy General Dynamics Corporation.*)

- incorrect orientation of laminae principal material directions
- damaged fibers
- wrinkles or ridges caused by improper compaction, winding, or layer alignment
- inclusion of foreign matter
- unacceptable joints in layers
- variation in thickness

Appreciation of the nature of the defects is heightened by direct observation of the manufacture of laminated fiber-reinforced composite materials.

1.6 CURRENT AND POTENTIAL ADVANTAGES OF FIBER-REINFORCED COMPOSITE MATERIALS

The advent of advanced fiber-reinforced composite materials has been called the biggest technical revolution since the jet engine (Ref. 1-4). This claim is very striking because the tremendous impact of the jet engine on military aircraft performance is readily apparent. The impact on commercial aviation is even more striking because the airlines switched from propeller-driven planes to all jet fleets within the span of just a few years.

The adjective "advanced" in advanced fiber-reinforced composite materials is used to distinguish composites with new ultrahigh strength and stiffness fibers such as boron and graphite from some of the more familiar fibers such as glass. The matrix material can be either a plastic such as epoxy or polyimide or a metal such as aluminum. Such advanced composites have two major types of advantages, among many others: improved strength and stiffness, especially when compared with other materials on a unit weight basis. For example, composites can be made that have the same strength and stiffness as high-strength steel, yet are 70 percent lighter! Other advanced composites are as much as three times as strong as aluminum, the common aircraft structural material, yet weigh only 60 percent as much! Moreover, as has already been noted, composite materials can be tailored to efficiently meet design requirements of strength, stiffness, and other parameters all in various directions. These advantages should lead to new aircraft and spacecraft designs that are radical departures from past efforts based on conventional materials. However, the aerospace industry was attracted to titanium in the 1950s for similar reasons, but found serious disadvantages after the investment of many millions of dollars in research, development, and tooling. That experience has caused a more cautious, yet more deliberately complete and well-balanced approach to composite materials development. The advantages of composite materials are so compelling that research and development is being conducted across broad fronts instead of down the most obvious paths. Whole organizations have sprung up to analyze, design, and fabricate parts made of composite materials.

The strength and stiffness advantages of advanced composite materials will be discussed in Sec. 1.6.1 and cost advantages in Sec. 1.6.2. Finally, examples of current usage of advanced composites are given in Sec. 1.6.3 along with some predictions of what the future holds.

1.6.1 Strength and Stiffness Advantages

One of the most common ways of expressing the effectiveness of strength or stiffness of a material is as a ratio of either of the quantities to the density, i.e., weight per unit volume. Such an index does not include the cost to achieve a certain strength or stiffness, but cost comparisons might not be valid by themselves since many factors influence cost.

A representation of the strength and stiffness of many materials is shown in Fig. 1-15 on the basis of effectiveness per unit weight. Common structural metals are denoted by squares. Various forms of advanced composite materials are denoted by three types of circles: fibers alone are represented by open circles; laminae with unidirectional fibers are shown by solid circles; and laminae with equal numbers of fibers in two perpendicular directions are shown by circles with a cross in them. Obviously, the most effective material lies in the upper right-hand corner of Fig. 1-15. Fibers alone are stiffer and stronger than when placed in a matrix. Also, unidirectional configurations are stiffer and

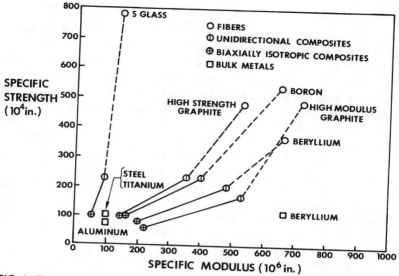

FIG. 1-15. Strength and stiffness of advanced composite materials.

stronger than biaxially isotropic configurations. Practical layups lie somewhere between unidirectional and biaxially isotropic configurations.

In Fig. 1-15, boron fibers exhibit the highest stiffness and strength efficiencies. When placed in a lamina as unidirectional fibers, the relative strength of boron drops significantly whereas the relative stiffness drops only a little. In a biaxially isotropic configuration, boron/epoxy is still stiffer than steel or titanium, although it is of the same relative strength. High strength graphite fibers and composites exhibit similar behavior. However, high modulus graphite fibers, although their stiffness is greater in all configurations than the other materials, have generally lower relative strengths (even lower than aluminum when placed in a biaxially isotropic configuration). S glass/epoxy in a unidirectional layup has about 2½ times the relative strength of steel or titanium, but is no stiffer (in fact, it is less stiff in a biaxially isotropic configuration than steel or titanium). Beryllium has about six times the relative stiffness of steel, titanium, or aluminum, but is no stronger. Beryllium wires are much stronger, but no stiffer than bulk beryllium. Beryllium wires in a matrix exhibit some of the same general characteristics as other composites.

Not all of the strength and stiffness advantages of fiber-reinforced composite materials can be transformed directly into structural advantages. Prominent among the reasons for this statement is the fact that the joints for members made of composite materials are more bulky than those for metal parts. These relative inefficiencies are being studied since they obviously affect the cost trade-offs for application of composite materials. Other limitations will be discussed subsequently.

1.6.2 Cost Advantages

Decreasing the cost of a material per pound depends on increasing manufacturing experience in a given process and on developing new, more effective manufacturing technology, among other things. Both these aspects are implicitly illustrated in Fig. 1-16. Graphite fibers are predicted to fall from several hundred dollars a pound in the early 1970s to $50 per pound in 1980 due to increased manufacturing experience and to the increased efficiencies of large-scale production. On the other hand, boron fibers, also several hundred dollars per pound in the early 1970s, are predicted to cost about a $100 per pound in 1980 because of inherent technological limitations. The estimated prices are based on boron that is deposited on a tungsten substrate. If a glass substrate can be used, one technological barrier would be overcome, and the cost of boron could be as low as that of graphite fibers. In addition, smaller fibers could be produced by the glass substrate process. One of the difficulties in working with boron is that it reacts chemically with many matrix materials, as does carbon to a lesser extent.

Cost per pound is but one aspect of the cost effectiveness of composite materials. A significant consideration is the scrappage in fabrication operations. Scrappage is the material that is trimmed or machined from the starting form of the material in achieving the final product. For most conventional materials, scrappage is returned to the material manufacturer for reprocessing.

However, scrappage should be less for composites than for conventional materials because composites are fabricated in as close to the final configuration as possible. For example, spars and longerons in airplane wings are beam elements that are usually tapered in both depth and width and have holes in their webs to decrease weight. The fabrication of such members from conventional materials such as aluminum or other alloys consists of hogging out (machining) a large blank of material that sometimes weighs as much as seven times the final spar weight. The scrappage is then 600 percent! On the other

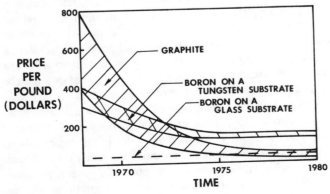

FIG. 1-16. Predicted cost of boron and graphite fibers.

hand, spars have been fabricated from composite materials with as little as 10 percent scrappage! This comparison may seem unfair in the light of other examples, but it actually is quite realistic. Composite materials are not claimed to be a cure-all for every application or even competitive with other materials. There are many instances in which composite materials are uniquely suited because of their peculiar fabrication process. Thus, this "special" case of a spar is not really special, but is actually a powerful example of the class of applications where composite materials offer significant advantages over conventional materials.

1.6.3 Current and Potential Usage of Composite Materials

Currently, almost every aerospace company is developing products made with fiber-reinforced composite materials. The usage of composite materials has progressed through several stages in the past ten years. First, *demonstration pieces* were built with the philosophy "let's see if we *can* build one." There may never have been any intention to put the part on an airplane and flight-test it. The second stage was *replacement pieces* where part of the objective was to flight-test a part that was designed to replace a metal part on an existing airplane. Examples of both stages are the boron/epoxy fuselage section and horizontal tail on the General Dynamics F-111 shown in Fig. 1- 17. Another example is the graphite/epoxy fuselage component for the Northrop F-5 made by General Dynamics shown in Fig. 1- 18. The third stage is actual *production pieces* where the plane is designed from the beginning to have various parts fabricated from fiber-reinforced composite materials. Examples are the boron/epoxy horizontal stabilizer in the Grumman F-14 shown in Fig. 1- 19 and the boron/epoxy horizontal and vertical stabilizers in the McDonnell-Douglas F-15 shown in Fig. 1- 20. Another example of a production piece is the graphite/epoxy horizontal and vertical stabilizers for the General Dynamics YF-16 in Fig. 1- 21. The final stage is the *all-composite airplane* that many people have dreamed of building for several years. This last goal has been approached in the deliberate, conservative, multistage fashion just outlined. A substantial composite materials technology and manufacturing base has been built and awaits further challenge.

The impact of composite materials use on jet engine performance is also very substantial. Currently, with various metal alloys, thrust-to-weight ratios of 5 to 1 are achieved. Reinforced plastics may lead to ratios as high as 16 to 1. Ultimately, with advanced graphite fiber composites, thrust-to-weight ratios on the order of 40 to 1 appear possible. An eightfold increase in the performance index of thrust-to-weight ratio should lead to drastically pyramided weight savings in an entire aircraft due to substantially lessened structural support requirements. However, the road to this goal can be perilous. For example, the

FIG. 1-17. Boron/epoxy fuselage section and horizontal tail for General Dynamics F-111. (*Courtesy General Dynamics Corp.*)

FIG. 1-18. Graphite/epoxy fuselage component made by General Dynamics for Northrop F-5. (*Courtesy General Dynamics Corp.*)

FIG. 1-19. Grumman F-14 with boron/epoxy horizontal stabilizers. (*Courtesy Grumman Aerospace Corp.*)

FIG. 1-20. McDonnell Douglas F-15 with boron/epoxy horizontal and vertical stabilizers. (*Courtesy McDonnell Douglas Corp.*)

FIG. 1-21. General Dynamics YF-16 with graphite/epoxy horizontal and vertical stabilizers. (*Courtesy General Dynamics Corp.*)

Rolls-Royce bankruptcy appears to be closely tied to a lost gamble on development of graphite/epoxy fan blades for the Lockheed L-1011 engines.

In the near future, aircraft will be built with a very high percentage of components made from composite materials. Only then will the full advantages of weight savings be realized since nearly all parts of a plane interact with or support other parts. Hence, weight reduction in one part of a plane pyramids over the entire plane. Weight reductions are well-motivated since the structure of a typical airplane might weigh 30 percent of the total weight with only about 10 percent being payload and the rest fuel, electronic gear, etc. Thus, if materials that are 50 percent more effective in stiffness and strength were used, then the weight would be reduced by the amount of the payload. The implications of such a reduction are manifold. The payload could be doubled, the range extended, operating efficiency improved, or some combination of these and other factors. Obviously, such benefits are welcome improvements, but there is a sometimes more significant benefit from weight savings. In the case of the ill-fated SST project, the possibility of carrying any payload at all was in doubt at one stage in the design. Similarly, the economic feasibility of VTOL craft may depend on the extensive use of composite materials. In all applications, improved fatigue life and reliability of composite materials are an added attraction.

REFERENCES

1-1 Dietz, Albert G. H.: "Composite Materials," 1965 Edgar Marburg Lecture, American Society for Testing and Materials, 1965.

1-2 Sutton, Willard H., B. Walter Rosen, and Donald G. Flom: Whisker-Reinforced Plastics for Space Applications, *SPE Journal*, November, 1964, pp. 1203-1209.

1-3 "Structural Design Guide for Advanced Composite Applications," vol. 1, Material Characterization, 2d ed., Advanced Composites Division, Air Force Materials Laboratory, January, 1971.

1-4 Judge, John F.: Composite Materials: The Coming Revolution, *Airline Management and Marketing*, September, 1969, pp. 85, 90, and 91.

Chapter 2
MACROMECHANICAL
BEHAVIOR OF A LAMINA

2.1 INTRODUCTION

A lamina is a flat or curved assemblage of fibers in a supporting matrix as defined in Sec. 1.4. Examples of two types of laminae are shown in Fig. 1-7. A lamina is the basic building block in a laminated fiber-reinforced composite. Thus, knowledge of the mechanical behavior of a lamina is essential to the understanding of laminated fiber-reinforced structures. The macromechanical behavior (i.e., the behavior of a lamina when only averaged apparent mechanical properties are considered, rather than the detailed interactions of the constituents of the composite) will be focused upon in this chapter. The basic restriction of the chapter is to linear elastic behavior as is the case in the next chapter on micromechanical behavior of a lamina.

In Sec. 2.2, the stress-strain relations (generalized Hooke's law) for anisotropic and orthotropic as well as isotropic materials are discussed. These relations have two commonly accepted manners of expression; compliances and stiffnesses as coefficients (elastic constants) of the stress-strain relations. The most attractive form of the stress-strain relations for orthotropic materials involves the engineering constants described in Sec. 2.3. The engineering constants are particularly helpful in describing composite behavior because they are defined by the use of very obvious and simple physical tests. Restrictions in the form of bounds are derived for the elastic constants in Sec. 2.4. These restrictions are useful in understanding the unusual behavior of composites relative to conventional isotropic materials. Attention is focused in Sec. 2.5 on stress-strain relations for an orthotropic material under plane stress conditions, the most common description of a composite lamina. These relations are transformed in Sec. 2.6 to coordinate systems that are not aligned with the principal material directions of the lamina. This transformation is necessary in order to describe the behavior of composite materials that have fibers running in directions other than the natural geometrical directions of the structural element (e.g., a helically wound circular cylindrical shell). The stress-strain relations for an orthotropic lamina with principal material directions that are not aligned with the obvious geometrical directions are further related to generalized engineering constants and anisotropic materials. The transformed reduced stiffnesses derived in Sec. 2.6 are shown in Sec. 2.7 to have certain combinations which are invariant with respect to rotation of coordinates in the plane of the lamina. The invariants are useful in design of laminated composites. Next, in Sec. 2.8, the important topic of lamina strength is addressed. There, the common approach for conventional isotropic materials of comparing the maximum principal stress with the maximum allowable stress is rejected for composite materials. Tests are described to obtain the stiffnesses and strengths of orthotropic composite laminae in principal material directions. The procedures for predicting the strength in nonprincipal material directions are discussed in Sec. 2.9. There, an energy-related failure criterion is seen to agree well with experimental data.

2.2 STRESS-STRAIN RELATIONS FOR ANISOTROPIC MATERIALS

The generalized Hooke's law relating stresses to strains can be written in contracted notation as

$$\sigma_i = C_{ij}\epsilon_j \qquad i,j = 1, \ldots, 6 \tag{2.1}$$

where σ_i are the stress components, C_{ij} is the stiffness matrix, and ϵ_j are the strain components. The contracted notation is defined in comparison to the usual tensor notation for three-dimensional stresses and strains in Table 2-1 for situations in which the stress and strain tensors are symmetric (the usual case when body forces are absent). Note that, by virtue of Table 2-1, the strains in contracted notation are therefore defined as

$$\epsilon_1 = \frac{\partial u}{\partial x} \qquad \epsilon_2 = \frac{\partial v}{\partial y} \qquad \epsilon_3 = \frac{\partial w}{\partial z}$$

$$\gamma_{23} = \frac{\partial v}{\partial z} + \frac{\partial w}{\partial y} \qquad \gamma_{31} = \frac{\partial w}{\partial x} + \frac{\partial u}{\partial z} \qquad \gamma_{12} = \frac{\partial u}{\partial y} + \frac{\partial v}{\partial x} \tag{2.2}$$

where u, v, and w are displacements in the x, y, and z directions.

The stiffness matrix, C_{ij}, has 36 constants in Eq. (2.1). However, less than 36 of the constants can be shown to be actually independent for elastic materials when the strain energy is considered. Elastic materials for which an elastic potential or strain energy density function exists have incremental work per unit volume of

$$dW = \sigma_i \, d\epsilon_i \tag{2.3}$$

when the stresses σ_i act through strains $d\epsilon_i$. However, because of the stress-strain relations, Eq. (2.1), the incremental work becomes

$$dW = C_{ij}\epsilon_j \, d\epsilon_i \tag{2.4}$$

Upon integration for all strains, the work per unit of volume is

TABLE 2-1. Comparison between tensor and contracted notation for stresses and strains

	Stresses		Strains	
	Tensor notation	Contracted notation	Tensor notation	Contracted notation
	σ_{11}	σ_1	ϵ_{11}	ϵ_1
	σ_{22}	σ_2	ϵ_{22}	ϵ_2
	σ_{33}	σ_3	ϵ_{33}	ϵ_3
	$\tau_{23} = \sigma_{23}$	σ_4	$\gamma_{23} = 2\epsilon_{23}$ *	ϵ_4
	$\tau_{31} = \sigma_{31}$	σ_5	$\gamma_{31} = 2\epsilon_{31}$	ϵ_5
	$\tau_{12} = \sigma_{12}$	σ_6	$\gamma_{12} = 2\epsilon_{12}$	ϵ_6

*Note that γ_{ij} ($i \neq j$) represents engineering shear strain whereas ϵ_{ij} ($i \neq j$) represents tensor shear strain.

$$W = \frac{1}{2} C_{ij} \epsilon_i \epsilon_j$$

$$(2.5)$$

However, Hooke's law, Eq. (2.1), can be derived from Eq. (2.5):

$$\frac{\partial W}{\partial \epsilon_i} = C_{ij} \epsilon_j$$

$$(2.6)$$

whereupon

$$\frac{\partial^2 W}{\partial \epsilon_i \partial \epsilon_j} = C_{ij}$$

$$(2.7)$$

Similarly,

$$\frac{\partial^2 W}{\partial \epsilon_j \partial \epsilon_i} = C_{ji}$$

$$(2.8)$$

But the order of differentiation of W is immaterial, so

$$C_{ij} = C_{ji}$$

$$(2.9)$$

Thus, the stiffness matrix is symmetric so only 21 of the constants are independent.

In a similar manner, we can show that

$$W = \frac{1}{2} S_{ij} \sigma_i \sigma_j$$

$$(2.10)$$

where S_{ij} is the compliance matrix defined by the inverse of the stress-strain relations, the strain-stress relations:

$$\epsilon_i = S_{ij} \sigma_j \quad i, j = 1, \ldots, 6$$

$$(2.11)$$

Reasoning similar to that in the preceding paragraph leads to the conclusion that

$$S_{ij} = S_{ji}$$

$$(2.12)$$

i.e., that the compliance matrix is symmetric and hence has only 21 independent constants. At this point, note that the stiffnesses and compliances are not described by mnemonic notation, but are unfortunately reversed in common usage. The stiffness and compliance components will be referred to as elastic constants.

With the foregoing reduction from 36 to 21 independent constants, the stress-strain relations are

$$
\begin{Bmatrix} \sigma_1 \\ \sigma_2 \\ \sigma_3 \\ \tau_{23} \\ \tau_{31} \\ \tau_{12} \end{Bmatrix} = \begin{bmatrix} C_{11} & C_{12} & C_{13} & C_{14} & C_{15} & C_{16} \\ C_{12} & C_{22} & C_{23} & C_{24} & C_{25} & C_{26} \\ C_{13} & C_{23} & C_{33} & C_{34} & C_{35} & C_{36} \\ C_{14} & C_{24} & C_{34} & C_{44} & C_{45} & C_{46} \\ C_{15} & C_{25} & C_{35} & C_{45} & C_{55} & C_{56} \\ C_{16} & C_{26} & C_{36} & C_{46} & C_{56} & C_{66} \end{bmatrix} \begin{Bmatrix} \epsilon_1 \\ \epsilon_2 \\ \epsilon_3 \\ \gamma_{23} \\ \gamma_{31} \\ \gamma_{12} \end{Bmatrix}
\tag{2.13}
$$

as the most general expression within the framework of linear elasticity. Actually, the relations in Eq. (2.13) are referred to as characterizing *anisotropic* materials since there are no planes of symmetry for the material properties. An alternative name for such an anisotropic material is a *triclinic* material. Materials with more property symmetry than anisotropic materials will be described in the next few paragraphs. Proof of the form of the stress-strain relations for the various cases of material property symmetry is given, for example, by Tsai (Ref. 2-1).

If there is one plane of material property symmetry, the stress-strain relations reduce to

$$
\begin{Bmatrix} \sigma_1 \\ \sigma_2 \\ \sigma_3 \\ \tau_{23} \\ \tau_{31} \\ \tau_{12} \end{Bmatrix} = \begin{bmatrix} C_{11} & C_{12} & C_{13} & 0 & 0 & C_{16} \\ C_{12} & C_{22} & C_{23} & 0 & 0 & C_{26} \\ C_{13} & C_{23} & C_{33} & 0 & 0 & C_{36} \\ 0 & 0 & 0 & C_{44} & C_{45} & 0 \\ 0 & 0 & 0 & C_{45} & C_{55} & 0 \\ C_{16} & C_{26} & C_{36} & 0 & 0 & C_{66} \end{bmatrix} \begin{Bmatrix} \epsilon_1 \\ \epsilon_2 \\ \epsilon_3 \\ \gamma_{23} \\ \gamma_{31} \\ \gamma_{12} \end{Bmatrix}
\tag{2.14}
$$

where the plane of symmetry is $z = 0$. Such a material is termed *monoclinic.* There are 13 independent elastic constants for monoclinic materials.

If there are two orthogonal planes of material property symmetry for a material, symmetry will exist relative to a third mutually orthogonal plane. The stress-strain relations in coordinates aligned with principal material directions[1] are

[1] Principal material directions are parallel to the intersections of the three orthogonal planes of material symmetry.

$$
\begin{Bmatrix} \sigma_1 \\ \sigma_2 \\ \sigma_3 \\ \tau_{23} \\ \tau_{31} \\ \tau_{12} \end{Bmatrix} = \begin{bmatrix} C_{11} & C_{12} & C_{13} & 0 & 0 & 0 \\ C_{12} & C_{22} & C_{23} & 0 & 0 & 0 \\ C_{13} & C_{23} & C_{33} & 0 & 0 & 0 \\ 0 & 0 & 0 & C_{44} & 0 & 0 \\ 0 & 0 & 0 & 0 & C_{55} & 0 \\ 0 & 0 & 0 & 0 & 0 & C_{66} \end{bmatrix} \begin{Bmatrix} \epsilon_1 \\ \epsilon_2 \\ \epsilon_3 \\ \gamma_{23} \\ \gamma_{31} \\ \gamma_{12} \end{Bmatrix}
\tag{2.15}
$$

and are said to define an *orthotropic* material. Note that there is no interaction between normal stresses σ_1, σ_2, σ_3 and shearing strains γ_{23}, γ_{31}, γ_{12} such as occurs in anisotropic materials (by virtue of the presence of, for example, C_{14}). Similarly, there is no interaction between shearing stresses and normal strains as well as none between shearing stresses and shearing strains in different planes. Note also that there are now only *nine* independent constants in the stiffness matrix.

If at every point of a material there is one plane in which the mechanical properties are equal in all directions, then the material is termed *transversely isotropic*. If, for example, the 1-2 plane is the special plane of isotropy, then the 1 and 2 subscripts on the stiffnesses are interchangeable. The stress-strain relations then have only *five* independent constants and are

$$
\begin{Bmatrix} \sigma_1 \\ \sigma_2 \\ \sigma_3 \\ \tau_{23} \\ \tau_{31} \\ \tau_{12} \end{Bmatrix} = \begin{bmatrix} C_{11} & C_{12} & C_{13} & 0 & 0 & 0 \\ C_{12} & C_{11} & C_{13} & 0 & 0 & 0 \\ C_{13} & C_{13} & C_{33} & 0 & 0 & 0 \\ 0 & 0 & 0 & C_{44} & 0 & 0 \\ 0 & 0 & 0 & 0 & C_{44} & 0 \\ 0 & 0 & 0 & 0 & 0 & (C_{11}-C_{12})/2 \end{bmatrix} \begin{Bmatrix} \epsilon_1 \\ \epsilon_2 \\ \epsilon_3 \\ \gamma_{23} \\ \gamma_{31} \\ \gamma_{12} \end{Bmatrix}
\tag{2.16}
$$

If there are an infinite number of planes of material property symmetry, then the foregoing relations simplify to the *isotropic* material case with only *two* independent constants in the stiffness matrix:

$$
\begin{Bmatrix} \sigma_1 \\ \sigma_2 \\ \sigma_3 \\ \tau_{23} \\ \tau_{31} \\ \tau_{12} \end{Bmatrix} = \begin{bmatrix} C_{11} & C_{12} & C_{12} & 0 & 0 & 0 \\ C_{12} & C_{11} & C_{12} & 0 & 0 & 0 \\ C_{12} & C_{12} & C_{11} & 0 & 0 & 0 \\ 0 & 0 & 0 & (C_{11}-C_{12})/2 & 0 & 0 \\ 0 & 0 & 0 & 0 & (C_{11}-C_{12})/2 & 0 \\ 0 & 0 & 0 & 0 & 0 & (C_{11}-C_{12})/2 \end{bmatrix} \begin{Bmatrix} \epsilon_1 \\ \epsilon_2 \\ \epsilon_3 \\ \gamma_{23} \\ \gamma_{31} \\ \gamma_{12} \end{Bmatrix}
$$

$$
\tag{2.17}
$$

The strain-stress relations for the five most common material property symmetry cases are shown in Eqs. (2.18), (2.19), (2.20), (2.21), and (2.22).

Anisotropic (21 independent constants)

$$
\begin{Bmatrix} \epsilon_1 \\ \epsilon_2 \\ \epsilon_3 \\ \gamma_{23} \\ \gamma_{31} \\ \gamma_{12} \end{Bmatrix}
=
\begin{bmatrix}
S_{11} & S_{12} & S_{13} & S_{14} & S_{15} & S_{16} \\
S_{12} & S_{22} & S_{23} & S_{24} & S_{25} & S_{26} \\
S_{13} & S_{23} & S_{33} & S_{34} & S_{35} & S_{36} \\
S_{14} & S_{24} & S_{34} & S_{44} & S_{45} & S_{46} \\
S_{15} & S_{25} & S_{35} & S_{45} & S_{55} & S_{56} \\
S_{16} & S_{26} & S_{36} & S_{46} & S_{56} & S_{66}
\end{bmatrix}
\begin{Bmatrix} \sigma_1 \\ \sigma_2 \\ \sigma_3 \\ \tau_{23} \\ \tau_{31} \\ \tau_{12} \end{Bmatrix}
\tag{2.18}
$$

Monoclinic (13 independent constants)
(for symmetry about z = 0)

$$
\begin{Bmatrix} \epsilon_1 \\ \epsilon_2 \\ \epsilon_3 \\ \gamma_{23} \\ \gamma_{31} \\ \gamma_{12} \end{Bmatrix}
=
\begin{bmatrix}
S_{11} & S_{12} & S_{13} & 0 & 0 & S_{16} \\
S_{12} & S_{22} & S_{23} & 0 & 0 & S_{26} \\
S_{13} & S_{23} & S_{33} & 0 & 0 & S_{36} \\
0 & 0 & 0 & S_{44} & S_{45} & 0 \\
0 & 0 & 0 & S_{45} & S_{55} & 0 \\
S_{16} & S_{26} & S_{36} & 0 & 0 & S_{66}
\end{bmatrix}
\begin{Bmatrix} \sigma_1 \\ \sigma_2 \\ \sigma_3 \\ \tau_{23} \\ \tau_{31} \\ \tau_{12} \end{Bmatrix}
\tag{2.19}
$$

Orthotropic (9 independent constants)

$$
\begin{Bmatrix} \epsilon_1 \\ \epsilon_2 \\ \epsilon_3 \\ \gamma_{23} \\ \gamma_{31} \\ \gamma_{12} \end{Bmatrix}
=
\begin{bmatrix}
S_{11} & S_{12} & S_{13} & 0 & 0 & 0 \\
S_{12} & S_{22} & S_{23} & 0 & 0 & 0 \\
S_{13} & S_{23} & S_{33} & 0 & 0 & 0 \\
0 & 0 & 0 & S_{44} & 0 & 0 \\
0 & 0 & 0 & 0 & S_{55} & 0 \\
0 & 0 & 0 & 0 & 0 & S_{66}
\end{bmatrix}
\begin{Bmatrix} \sigma_1 \\ \sigma_2 \\ \sigma_3 \\ \tau_{23} \\ \tau_{31} \\ \tau_{12} \end{Bmatrix}
\tag{2.20}
$$

Transversely isotropic (5 independent constants)
(for a 1-2 symmetry plane)

$$
\begin{Bmatrix} \epsilon_1 \\ \epsilon_2 \\ \epsilon_3 \\ \gamma_{23} \\ \gamma_{31} \\ \gamma_{12} \end{Bmatrix} =
\begin{bmatrix}
S_{11} & S_{12} & S_{13} & 0 & 0 & 0 \\
S_{12} & S_{11} & S_{13} & 0 & 0 & 0 \\
S_{13} & S_{13} & S_{33} & 0 & 0 & 0 \\
0 & 0 & 0 & S_{44} & 0 & 0 \\
0 & 0 & 0 & 0 & S_{44} & 0 \\
0 & 0 & 0 & 0 & 0 & 2(S_{11}-S_{12})
\end{bmatrix}
\begin{Bmatrix} \sigma_1 \\ \sigma_2 \\ \sigma_3 \\ \tau_{23} \\ \tau_{31} \\ \tau_{12} \end{Bmatrix}
\qquad (2.21)
$$

Isotropic (2 independent constants)

$$
\begin{Bmatrix} \epsilon_1 \\ \epsilon_2 \\ \epsilon_3 \\ \gamma_{23} \\ \gamma_{31} \\ \gamma_{12} \end{Bmatrix} =
\begin{bmatrix}
S_{11} & S_{12} & S_{12} & 0 & 0 & 0 \\
S_{12} & S_{11} & S_{12} & 0 & 0 & 0 \\
S_{12} & S_{12} & S_{11} & 0 & 0 & 0 \\
0 & 0 & 0 & 2(S_{11}-S_{12}) & 0 & 0 \\
0 & 0 & 0 & 0 & 2(S_{11}-S_{12}) & 0 \\
0 & 0 & 0 & 0 & 0 & 2(S_{11}-S_{12})
\end{bmatrix}
\begin{Bmatrix} \sigma_1 \\ \sigma_2 \\ \sigma_3 \\ \tau_{23} \\ \tau_{31} \\ \tau_{12} \end{Bmatrix}
\qquad (2.22)
$$

2.3 ENGINEERING CONSTANTS FOR ORTHOTROPIC MATERIALS

Engineering constants (also known as technical constants) are generalized Young's moduli, Poisson's ratios, and shear moduli as well as some other behavioral constants that will be discussed in Sec. 2.6. These constants are measured in simple tests such as uniaxial tension or pure shear tests. Thus, these constants with their obvious physical interpretation have more direct meaning than the components of the relatively abstract compliance and stiffness matrices used in Sec. 2.2.

Most simple tests are performed with a known load or stress. The resulting displacement or strain is then measured. Thus, the components of the compliance (S_{ij}) matrix are determined more directly than those of the stiffness (C_{ij}) matrix. For an orthotropic material, the compliance matrix components in terms of the engineering constants are

$$[S_{ij}] = \begin{bmatrix} \dfrac{1}{E_1} & -\dfrac{\nu_{21}}{E_2} & -\dfrac{\nu_{31}}{E_3} & 0 & 0 & 0 \\[2ex] -\dfrac{\nu_{12}}{E_1} & \dfrac{1}{E_2} & -\dfrac{\nu_{32}}{E_3} & 0 & 0 & 0 \\[2ex] -\dfrac{\nu_{13}}{E_1} & -\dfrac{\nu_{23}}{E_2} & \dfrac{1}{E_3} & 0 & 0 & 0 \\[2ex] 0 & 0 & 0 & \dfrac{1}{G_{23}} & 0 & 0 \\[2ex] 0 & 0 & 0 & 0 & \dfrac{1}{G_{31}} & 0 \\[2ex] 0 & 0 & 0 & 0 & 0 & \dfrac{1}{G_{12}} \end{bmatrix} \qquad (2.23)$$

where

E_1, E_2, E_3 = Young's moduli in 1, 2, and 3 directions, respectively.

ν_{ij} = Poisson's ratio for transverse strain in the j-direction when stressed in the i-direction, that is,

$$\nu_{ij} = -\frac{\epsilon_j}{\epsilon_i} \qquad (2.24)$$

for $\sigma_i = \sigma$ and all other stresses are zero.

G_{23}, G_{31}, G_{12} = shear moduli in the 2-3, 3-1, and 1-2 planes, respectively.

Recall that for an orthotropic material there are nine independent constants because

$$S_{ij} = S_{ji} \qquad (2.25)$$

since the compliance matrix is the inverse of the stiffness (C_{ij}) matrix that was shown to be symmetric in Eq. (2.9). When engineering constants are substituted in Eq. (2.25),

$$\frac{\nu_{ij}}{E_i} = \frac{\nu_{ji}}{E_j} \quad i, j = 1, 2, 3 \qquad (2.26)$$

Thus, there are three reciprocal relations that must be satisfied for an orthotropic material. Moreover, only ν_{12}, ν_{13}, and ν_{23} need be further considered since ν_{21}, ν_{31}, and ν_{32} can be expressed in terms of the first-mentioned Poisson's ratios and the Young's moduli. The latter Poisson's ratios should not be forgotten, however, because for some tests they are what is actually being measured.

The difference between ν_{12} and ν_{21} for an orthotropic material is emphasized with the aid of Fig. 2-1 where two cases of uniaxial stress are shown for a square element. First, a stress is applied in the 1-direction in Fig. 2-1. Then, from Eqs. (2.20) and (2.23), the strains are

$$^1\epsilon_1 = \frac{\sigma}{E_1} \qquad ^1\epsilon_2 = \left(-\frac{\nu_{12}}{E_1}\right)\sigma \qquad\qquad (2.27)$$

so the deformations are

$$^1\Delta_1 = \frac{\sigma L}{E_1} \qquad ^1\Delta_2 = \left(\frac{\nu_{12}}{E_1}\right)\sigma L \qquad\qquad (2.28)$$

where the direction of loading is denoted by the superscript. Second, the same value of stress is applied in the 2-direction in Fig. 2-1. The strains are

$$^2\epsilon_1 = \left(-\frac{\nu_{21}}{E_2}\right)\sigma \qquad ^2\epsilon_2 = \frac{\sigma}{E_2} \qquad\qquad (2.29)$$

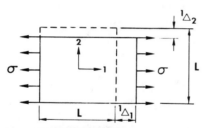

STRESS IN 1-DIRECTION

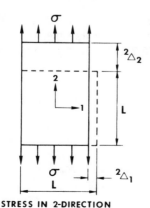

STRESS IN 2-DIRECTION

FIG. 2-1. Distinction between ν_{12} and ν_{21}.

and the deformations are

$$^2\Delta_1 = \left(\frac{\nu_{21}}{E_2}\right)\sigma L \qquad ^2\Delta_2 = \frac{\sigma L}{E_2} \tag{2.30}$$

Obviously, if $E_1 > E_2$ then $^1\Delta_1 < {}^2\Delta_2$ as we would expect. However, because of the reciprocal relations, irrespective of the values of E_1 and E_2,

$$^1\Delta_2 = {}^2\Delta_1 \tag{2.31}$$

which is an obvious generalization of Betti's law to the treatment of anisotropic materials. That is, the transverse deformation (and transverse strain) is the same when the stress is applied in the 2-direction as when it is applied in the 1-direction.

   Since the stiffness and compliance matrices are mutually inverse, it follows by matrix algebra that their components are related as follows for orthotropic materials:

$$C_{11} = \frac{S_{22}S_{33} - S_{23}^2}{S} \qquad\qquad C_{12} = \frac{S_{13}S_{23} - S_{12}S_{33}}{S}$$

$$C_{22} = \frac{S_{33}S_{11} - S_{13}^2}{S} \qquad\qquad C_{13} = \frac{S_{12}S_{23} - S_{13}S_{22}}{S}$$

$$\tag{2.32}$$

$$C_{33} = \frac{S_{11}S_{22} - S_{12}^2}{S} \qquad\qquad C_{23} = \frac{S_{12}S_{13} - S_{23}S_{11}}{S}$$

$$C_{44} = \frac{1}{S_{44}} \qquad C_{55} = \frac{1}{S_{55}} \qquad C_{66} = \frac{1}{S_{66}}$$

where

$$S = S_{11}S_{22}S_{33} - S_{11}S_{23}^2 - S_{22}S_{13}^2 - S_{33}S_{12}^2 + 2S_{12}S_{23}S_{13} \tag{2.33}$$

In Eq. (2.32), the symbols C and S can be interchanged everywhere to provide the converse relationship.

   The stiffness matrix, C_{ij}, for an orthotropic material in terms of the engineering constants is obtained by inversion of the compliance matrix, S_{ij}, in Eq. (2.23) or by substitution in Eqs. (2.32) and (2.33). The nonzero stiffnesses in Eq. (2.15) are

$$C_{11} = \frac{1 - \nu_{23}\nu_{32}}{E_2 E_3 \Delta}$$

$$C_{12} = \frac{\nu_{21} + \nu_{31}\nu_{23}}{E_2 E_3 \Delta} = \frac{\nu_{12} + \nu_{32}\nu_{13}}{E_1 E_3 \Delta}$$

$$C_{13} = \frac{\nu_{31} + \nu_{21}\nu_{32}}{E_2 E_3 \Delta} = \frac{\nu_{13} + \nu_{12}\nu_{23}}{E_1 E_2 \Delta} \tag{2.34}$$

$$C_{22} = \frac{1 - \nu_{13}\nu_{31}}{E_1 E_3 \Delta}$$

$$C_{23} = \frac{\nu_{32} + \nu_{12}\nu_{31}}{E_1 E_3 \Delta} = \frac{\nu_{23} + \nu_{21}\nu_{13}}{E_1 E_2 \Delta}$$

$$C_{33} = \frac{1 - \nu_{12}\nu_{21}}{E_1 E_2 \Delta}$$

$$C_{44} = G_{23}$$

$$C_{55} = G_{31}$$

$$C_{66} = G_{12}$$

(2.34)
(cont'd.)

where

$$\Delta = \frac{1 - \nu_{12}\nu_{21} - \nu_{23}\nu_{32} - \nu_{31}\nu_{13} - 2\nu_{21}\nu_{32}\nu_{13}}{E_1 E_2 E_3} \qquad (2.35)$$

Note especially that if a material is suspected to be orthotropic, mechanical tests at various angles will reveal whether there are directions for which shear coupling does not exist. Hence, the orthotropy, isotropy, or lack thereof can be determined, although at a sometimes significant cost. The easiest way to determine principal material directions is visual observation. However, for visual observation to work, the characteristics of the material must obviously be readily seen by the naked eye. For example, in a fiber-reinforced lamina made from the boron/epoxy tape in Fig. 1-11, the longitudinal direction is readily determined (and defined) to be the 1-direction. Similarly, the 2-direction is in the plane of the tape transverse to the longitudinal direction. Finally, the 3-direction is defined to be perpendicular to the plane of the tape.

2.4 RESTRICTIONS ON ELASTIC CONSTANTS

2.4.1 Isotropic Materials

For isotropic materials, certain relations between the elastic constants must be satisfied. For example, the shear modulus is defined in terms of the elastic modulus, E, and Poisson's ratio, ν, as

$$G = \frac{E}{2(1 + \nu)} \qquad (2.36)$$

Thus, in order that E and G always be positive, i.e., that a positive normal stress or shear stress times the respective normal strain or shear strain yield *positive* work,

$$\nu > -1 \qquad (2.37)$$

In a similar manner, if an isotropic body is subjected to hydrostatic pressure, p, then the volumetric strain, the sum of the three normal or extensional strains, is defined by

$$\theta = \epsilon_x + \epsilon_y + \epsilon_z = \frac{p}{E/3(1 - 2\nu)} = \frac{p}{K} \tag{2.38}$$

Then the bulk modulus,

$$K = \frac{E}{3(1 - 2\nu)} \tag{2.39}$$

is positive only if E is positive and

$$\nu < \tfrac{1}{2} \tag{2.40}$$

If the bulk modulus were negative, a hydrostatic pressure would cause expansion of a cube of isotropic material. Thus, in isotropic materials, Poisson's ratio is restricted to the range

$$-1 < \nu < \tfrac{1}{2} \tag{2.41}$$

in order that shear or hydrostatic loading not produce negative strain energy.

2.4.2 Orthotropic Materials

For orthotropic materials, the relations between elastic constants are more complex. Those relations must be investigated with rigor in order to avoid the pitfalls of an intuition built up on the basis of working with isotropic materials. First, the product of a stress component and the corresponding strain component represents work done by the stress. The sum of the work done by all stress components must be positive in order to avoid the creation of energy. This latter condition provides a *thermodynamic constraint* on the values of the elastic constants. What was previously accomplished for isotropic materials is, in reality, a consequence of such a constraint. The constraint was generalized to include orthotropic materials by Lempriere (Ref. 2-2). Formally, he required the matrices relating stress to strain to be positive-definite, i.e., to have positive principal values or invariants. Thus, both the stiffness and compliance matrices must be positive-definite.

This mathematical condition can be replaced by the following physical argument. If only one normal stress is applied at a time, the corresponding strain is determined by the diagonal elements of the compliance matrix. Thus, those elements must be positive, that is,

$$S_{11}, S_{22}, S_{33}, S_{44}, S_{55}, S_{66} > 0 \tag{2.42}$$

or, in terms of the engineering constants,

$$E_1, E_2, E_3, G_{23}, G_{31}, G_{12} > 0 \tag{2.43}$$

Similarly, under suitable constraints, deformation is possible in which only one extensional strain arises. Again, work is produced by the corresponding stress alone. Thus, since the work done is determined by the diagonal elements of the stiffness matrix, those elements must be positive, that is,

$$C_{11}, C_{22}, C_{33}, C_{44}, C_{55}, C_{66} > 0 \tag{2.44}$$

whereupon from Eq. (2.34)

$$(1 - \nu_{23}\nu_{32}), (1 - \nu_{13}\nu_{31}), (1 - \nu_{12}\nu_{21}) > 0 \tag{2.45}$$

and

$$\bar{\Delta} = 1 - \nu_{12}\nu_{21} - \nu_{23}\nu_{32} - \nu_{31}\nu_{13} - 2\nu_{21}\nu_{32}\nu_{13} > 0 \tag{2.46}$$

since the determinant of a matrix must be positive for positive definiteness. Also, from Eq. (2.32), the positiveness of the stiffnesses leads to

$$|S_{23}| < (S_{22}S_{33})^{\frac{1}{2}}$$
$$|S_{13}| < (S_{11}S_{33})^{\frac{1}{2}} \tag{2.47}$$
$$|S_{12}| < (S_{11}S_{22})^{\frac{1}{2}}$$

By use of the condition of symmetry of the compliances, Eq. (2.12), in the form

$$\frac{\nu_{ij}}{E_i} = \frac{\nu_{ji}}{E_j} \quad i,j = 1,2,3 \tag{2.48}$$

the conditions of Eq. (2.45) can be written as

$$|\nu_{21}| < \left(\frac{E_2}{E_1}\right)^{\frac{1}{2}} \quad |\nu_{12}| < \left(\frac{E_1}{E_2}\right)^{\frac{1}{2}}$$
$$|\nu_{32}| < \left(\frac{E_3}{E_2}\right)^{\frac{1}{2}} \quad |\nu_{23}| < \left(\frac{E_2}{E_3}\right)^{\frac{1}{2}} \tag{2.49}$$
$$|\nu_{13}| < \left(\frac{E_1}{E_3}\right)^{\frac{1}{2}} \quad |\nu_{31}| < \left(\frac{E_3}{E_1}\right)^{\frac{1}{2}}$$

Equations (2.49) can also be obtained from Eqs. (2.47) if the definitions for S_{ij} in terms of the engineering constants are substituted. Similarly, Eq. (2.46) can be expressed as

$$\nu_{21}\nu_{32}\nu_{13} < \frac{1 - \nu_{21}^2\left(\frac{E_1}{E_2}\right) - \nu_{32}^2\left(\frac{E_2}{E_3}\right) - \nu_{13}^2\left(\frac{E_3}{E_1}\right)}{2} < \frac{1}{2} \tag{2.50}$$

and can be regrouped to read

$$\left[1 - \nu_{32}^2\left(\frac{E_2}{E_3}\right)\right]\left[1 - \nu_{13}^2\left(\frac{E_3}{E_1}\right)\right] - \left[\nu_{21}\left(\frac{E_1}{E_2}\right)^{\frac{1}{2}} + \nu_{32}\nu_{13}\left(\frac{E_2}{E_1}\right)^{\frac{1}{2}}\right]^2 > 0 \qquad (2.51)$$

In order to obtain a constraint on one Poisson's ratio, ν_{21}, in terms of two others, ν_{32} and ν_{13}, Eq. (2.51) can be further rearranged as

$$-\left\{\nu_{32}\nu_{13}\left(\frac{E_2}{E_1}\right) + \left[1 - \nu_{32}^2\left(\frac{E_2}{E_3}\right)\right]^{\frac{1}{2}}\left[1 - \nu_{13}^2\left(\frac{E_3}{E_1}\right)\right]^{\frac{1}{2}}\left(\frac{E_2}{E_1}\right)^{\frac{1}{2}}\right\}$$

$$< \nu_{21} < \qquad (2.52)$$

$$-\left\{\nu_{32}\nu_{13}\left(\frac{E_2}{E_1}\right) - \left[1 - \nu_{32}^2\left(\frac{E_2}{E_3}\right)\right]^{\frac{1}{2}}\left[1 - \nu_{13}^2\left(\frac{E_3}{E_1}\right)\right]^{\frac{1}{2}}\left(\frac{E_2}{E_1}\right)^{\frac{1}{2}}\right\}$$

Similar expressions can be obtained for ν_{32} and ν_{13}.

The preceding restrictions on engineering constants for orthotropic materials are used to examine experimental data to see if they are physically consistent within the framework of the mathematical elasticity model. In testing boron/epoxy composite materials, Dickerson and DiMartino (Ref. 2-3) reported Poisson's ratios as high as 1.97 for strain in the 2-direction due to loading in the 1-direction (ν_{12}). The reported values of the Young's moduli for the two directions are $E_1 = 11.86 \times 10^6$ psi and $E_2 = 1.33 \times 10^6$ psi. Thus,

$$\left(\frac{E_1}{E_2}\right)^{\frac{1}{2}} = 2.99 \qquad (2.53)$$

and the condition

$$|\nu_{12}| < \left(\frac{E_1}{E_2}\right)^{\frac{1}{2}} \qquad (2.54)$$

is satisfied. Accordingly, $\nu_{12} = 1.97$ is a reasonable number even though our intuition based on isotropic materials rejects such a large number. Insufficient data were reported to verify the determinant condition, Eq. (2.46), that may be more stringent. Also, the "converse" Poisson's ratio, ν_{21}, was reported as .22. This value satisfies the symmetry condition or reciprocal relations in Eq. (2.48).

Should the measured material properties satisfy the constraints, we can proceed with confidence to design structures with the material. Otherwise, we have reason to doubt either the material model or the experimental data or both!

The restrictions on engineering constants can also be used in the solution of practical engineering analysis problems. For example, consider a differential equation which has several solutions depending on the relative values of the coefficients in the differential equation. Those coefficients in a physical problem of deformation of a body involve the elastic constants. The restrictions on elastic constants can then be used to determine which solution to the differential equation is applicable.

Problem Set 2.4

Exercise 2.4.1 Show that the determinant inequality in Eq. (2.46) for orthotropic materials correctly reduces to $\nu < \frac{1}{2}$ for isotropic materials.

Exercise 2.4.2 Derive Eq. (2.50) from the determinant inequality in Eq. (2.46).

Exercise 2.4.3 Derive Eq. (2.51) from Eq. (2.50).

Exercise 2.4.4 Derive Eq. (2.52) from Eq. (2.51).

Exercise 2.4.5 Show that Eq. (2.52) reduces for isotropic materials to the known bounds on ν.

2.5 STRESS–STRAIN RELATIONS FOR PLANE STRESS IN AN ORTHOTROPIC MATERIAL

For a lamina in the 1-2 plane as shown in Fig. 2-2, a plane stress state is defined by setting

$$\sigma_3 = 0 \qquad \tau_{23} = 0 \qquad \tau_{31} = 0 \tag{2.55}$$

in the three-dimensional stress-strain relations given in Eqs. (2.18)–(2.22) for anisotropic, monoclinic, orthotropic, transversely isotropic, or isotropic materials. For orthotropic materials, such a procedure results in implied strains of

$$\epsilon_3 = S_{13}\sigma_1 + S_{23}\sigma_2$$
$$\gamma_{23} = 0 \qquad \gamma_{31} = 0 \tag{2.56}$$

Moreover, the strain-stress relations in Eq. (2.20) reduce to

$$\begin{Bmatrix} \epsilon_1 \\ \epsilon_2 \\ \gamma_{12} \end{Bmatrix} = \begin{bmatrix} S_{11} & S_{12} & 0 \\ S_{12} & S_{22} & 0 \\ 0 & 0 & S_{66} \end{bmatrix} \begin{Bmatrix} \sigma_1 \\ \sigma_2 \\ \tau_{12} \end{Bmatrix} \tag{2.57}$$

supplemented by Eq. (2.56) where

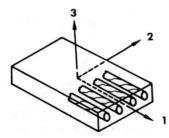

FIG. 2.2. Unidirectionally reinforced lamina.

$$S_{11} = \frac{1}{E_1}$$

$$S_{12} = -\frac{\nu_{12}}{E_1} = -\frac{\nu_{21}}{E_2}$$

$$S_{22} = \frac{1}{E_2}$$ (2.58)

$$S_{66} = \frac{1}{G_{12}}$$

Note that in order to determine ϵ_3 in Eq. (2.56), ν_{13} and ν_{23} must be known in addition to those engineering constants shown in Eq. (2.58). That is, ν_{13} and ν_{23} arise from S_{13} and S_{23} in Eq. (2.56).

The strain-stress relations in Eq. (2.57) can be inverted to obtain the stress-strain relations:

$$\begin{Bmatrix} \sigma_1 \\ \sigma_2 \\ \tau_{12} \end{Bmatrix} = \begin{bmatrix} Q_{11} & Q_{12} & 0 \\ Q_{12} & Q_{22} & 0 \\ 0 & 0 & Q_{66} \end{bmatrix} \begin{Bmatrix} \epsilon_1 \\ \epsilon_2 \\ \gamma_{12} \end{Bmatrix}$$ (2.59)

where the Q_{ij}, the so-called reduced stiffnesses, are

$$Q_{11} = \frac{S_{22}}{S_{11}S_{22} - S_{12}^2}$$

$$Q_{12} = -\frac{S_{12}}{S_{11}S_{22} - S_{12}^2}$$

$$Q_{22} = \frac{S_{11}}{S_{11}S_{22} - S_{12}^2}$$ (2.60)

$$Q_{66} = \frac{1}{S_{66}}$$

or, in terms of the engineering constants,

$$Q_{11} = \frac{E_1}{1 - \nu_{12}\nu_{21}}$$

$$Q_{12} = \frac{\nu_{12}E_2}{1 - \nu_{12}\nu_{21}} = \frac{\nu_{21}E_1}{1 - \nu_{12}\nu_{21}}$$ (2.61)

$$Q_{22} = \frac{E_2}{1 - \nu_{12}\nu_{21}}$$

$$Q_{66} = G_{12}$$

The preceding stress-strain and strain-stress relations are the basis for the stiffness and stress analysis of an individual lamina subjected to forces in its own plane. The relations are therefore indispensible in the analysis of laminates.

Note that there are four independent material properties, E_1, E_2, ν_{12} and G_{12}, in Eqs. (2.57) and (2.59) when Eqs. (2.58) and (2.61) are considered in addition to the reciprocal relation

$$\frac{\nu_{12}}{E_1} = \frac{\nu_{21}}{E_2} \tag{2.62}$$

For isotropic materials under plane stress, the strain-stress relations are

$$\begin{Bmatrix} \epsilon_1 \\ \epsilon_2 \\ \gamma_{12} \end{Bmatrix} = \begin{bmatrix} S_{11} & S_{12} & 0 \\ S_{12} & S_{11} & 0 \\ 0 & 0 & 2(S_{11} - S_{12}) \end{bmatrix} \begin{Bmatrix} \sigma_1 \\ \sigma_2 \\ \tau_{12} \end{Bmatrix} \tag{2.63}$$

where

$$S_{11} = \frac{1}{E}$$
$$S_{12} = -\frac{\nu}{E} \tag{2.64}$$

and the stress-strain relations are

$$\begin{Bmatrix} \sigma_1 \\ \sigma_2 \\ \tau_{12} \end{Bmatrix} = \begin{bmatrix} Q_{11} & Q_{12} & 0 \\ Q_{12} & Q_{11} & 0 \\ 0 & 0 & Q_{66} \end{bmatrix} \begin{Bmatrix} \epsilon_1 \\ \epsilon_2 \\ \gamma_{12} \end{Bmatrix} \tag{2.65}$$

where

$$Q_{11} = \frac{E}{1 - \nu^2}$$
$$Q_{12} = \frac{\nu E}{1 - \nu^2} \tag{2.66}$$
$$Q_{66} = \frac{E}{2(1 + \nu)} = G$$

The preceding isotropic relations can be obtained either from the orthotropic relations by equating E_1 to E_2 and G_{12} to G or by the same manner as the orthotropic relations were obtained.

2.6 STRESS-STRAIN RELATIONS FOR A LAMINA OF ARBITRARY ORIENTATION

In Sec. 2.5, the stresses and strains were defined in the principal material directions for an orthotropic material. However, the principal directions of orthotropy often do not coincide with coordinate directions that are geometrically natural to the solution of the problem. For example, consider the helically wound fiberglass-reinforced circular cylindrical shell in Fig. 2-3. There, the coordinates natural to the solution of the shell problem are the shell coordinates

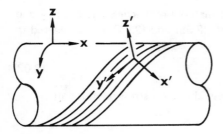

FIG. 2-3. Helically wound fiber-reinforced circular cylindrical shell.

x, y, z whereas the principal material coordinates are x', y', z'. The wrap angle is defined by cos (y', y) = cos α; also, $z' = z$. Other examples include laminated plates with different laminae at different orientations. Thus, a relation is needed between the stresses and strains in the principal material directions and those in the body coordinates. Then, a method of transforming stress-strain relations from one coordinate system to another is also needed.

At this point, we recall from elementary mechanics of materials the transformation equations for expressing stresses in an x-y coordinate system in terms of stresses in a 1-2 coordinate system,

$$\begin{Bmatrix} \sigma_x \\ \sigma_y \\ \tau_{xy} \end{Bmatrix} = \begin{bmatrix} \cos^2\theta & \sin^2\theta & -2\sin\theta\cos\theta \\ \sin^2\theta & \cos^2\theta & 2\sin\theta\cos\theta \\ \sin\theta\cos\theta & -\sin\theta\cos\theta & \cos^2\theta - \sin^2\theta \end{bmatrix} \begin{Bmatrix} \sigma_1 \\ \sigma_2 \\ \tau_{12} \end{Bmatrix} \quad (2.67)$$

where θ is the angle *from* the x-axis *to* the 1-axis (see Fig. 2-4). Note especially that the transformation has nothing to do with the material properties but is merely a rotation of stresses.

Similarly, the strain transformation equations are

$$\begin{Bmatrix} \epsilon_x \\ \epsilon_y \\ \dfrac{\gamma_{xy}}{2} \end{Bmatrix} = \begin{bmatrix} \cos^2\theta & \sin^2\theta & -2\sin\theta\cos\theta \\ \sin^2\theta & \cos^2\theta & 2\sin\theta\cos\theta \\ \sin\theta\cos\theta & -\sin\theta\cos\theta & \cos^2\theta - \sin^2\theta \end{bmatrix} \begin{Bmatrix} \epsilon_1 \\ \epsilon_2 \\ \dfrac{\gamma_{12}}{2} \end{Bmatrix} \quad (2.68)$$

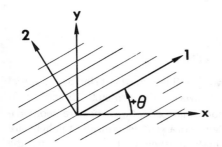

FIG. 2-4. Positive rotation of principal material axes from arbitrary xy axes.

where we observe that strains do transform with the same transformation as stresses if the tensor definition of shear strain is used (which is equivalent to dividing the engineering shear strain by two).

The transformations are commonly written as

$$\left\{\begin{array}{c} \sigma_x \\ \sigma_y \\ \tau_{xy} \end{array}\right\} = [T]^{-1} \left\{\begin{array}{c} \sigma_1 \\ \sigma_2 \\ \tau_{12} \end{array}\right\} \tag{2.69}$$

$$\left\{\begin{array}{c} \epsilon_x \\ \epsilon_y \\ \dfrac{\gamma_{xy}}{2} \end{array}\right\} = [T]^{-1} \left\{\begin{array}{c} \epsilon_1 \\ \epsilon_2 \\ \dfrac{\gamma_{12}}{2} \end{array}\right\} \tag{2.70}$$

where the superscript -1 denotes the matrix inverse and

$$[T] = \begin{bmatrix} \cos^2\theta & \sin^2\theta & 2\sin\theta\cos\theta \\ \sin^2\theta & \cos^2\theta & -2\sin\theta\cos\theta \\ -\sin\theta\cos\theta & \sin\theta\cos\theta & \cos^2\theta - \sin^2\theta \end{bmatrix} \tag{2.71}$$

However, if the matrix

$$[R] = \begin{bmatrix} 1 & 0 & 0 \\ 0 & 1 & 0 \\ 0 & 0 & 2 \end{bmatrix} \tag{2.72}$$

due to Reuter (Ref. 2-4) is introduced, the more natural strain vectors

$$\left\{\begin{array}{c} \epsilon_1 \\ \epsilon_2 \\ \gamma_{12} \end{array}\right\} = [R] \left\{\begin{array}{c} \epsilon_1 \\ \epsilon_2 \\ \dfrac{\gamma_{12}}{2} \end{array}\right\} \tag{2.73}$$

$$\left\{\begin{array}{c} \epsilon_x \\ \epsilon_y \\ \gamma_{xy} \end{array}\right\} = [R] \left\{\begin{array}{c} \epsilon_x \\ \epsilon_y \\ \dfrac{\gamma_{xy}}{2} \end{array}\right\} \tag{2.74}$$

can be used instead of the modified strain vectors in the strain transformations as well as in stress-strain law transformations. The beauty of Reuter's transformation is that concise matrix notation can then be used. As a result, the ordinary expressions for stiffness and compliance matrices with awkward factors of 1/2 and 2 in various rows and columns are avoided.

A so-called *specially orthotropic lamina* is one whose principal material axes are aligned with the natural body axes for the problem, for example,

$$\begin{Bmatrix} \sigma_x \\ \sigma_y \\ \tau_{xy} \end{Bmatrix} = \begin{Bmatrix} \sigma_1 \\ \sigma_2 \\ \tau_{12} \end{Bmatrix} = \begin{bmatrix} Q_{11} & Q_{12} & 0 \\ Q_{12} & Q_{22} & 0 \\ 0 & 0 & Q_{66} \end{bmatrix} \begin{Bmatrix} \epsilon_1 \\ \epsilon_2 \\ \gamma_{12} \end{Bmatrix} \tag{2.75}$$

where the principal material axes are shown in Fig. 2-2. These stress-strain relations were introduced in Sec. 2.5 and apply when the principal material directions of an orthotropic lamina are used as coordinates.

However, as mentioned previously, orthotropic laminae are often constructed in such a manner that the principal material directions do not coincide with the natural directions of the body. This statement is not to be interpreted as meaning that the material itself is no longer orthotropic; rather, we are just looking at an orthotropic material in an unnatural manner, i.e., in a coordinate system that is oriented at some finite angle to the principal material coordinate system. Then, basically, the question is: Given the stress-strain relations in the principal material directions, what are the stress-strain relations in the *xy* coordinate directions?

Accordingly, we use the stress and strain transformations of Eqs. (2.69) and (2.70) along with Reuter's matrix, Eq. (2.72), after abbreviating Eq. (2.75) as

$$\begin{Bmatrix} \sigma_1 \\ \sigma_2 \\ \tau_{12} \end{Bmatrix} = [Q] \begin{Bmatrix} \epsilon_1 \\ \epsilon_2 \\ \gamma_{12} \end{Bmatrix} \tag{2.76}$$

to obtain

$$\begin{Bmatrix} \sigma_x \\ \sigma_y \\ \tau_{xy} \end{Bmatrix} = [T]^{-1} \begin{Bmatrix} \sigma_1 \\ \sigma_2 \\ \tau_{12} \end{Bmatrix} = [T]^{-1} [Q] [R] [T] [R]^{-1} \begin{Bmatrix} \epsilon_x \\ \epsilon_y \\ \gamma_{xy} \end{Bmatrix} \tag{2.77}$$

However, $[R] [T] [R]^{-1}$ can be shown to be $[T]^{-T}$ where the superscript T denotes the matrix transpose. Then, if we use the abbreviation

$$[\bar{Q}] = [T]^{-1} [Q] [T]^{-T} \tag{2.78}$$

the stress-strain relations in *xy* coordinates are

$$
\left\{\begin{array}{c} \sigma_x \\ \sigma_y \\ \tau_{xy} \end{array}\right\} = [\bar{Q}] \left\{\begin{array}{c} \epsilon_x \\ \epsilon_y \\ \gamma_{xy} \end{array}\right\} = \left[\begin{array}{ccc} \bar{Q}_{11} & \bar{Q}_{12} & \bar{Q}_{16} \\ \bar{Q}_{12} & \bar{Q}_{22} & \bar{Q}_{26} \\ \bar{Q}_{16} & \bar{Q}_{26} & \bar{Q}_{66} \end{array}\right] \left\{\begin{array}{c} \epsilon_x \\ \epsilon_y \\ \gamma_{xy} \end{array}\right\}
$$

(2.79)

in which

$$\bar{Q}_{11} = Q_{11} \cos^4\theta + 2(Q_{12} + 2Q_{66}) \sin^2\theta \cos^2\theta + Q_{22} \sin^4\theta$$

$$\bar{Q}_{12} = (Q_{11} + Q_{22} - 4Q_{66}) \sin^2\theta \cos^2\theta + Q_{12}(\sin^4\theta + \cos^4\theta)$$

$$\bar{Q}_{22} = Q_{11} \sin^4\theta + 2(Q_{12} + 2Q_{66}) \sin^2\theta \cos^2\theta + Q_{22} \cos^4\theta$$

$$\bar{Q}_{16} = (Q_{11} - Q_{12} - 2Q_{66}) \sin\theta \cos^3\theta + (Q_{12} - Q_{22} + 2Q_{66}) \sin^3\theta \cos\theta$$

$$\bar{Q}_{26} = (Q_{11} - Q_{12} - 2Q_{66}) \sin^3\theta \cos\theta + (Q_{12} - Q_{22} + 2Q_{66}) \sin\theta \cos^3\theta$$

$$\bar{Q}_{66} = (Q_{11} + Q_{22} - 2Q_{12} - 2Q_{66}) \sin^2\theta \cos^2\theta + Q_{66}(\sin^4\theta + \cos^4\theta)$$

(2.80)

where the bar over the $\bar{Q}_{ij}$ matrix denotes that we are dealing with the transformed reduced stiffnesses instead of the reduced stiffnesses, Q_{ij}.

Note that the transformed reduced stiffness matrix $\bar{Q}_{ij}$ has terms in all nine positions in contrast to the presence of zeros in the reduced stiffness matrix Q_{ij}. However, there are still only *four* independent material constants since the lamina is orthotropic. In the general case with body coordinates x and y, there is coupling between shear strain and normal stresses and between shear stress and normal strains. Thus, in body coordinates, even an orthotropic lamina appears to be anisotropic. However, because such a lamina does have orthotropic character-istics in principal material directions, it is called a *generally orthotropic lamina* because it can be represented by the stress-strain relations in Eq. (2.79).

The only advantage associated with generally orthotropic laminae as opposed to anisotropic laminae is that generally orthotropic laminae are easier to characterize experimentally. However, if we do not realize that principal material axes exist, then a generally orthotropic lamina is indistinguishable from an anisotropic lamina. That is, we cannot take away the inherent orthotropic character of a lamina, but we can orient the lamina in such a manner as to make that character difficult to recognize.

As an alternative to the foregoing procedure, we can express the strains in terms of the stresses in body coordinates by either (1) inversion of the stress-strain relations in Eq. (2.79) or (2) transformation of the strain-stress relations in principal material directions from Eq. (2.57),

$$
\left\{\begin{array}{c} \epsilon_1 \\ \epsilon_2 \\ \gamma_{12} \end{array}\right\} = \left[\begin{array}{ccc} S_{11} & S_{12} & 0 \\ S_{12} & S_{22} & 0 \\ 0 & 0 & S_{66} \end{array}\right] \left\{\begin{array}{c} \sigma_1 \\ \sigma_2 \\ \tau_{12} \end{array}\right\}
$$

(2.81)

to body coordinates. We choose the second approach and apply the transformations of Eqs. (2.69) and (2.70) along with Reuter's matrix, Eq. (2.72), to obtain

$$\begin{Bmatrix} \epsilon_x \\ \epsilon_y \\ \gamma_{xy} \end{Bmatrix} = [T]^T [S] [T] \begin{Bmatrix} \sigma_x \\ \sigma_y \\ \tau_{xy} \end{Bmatrix} = \begin{bmatrix} \bar{S}_{11} & \bar{S}_{12} & \bar{S}_{16} \\ \bar{S}_{12} & \bar{S}_{22} & \bar{S}_{26} \\ \bar{S}_{16} & \bar{S}_{26} & \bar{S}_{66} \end{bmatrix} \begin{Bmatrix} \sigma_x \\ \sigma_y \\ \tau_{xy} \end{Bmatrix} \tag{2.82}$$

where $[R] [T]^{-1} [R]^{-1}$ was found to be $[T]^T$ and

$$\bar{S}_{11} = S_{11} \cos^4\theta + (2S_{12} + S_{66}) \sin^2\theta \cos^2\theta + S_{22} \sin^4\theta$$

$$\bar{S}_{12} = S_{12} (\sin^4\theta + \cos^4\theta) + (S_{11} + S_{22} - S_{66}) \sin^2\theta \cos^2\theta$$

$$\bar{S}_{22} = S_{11} \sin^4\theta + (2S_{12} + S_{66}) \sin^2\theta \cos^2\theta + S_{22} \cos^4\theta \tag{2.83}$$

$$\bar{S}_{16} = (2S_{11} - 2S_{12} - S_{66}) \sin\theta \cos^3\theta - (2S_{22} - 2S_{12} - S_{66}) \sin^3\theta \cos\theta$$

$$\bar{S}_{26} = (2S_{11} - 2S_{12} - S_{66}) \sin^3\theta \cos\theta - (2S_{22} - 2S_{12} - S_{66}) \sin\theta \cos^3\theta$$

$$\bar{S}_{66} = 2(2S_{11} + 2S_{22} - 4S_{12} - S_{66}) \sin^2\theta \cos^2\theta + S_{66} (\sin^4\theta + \cos^4\theta)$$

Recall that the S_{ij} are defined in terms of the engineering constants in Eq. (2.58).

Because of the presence of $\bar{Q}_{16}$ and $\bar{Q}_{26}$ in Eq. (2.79) and of S_{16} and S_{26} in Eq. (2.82), the solution of problems involving so-called generally orthotropic laminae is more difficult than problems with so-called specially orthotropic laminae. As a matter of fact, there is no difference between solutions for generally orthotropic laminae and those for anisotropic laminae whose stress-strain relations, under conditions of plane stress, can be written as

$$\begin{Bmatrix} \sigma_1 \\ \sigma_2 \\ \tau_{12} \end{Bmatrix} = \begin{bmatrix} Q_{11} & Q_{12} & Q_{16} \\ Q_{12} & Q_{22} & Q_{26} \\ Q_{16} & Q_{26} & Q_{66} \end{bmatrix} \begin{Bmatrix} \epsilon_1 \\ \epsilon_2 \\ \gamma_{12} \end{Bmatrix} \tag{2.84}$$

or in inverted form as

$$\begin{Bmatrix} \epsilon_1 \\ \epsilon_2 \\ \gamma_{12} \end{Bmatrix} = \begin{bmatrix} S_{11} & S_{12} & S_{16} \\ S_{12} & S_{22} & S_{26} \\ S_{16} & S_{26} & S_{66} \end{bmatrix} \begin{Bmatrix} \sigma_1 \\ \sigma_2 \\ \tau_{12} \end{Bmatrix} \tag{2.85}$$

where, in terms of the engineering constants, the anisotropic compliances are

$$S_{11} = \frac{1}{E_1}$$

$$S_{12} = - \frac{\nu_{12}}{E_1} = - \frac{\nu_{21}}{E_2}$$

$$S_{22} = \frac{1}{E_2} \tag{2.86}$$

$$S_{66} = \frac{1}{G_{12}}$$

$$S_{16} = \frac{\eta_{12,1}}{E_1} = \frac{\eta_{1,12}}{G_{12}}$$

(2.86)
(cont'd.)

$$S_{26} = \frac{\eta_{12,2}}{E_2} = \frac{\eta_{2,12}}{G_{12}}$$

Note that some new engineering constants have been used. The new constants are called *coefficients of mutual influence* by Lekhnitski (Ref. 2-5) and are defined as

$\eta_{i,ij}$ = coefficient of mutual influence of the first kind which characterizes stretching in the *i*-direction caused by shear in the *ij*-plane, that is,

$$\eta_{i,ij} = \frac{\epsilon_i}{\gamma_{ij}}$$

(2.87)

for $\tau_{ij} = \tau$ and all other stresses are zero.

$\eta_{ij,i}$ = coefficient of mutual influence of the second kind which characterizes shearing in the *ij*-plane caused by a normal stress in the *i*-direction, that is,

$$\eta_{ij,i} = \frac{\gamma_{ij}}{\epsilon_i}$$

(2.88)

for $\sigma_i = \sigma$ and all other stresses are zero.

Lekhnitski defines the coefficients of mutual influence and the Poisson's ratios with subscripts that are reversed from the present notation.

Other anisotropic elasticity relations are used to define *Chentsov coefficients* which are to shearing stresses and shearing strains what Poisson's ratios are to normal stresses and normal strains. However, the Chentsov coefficients do not affect the in-plane behavior of laminae under plane stress since the coefficients are related to S_{45}, S_{46}, S_{56} in Eq. (2.18).

The Chentsov coefficients are defined as

$\mu_{ij,kl}$ = Chentsov coefficient which characterizes the shearing strain in the *ij*-plane due to shearing stress in the *kl*-plane, that is,

$$\mu_{ij,kl} = \frac{\gamma_{ij}}{\gamma_{kl}}$$

(2.89)

for $\tau_{kl} = \tau$ and all other stresses are zero.

The Chentsov coefficients are subject to the reciprocal relations

$$\frac{\mu_{ij,kl}}{G_{kl}} = \frac{\mu_{kl,ij}}{G_{ij}}$$

(2.90)

Thus, the out-of-plane shearing strains of a lamina due to in-plane shearing stress and normal stresses are

$$\gamma_{13} = \frac{\eta_{1,13}\sigma_1 + \eta_{2,13}\sigma_2 + \mu_{12,13}\tau_{12}}{G_{13}}$$

$$\gamma_{23} = \frac{\eta_{1,23}\sigma_1 + \eta_{2,23}\sigma_2 + \mu_{12,23}\tau_{12}}{G_{23}}$$

$$(2.91)$$

wherein both the Chentsov coefficients and the coefficients of mutual influence of the first kind are required. Note that neither of these shear strains arise in an orthotropic material unless it is stressed in directions other than the principal material directions. In such cases, the Chentsov coefficients and the coefficients of mutual influence would be obtained from the transformed compliances as in the following paragraph.

Compare the transformed orthotropic compliances in Eq. (2.83) with the anisotropic compliances in terms of engineering constants in Eq. (2.86). It is obvious from the comparison that an apparent coefficient of mutual influence results when an orthotropic lamina is stressed in nonprincipal material directions. Redesignate the coordinates 1 and 2 in Eq. (2.85) as x and y since, by definition, an anisotropic material has *no* principal material directions. Then, substitute the redesignated S_{ij} from Eq. (2.86) in Eq. (2.83) along with the orthotropic compliances in Eq. (2.58). Finally, the apparent engineering constants for an orthotropic lamina that is stressed in nonprincipal xy coordinates are:

$$\frac{1}{E_x} = \frac{1}{E_1}\cos^4\theta + \left(\frac{1}{G_{12}} - \frac{2\nu_{12}}{E_1}\right)\sin^2\theta\,\cos^2\theta + \frac{1}{E_2}\sin^4\theta$$

$$\nu_{xy} = E_x\left[\frac{\nu_{12}}{E_1}(\sin^4\theta + \cos^4\theta) - \left(\frac{1}{E_1} + \frac{1}{E_2} - \frac{1}{G_{12}}\right)\sin^2\theta\,\cos^2\theta\right]$$

$$\frac{1}{E_y} = \frac{1}{E_1}\sin^4\theta + \left(\frac{1}{G_{12}} - \frac{2\nu_{12}}{E_1}\right)\sin^2\theta\,\cos^2\theta + \frac{1}{E_2}\cos^4\theta$$

$$\frac{1}{G_{xy}} = 2\left(\frac{2}{E_1} + \frac{2}{E_2} + \frac{4\nu_{12}}{E_1} - \frac{1}{G_{12}}\right)\sin^2\theta\,\cos^2\theta + \frac{1}{G_{12}}(\sin^4\theta + \cos^4\theta)$$

$$\eta_{xy,x} = E_x\left[\left(\frac{2}{E_1} + \frac{2\nu_{12}}{E_1} - \frac{1}{G_{12}}\right)\sin\theta\,\cos^3\theta - \left(\frac{2}{E_2} + \frac{2\nu_{12}}{E_1} - \frac{1}{G_{12}}\right)\sin^3\theta\,\cos\theta\right]$$

$$\eta_{xy,y} = E_y\left[\left(\frac{2}{E_1} + \frac{2\nu_{12}}{E_1} - \frac{1}{G_{12}}\right)\sin^3\theta\,\cos\theta - \left(\frac{2}{E_2} + \frac{2\nu_{12}}{E_1} - \frac{1}{G_{12}}\right)\sin\theta\,\cos^3\theta\right]$$

$$(2.92)$$

An important implication of the presence of the coefficient of mutual influence is that off-axis (nonprincipal material direction) tensile tests for composites

result in shear deformation as well as axial extension. This subject is investigated
further in Sec. 2.8. At this point, recognize that Eq. (2.92) is a quantification of
the foregoing implication for tensile tests and of the qualitative observations
made in Sec. 1.2.

The apparent anisotropic moduli for an orthotropic lamina stressed at an
angle θ to the principal material directions vary with θ as in Eq. (2.92). To gain
an appreciation for how the moduli vary, they can be plotted for specific
materials. Values typical of a glass/epoxy composite are plotted from Eq. (2.92)
in Fig. 2-5. Similarly, values for a boron/epoxy composite are plotted in Fig. 2-6.
In both figures, E_x is divided by E_2 and G_{xy} is divided by G_{12}. This
normalization is done to permit a convenient comparison of most of the moduli
in a single figure. Note in both figures that G_{xy} is largest at $\theta = 45°$. The
coefficient of mutual influence $\eta_{xy,x}$ is, of course, zero at $\theta = 0°$ and $\theta = 90°$,
but achieves large values compared to ν_{xy} for intermediate angles. The modulus
E_y behaves essentially like E_x, except E_y is, of course, small for θ near $0°$ and
large when θ is near $90°$. Similar comments could be made for ν_{yx} and $\eta_{xy,y}$.

The values in Figs. 2-5 and 2-6 are not entirely typical of all composite
materials. For example, follow the hints in Exercise 2.6.7 to demonstrate that
E_x can actually exceed both E_1 and E_2 for some orthotropic laminae. Similarly,
E_x can be shown to be smaller than both E_1 and E_2 (note that for boron/

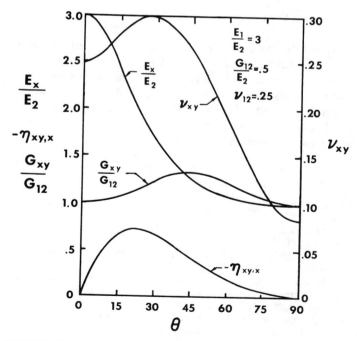

FIG. 2-5. Normalized moduli for glass/epoxy.

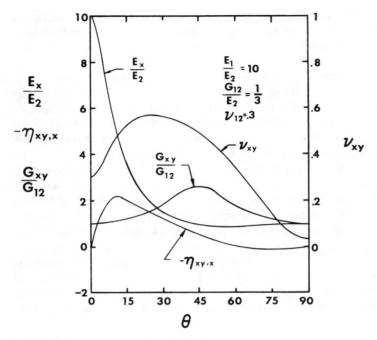

FIG. 2-6. Normalized moduli for boron/epoxy.

epoxy in Fig. 2-6 E_x is slightly smaller than E_2 in the neighborhood of $\theta = 60°$). These results were summarized by Jones (Ref. 2-6) as a simple theorem: the extremum (largest and smallest) material properties do not necessarily occur in principal material directions. The moduli G_{xy}, ν_{xy}, and $\eta_{xy,x}$ can exhibit similar peculiarities within the scope of Eq. (2.92). Nothing should, therefore, be taken for granted with a new composite material: its moduli as a function of θ should be examined to truly understand its character.

Problem Set 2.6

Exercise 2.6.1 Derive Eq. (2.77).

Exercise 2.6.2 Prove $[R] [T] [R]^{-1} = [T]^{-T}$.

Exercise 2.6.3 Derive Eq. (2.82).

Exercise 2.6.4 Prove $[R] [T]^{-1} [R]^{-1} = [T]^{T}$.

Exercise 2.6.5 Derive Eq. (2.92) starting from Eq. (2.83) supplemented by the redesignated Eq. (2.85) as well as Eqs. (2.86) and (2.58).

Exercise 2.6.6 Plot E_x, E_y, G_{xy}, ν_{xy}, $\eta_{xy,x}$ and $\eta_{xy,y}$ as functions of θ from $\theta = 0°$ to $\theta = 90°$ for high modulus graphite/epoxy, an orthotropic material with $E_1 = 30 \times 10^6$ psi, $E_2 = .75 \times 10^6$ psi, $G_{12} = .375 \times 10^6$ psi and $\nu_{12} = .25$.

Exercise 2.6.7 Show that the apparent direct modulus of an orthotropic material as a function of θ [the first equation in Eqs. (2.92)] can be written as

$$\frac{E_1}{E_x} = (1 + a - 4b)\cos^4\theta + (4b - 2a)\cos^2\theta + a$$

where $a = E_1/E_2$ and $b = \frac{1}{4}(E_1/G_{12} - 2\nu_{12})$. Use the derivatives of E_x to find its maxima and minima in the manner of Appendix B. Hence, show that E_x is greater than both E_1 and E_2 for some values of θ if

$$G_{12} > \frac{E_1}{2(1 + \nu_{12})}$$

and that E_x is less than both E_1 and E_2 for some values of θ if

$$G_{12} < \frac{E_1}{2[(E_1/E_2) + \nu_{12}]}$$

That is, show that an orthotropic material can have an apparent Young's modulus that either exceeds or is less than the Young's moduli in both principal material directions. Plot E_x/E_1 for some contrived materials that exemplify the above relations.

2.7 INVARIANT PROPERTIES OF AN ORTHOTROPIC LAMINA

The transformed reduced stiffnesses in Eq. (2.80) are obviously very complicated functions of the four independent material properties E_1, E_2, ν_{12}, and G_{12} as well as the angle of rotation, θ. To understand the physical implications of the various rotations that occur in actual laminates would require considerable practical experience. Matching up the highest E of E_1 and E_2 with the laminate direction requiring the highest stiffness is easy. However, if the design situation includes requirements for various stiffnesses in several directions, then we must have a rationale for deciding the orientation of the lamina that make up a laminate. Obviously, then, we must understand how an individual lamina changes stiffness as it is reoriented at different angles to the reference direction. However, the present form of the transformation relations is not conducive to understanding their physical significance.

Tsai and Pagano (Ref. 2-7) accomplished an ingenious recasting of the stiffness transformation equations that enables ready understanding of the consequences of rotating a lamina in a laminate. By the use of various trigonometric identities, the transformed reduced stiffnesses, Eq. (2.80), can be written as

$$\bar{Q}_{11} = U_1 + U_2 \cos 2\theta + U_3 \cos 4\theta$$

$$\bar{Q}_{12} = U_4 - U_3 \cos 4\theta$$

$$\bar{Q}_{22} = U_1 - U_2 \cos 2\theta + U_3 \cos 4\theta$$

$$\bar{Q}_{16} = -\frac{1}{2}U_2 \sin 2\theta - U_3 \sin 4\theta \qquad\qquad (2.93)$$

$$\bar{Q}_{26} = -\frac{1}{2}U_2 \sin 2\theta + U_3 \sin 4\theta$$

$$\bar{Q}_{66} = U_5 - U_3 \cos 4\theta$$

in which

$$U_1 = \frac{3Q_{11} + 3Q_{22} + 2Q_{12} + 4Q_{66}}{8} \qquad\qquad (2.94)$$

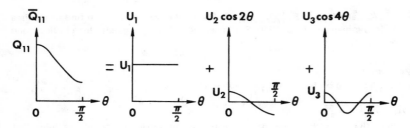

FIG. 2-7. Decomposition of $\bar{Q}_{11}$ into components.

$$U_2 = \frac{Q_{11} - Q_{22}}{2}$$

$$U_3 = \frac{Q_{11} + Q_{22} - 2Q_{12} - 4Q_{66}}{8}$$

$$U_4 = \frac{Q_{11} + Q_{22} + 6Q_{12} - 4Q_{66}}{8}$$

$$U_5 = \frac{Q_{11} + Q_{22} - 2Q_{12} + 4Q_{66}}{8}$$

(2.94)
(cont'd.)

The advantage of writing the transformation equations in the form of Eq. (2.93) is that the parts of $\bar{Q}_{11}$, $\bar{Q}_{12}$, $\bar{Q}_{22}$, and $\bar{Q}_{66}$ are then obviously invariant under rotations about the z axis (perpendicular to the lamina). This concept of invariance is useful when examining the prospect of orienting a lamina at various angles to achieve a certain stiffness profile. For example,

$$\bar{Q}_{11} = U_1 + U_2 \cos 2\theta + U_3 \cos 4\theta \tag{2.95}$$

can be decomposed into its components in the graphical manner of Fig. 2-7. There, we see that the value of $\bar{Q}_{11}$ is determined by a fixed constant, U_1, plus a quantity of low frequency variation with θ plus another quantity of higher frequency variation with θ. Thus, U_1 is an effective measure of lamina stiffness in a design application since it is not affected by orientation. The concept of

TABLE 2-2. Transformation equations for $\bar{Q}_{ij}$ and $\bar{Q}'_{ij}$*

	Constant	cos 2θ	sin 2θ	cos 4θ	sin 4θ
$\bar{Q}'_{11}$	U_1	U_2	$2U_6$	U_3	U_7
$\bar{Q}'_{22}$	U_1	$-U_2$	$-2U_6$	U_3	U_7
$\bar{Q}'_{12}$	U_4	0	0	$-U_3$	$-U_7$
$\bar{Q}'_{66}$	U_5	0	0	$-U_3$	$-U_7$
$2\bar{Q}'_{16}$	0	$2U_6$	$-U_2$	$2U_7$	$-2U_3$
$2\bar{Q}'_{26}$	0	$2U_6$	$-U_2$	$-2U_7$	$2U_3$

*$\bar{Q}'_{ij}$ are for anisotropic materials. The $\bar{Q}_{ij}$ for orthotropic materials are obtained by deleting U_6 and U_7 from the definitions of $\bar{Q}'_{ij}$.

invariance will be more useful in the study of laminates because laminates are made of laminae at various orientations to achieve a certain stiffness. Such tailoring of the material and structural configuration, however, comes at the expense of capabilities in other directions. For example, from observation of the variable nature of $\overline{Q}_{11}$, apparently to try to meet a required stiffness in one direction leads to a lower stiffness in some other direction unless the requirement is as low as Q_{22} in this example.

Similar invariance concepts for anisotropic materials were also developed by Tsai and Pagano (Ref. 2-7). For anisotropy, the following definitions

$$U_6 = \frac{Q_{16} + Q_{26}}{2}$$

$$U_7 = \frac{Q_{16} - Q_{26}}{2}$$

(2.96)

must be appended to Eq. (2.94) and worked into Eq. (2.93) in such a manner that Table 2-2 results from which the transformation equations can be written.

The actual invariants in "invariant properties of a lamina" include not only U_1, U_4, and U_5 since they are the constant terms in Eq. (2.93) but functions related to U_1, U_4, and U_5 as shown in the exercises. Utilization of invariance concepts will be deferred until after the development of lamination concepts in Chap. 4.

Problem Set 2.7

Exercise 2.7.1 Show that $Q_{11} + Q_{22} + 2Q_{12}$ *is invariant under rotation about the z axis, that is, that*

$$\overline{Q}_{11} + \overline{Q}_{22} + 2\overline{Q}_{12} = Q_{11} + Q_{22} + 2Q_{12}$$

Exercise 2.7.2 Show that $Q_{66} - Q_{12}$ *is invariant under rotation about the z axis, that is, that*

$$\overline{Q}_{66} - \overline{Q}_{12} = Q_{66} - Q_{12}$$

Exercise 2.7.3 Show that $U_5 = (U_1 - U_4)/2$, *that is, that the quantities* U_1, U_4, *and* U_5 *are related and only two are independent since one can be expressed in terms of the other two.*

2.8 STRENGTH OF AN ORTHOTROPIC LAMINA

2.8.1 Strength Concepts

The strength characteristics of an orthotropic lamina are just as important a building block in the description of laminates as the stiffness characteristics. Since it is physically impossible to obtain the strength characteristics of a lamina at all possible orientations, a means must be determined of obtaining the characteristics at any orientation in terms of characteristics in the principal material directions. In such an extension of the information obtained in principal material directions, the well-known concepts of principal stresses and principal strains are of no value. As review, the central issue here is that principal stresses and strains are the largest values irrespective of direction or orientation; however, direction of stress or strain has absolutely no significance for isotropic materials by definition. Because of orthotropy, the axes of principal stress do not coincide with the axes of principal strain. Moreover, because the strength is lower in one direction than another, the highest stress may not be the stress governing the design. A rational comparison of the actual stress field with the allowable stress field is required.

What has been accomplished in preceding sections on stiffness relationships serves as the basis for determination of the actual stress field; what remains is the definition of the allowable stress field. The first step in such a definition is the establishment of allowable stresses or strengths in the principal material directions. Such information is basic to the study of strength of an orthotropic lamina.

For a lamina stressed in its own plane, there are three fundamental strengths if the lamina has equal strengths in tension and compression:

X = axial or longitudinal strength

Y = transverse strength

S = shear strength

(The units are force/area, that is, allowable stresses). The directions of each of these strengths are shown in Fig. 2-8; obviously, the strengths result from independent application of the respective stresses, σ_1, σ_2, τ_{12} .

That the principal stresses are not of interest in determining the strength of an orthotropic lamina is illustrated by the following example. Consider the lamina with unidirectional fibers shown in Fig. 2-8. Say that the hypothetical strengths of the lamina in the 1-2 plane are:

X = 50,000 psi

Y =   1,000 psi

S =   2,000 psi

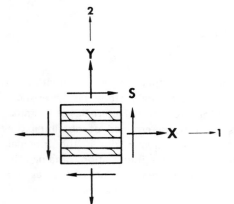

FIG. 2-8. Definitions of fundamental strengths for a unidirectionally reinforced lamina.

The stiffness would also be high in the 1-direction and low in the 2-direction as is easily imagined on the basis of the fiber orientation. Imagine that, in the 1-2 plane, the stresses are:

$$\sigma_1 = 45,000 \text{ psi}$$
$$\sigma_2 = 2,000 \text{ psi}$$
$$\tau_{12} = 1,000 \text{ psi}$$

Then obviously the maximum principal stress is lower than the largest strength. However, σ_2 is greater than Y so the lamina must fail under the imposed stresses. The key observation is that strength is a function of *orientation* of stresses in an orthotropic lamina. In contrast, for an isotropic material, strength is independent of body orientation relative to the imposed stresses.

If the material has unequal properties in tension and compression as do most composite materials, then the following strengths are required:

X_t = axial or longitudinal strength in tension

X_c = axial or longitudinal strength in compression

Y_t = transverse strength in tension

Y_c = transverse strength in compression

S = shear strength

Remember that the preceding strengths must be defined in principal material directions.

The shear strength in the principal material directions is seen to be independent of differences in tensile and compressive behavior as it must be by definition of a pure shear stress. That is, the shear stress whether "positive" or "negative" has the same maximum value for materials that exhibit different behavior in tension than in compression. This statement is rationalized by observation of Fig. 2-9 wherein positive and negative shear stresses are applied to a unidirectionally reinforced lamina. The convention of which shear stress is positive is consistent with Pagano and Chou's convention (Ref. 2-8). Note in Fig. 2-9 that there is no difference between the stress fields labeled positive and negative shear stress. The two stress fields are mirror images of each other, even when the principal stresses are examined as in the lower half of Fig. 2-9. Thus, the maximum value of shear stress is the same in both cases.

However, the maximum value of shear stress in other than principal material directions depends on the sign of the shear stress. For example, at 45° to the principal material axes, positive and negative shear stresses result in normal stresses of opposite signs on the fibers as in Fig. 2-10. There, for positive shear stress, tensile stresses result in the fiber direction, and compressive stresses arise perpendicular to the fibers. For negative shear stress, compressive stresses exist in the fiber direction and tensile stresses transverse to the fibers. However, both the normal strengths and normal stiffnesses for the material are different under tension than under compression. Thus, the apparent shear strengths and

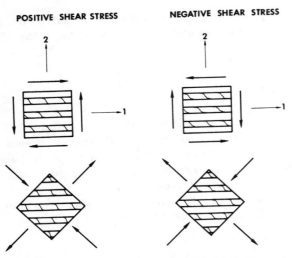

FIG. 2-9. Shear stress in principal material directions.

shear stiffnesses are different for positive and negative shear stresses applied at 45° to the principal material directions. This rationale can readily be extended from the simple unidirectionally reinforced lamina to woven materials.

The foregoing example is but one of the difficulties encountered in analysis of orthotropic materials with different properties in tension and compression. Moreover, the example is included to illustrate how basic information in the principal material directions can be transformed to other useful coordinate directions depending on the stress field under considerations. Such transforma-

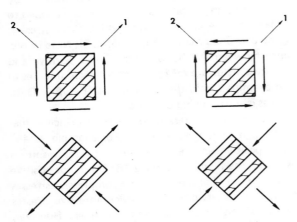

FIG. 2-10. Shear stress at 45° to principal material directions.

tions are simply indications that the basic information, whether strengths or stiffnesses, is in tensor form and therefore is subject to the usual rules governing tensor transformations given in Appendix A.

The topic of materials with different strengths and stiffnesses in tension than in compression will not be covered further in much depth (except to report different strengths) because research on such materials is still in its infancy. However, the topic is very important for the general class of composite materials, if not fiber-reinforced laminated composites. Ambartsumyan and his associates first reported research on this topic in 1965 (see, e.g., Ref. 2-9 and the list of references in Ref. 2-10). A few Americans have also investigated some aspects of the mechanics of these materials (see Jones, Ref. 2-10).

2.8.2 Experimental Determination of Strength and Stiffness

For orthotropic materials with equal properties in tension and compression, certain basic experiments can be performed to obtain the properties in the principal material directions. The experiments, if conducted properly, generally reveal both the strength and stiffness characteristics of the material. To reiterate, the stiffness characteristics are

$E_1 = $ Young's modulus in the 1-direction

$E_2 = $ Young's modulus in the 2-direction

$\nu_{12} = -\dfrac{\epsilon_2}{\epsilon_1}$ for $\sigma_1 = \sigma$ and all other stresses are zero

$\nu_{21} = -\dfrac{\epsilon_1}{\epsilon_2}$ for $\sigma_2 = \sigma$ and all other stresses are zero

$G_{12} = $ shear modulus in the 1-2 plane

where only three of E_1, E_2, ν_{12}, ν_{21} are independent and the strength characteristics are

$X = $ axial or longitudinal strength (1-direction)

$Y = $ transverse strength (2-direction)

$S = $ shear strength (1-2 plane)

Several experiments will now be described from which the foregoing basic stiffness and strength information can be obtained. The basic tenet of the experiments is that the stress-strain behavior of the materials is linear from zero load to the ultimate or fracture load. Such linear behavior is typical for glass/epoxy composites and is quite reasonable for boron/epoxy composites except for the shear behavior which is very nonlinear to fracture. This characteristic of linear elastic behavior to fracture is quite similar to the analysis of bodies that exhibit linear elastic behavior up until the onset of plasticity. Thus, certain concepts of the theory of plasticity such as yield functions are useful analogues for the strength theories that will be discussed later.

A key element in the experimental determination of the stiffness and strength characteristics of a lamina is the imposition of a uniform stress state in the specimen. Such loading is relatively easy for isotropic materials. However, for composite materials, the orthotropy often introduces coupling between

(1)   normal stresses and shear strains
(2)   shear stresses and normal strains
(3)   normal stresses and bending curvatures
(4)   bending stresses and normal strains

etc. when loaded in nonprincipal material directions for which the stress-strain relations are given in Eq. (2.82). Thus, special care must be taken to assure obtaining the desired information. This care typifies the knowledge required to treat composites.

First, consider a uniaxial tension test in the 1-direction on a flat piece of unidirectionally reinforced lamina as shown in Fig. 2-11. The strains ϵ_1 and ϵ_2 are measured in the test whereupon, by definition,

$$\sigma_1 = \frac{P}{A}$$

$$E_1 = \frac{\sigma_1}{\epsilon_1}$$

$$\nu_{12} = -\frac{\epsilon_2}{\epsilon_1} \tag{2.97}$$

$$X = \frac{P_{ult}}{A}$$

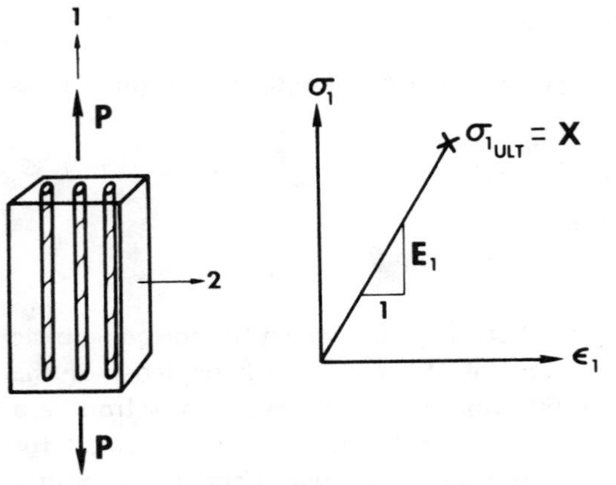

FIG. 2-11. Uniaxial loading in the 1-direction.

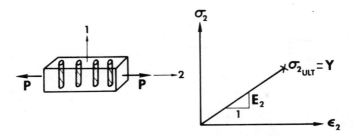

FIG. 2-12. Uniaxial loading in the 2-direction.

where A is the cross-sectional area of the specimen end perpendicular to the applied load.

Second, consider a uniaxial tension test in the 2-direction on a flat piece of unidirectionally reinforced lamina as in Fig. 2-12. As in the first experiment, ϵ_1 and ϵ_2 are measured so

$$\sigma_2 = \frac{P}{A}$$

$$E_2 = \frac{\sigma_2}{\epsilon_2}$$ (2.98)

$$\nu_{21} = -\frac{\epsilon_1}{\epsilon_2}$$

$$Y = \frac{P_{ult}}{A}$$

where again A is the cross-sectional area of the specimen end.

At this point, the stiffness properties should satisfy the reciprocal relations

$$\frac{\nu_{12}}{E_1} = \frac{\nu_{21}}{E_2}$$ (2.99)

or else one of three possibilities exists:

    (1)   The data were measured incorrectly.
    (2)   The calculations were performed incorrectly.
    (3)   The material cannot be described by linear elastic stress-strain relations.

Third, consider a uniaxial tension test at $45°$ to the 1-direction on a flat piece of lamina as shown in Fig. 2-13. By measurement of ϵ_x alone, obviously

$$E_x = \frac{P/A}{\epsilon_x}$$ (2.100)

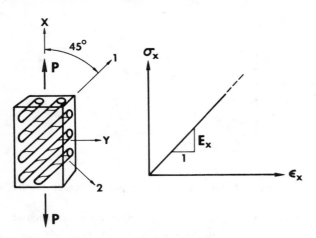

FIG. 2-13. Uniaxial loading at 45° to the 1-direction.

Then, by use of the transformation relations in Eq. (2.92)

$$\frac{1}{E_x} = \frac{1}{4}\left(\frac{1}{E_1} - \frac{2\nu_{12}}{E_1} + \frac{1}{G_{12}} + \frac{1}{E_2}\right) \tag{2.101}$$

wherein G_{12} is the only unknown. Thus,

$$G_{12} = \frac{1}{\left(\dfrac{4}{E_x} - \dfrac{1}{E_1} - \dfrac{1}{E_2} + \dfrac{2\nu_{12}}{E_1}\right)} \tag{2.102}$$

A relationship such as Eq. (2.101) does not exist for strengths since strengths do not necessarily transform like stiffnesses. Thus, this experiment cannot be relied upon to determine S, the ultimate shear stress since a pure shear deformation mode has not been excited with accompanying failure in shear. Accordingly, other approaches to obtaining S must be used.

Before turning to the other approaches to the determination of the shear strength, however, it is appropriate to comment on the ease of performing this third test. From Eq. (2.82), it is apparent that because of the presence of $\bar{S}_{16}$ there is coupling between the normal stress σ_x and shear strain γ_{xy}. Thus, although just a force P is indicated in Fig. 2-13, the experiment cannot be properly conducted unless the force is applied uniformly across the end and, in addition, the ends of the lamina are free to deform in the manner shown on the left of Fig. 2-14. Otherwise, if for example the end edges of the lamina were clamped in a testing machine and a resultant force P were applied, then the lamina would be restrained from shearing deformation so it would twist in the fashion shown on the right in Fig. 2-14 (Ref. 2-11). In the center of such a specimen, if it is long enough as compared to its width, the deformation is similar to the shearing and extension of the unrestrained lamina in Fig. 2-14. That is, away from St. Venant end effects, the type of test doesn't matter.

However, normally we don't choose to use enough material to have a useful gage section.

An additional characteristic of the off-axis test displayed in Figs. 2-13 and 2-14 is that the modulus E_x is not actually measured. Instead, the transformed reduced stiffness $\overline{Q}_{11}$ is measured unless the specimen has a high length to width ratio. The reason for this discrepancy is that the geometrically admissible state of strain in the specimen depends strongly on the geometry. If the specimen is long and slender, the boundary conditions at the specimen end grips are of no consequence à la St. Venant. Accordingly, a pure uniaxial strain is obtained and

$$\sigma_x = E_x \epsilon_x \qquad (2.103)$$

However, for a short fat specimen, the end restraint of $\sigma_x \neq 0$, $\epsilon_y = \gamma_{xy} = 0$ leads to a stress-strain relation

$$\sigma_x = \overline{Q}_{11} \epsilon_x \qquad (2.104)$$

that is consistent with Eq. (2.79). The reader should verify Eqs. (2.103) and (2.104) by imposing the stated conditions and deriving the relation for σ_x. That the difference between E_x in Eq. (2.103) and $\overline{Q}_{11}$ in Eq. (2.104) is significant is best illustrated by Fig. 2-15 for graphite/epoxy specimens. There, for an off-axis test at $30°$ to the fiber direction, the value of $\overline{Q}_{11}$ is 10.4 times as great as E_x. Similar differences exist for $\overline{Q}_{66}$ versus G_{xy}. For materials with lower values of E_1/E_2, the difference between $\overline{Q}_{11}$ and E_x is smaller. The practical significance of the difference between $\overline{Q}_{11}$ and E_x is that the length-to-width ratio of off-axis specimens must be large enough to assure that we are measuring E_x and not $\overline{Q}_{11}$.

The last test to be discussed actually consists of a collection of tests to determine the shear modulus and strength. Several tests are discussed because

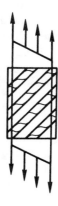

NO END EFFECT   **RESTRAINED ENDS**

FIG. 2-14. Deformation of a unidirectionally reinforced lamina loaded at $45°$ to the fibers.

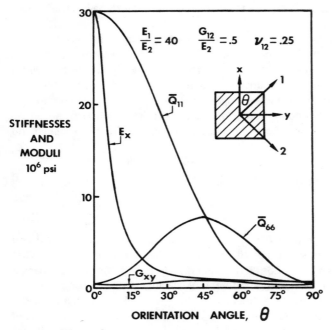

FIG. 2-15. Stiffnesses $\bar{Q}_{11}$ and $\bar{Q}_{66}$ versus moduli E_x and G_{xy}.

each has its faults as will be seen and because, to some extent, there has been no universal agreement on the best test for shear properties.

The torsion-tube test as described by Whitney, Pagano, and Pipes (Ref. 2-12) is depicted schematically in Fig. 2-16. There, a thin circular tube is subjected to a torque, T at the ends. The tube is made of multiple laminae with their fibers aligned either all parallel to the tube axis or all circumferentially. Reasonable assurance of a constant stress state through the tube thickness exists if the tube is only a few laminae thick. However, then serious end grip difficulties can arise because of the flimsy nature of the tube. Usually, the thickness of the tube ends must be built up by bonding on additional layers in order to introduce the load such that failure occurs in the central uniformly stressed portion of the

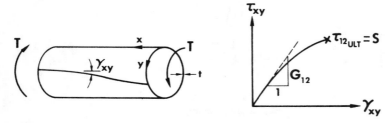

FIG. 2-16. Torsion tube test.

tube. Torsion tubes are expensive to fabricate and require relatively sophisticated instrumentation. If the shearing strain γ_{12} is measured under imposed shear stress τ_{12}, then

$$\tau_{12} = \frac{T}{2\pi r^2 t} \tag{2.105}$$

$$S = \tau_{12ult} = \frac{T_{ult}}{2\pi r^2 t} \tag{2.106}$$

Also, the shear modulus is

$$G_{12} = \frac{\tau_{12}}{\gamma_{12}} \tag{2.107}$$

for the linear portion of the stress-strain curve. However, a typical shear stress-shear strain curve is quite nonlinear as shown in Fig. 2-16. Accordingly, the whole stress-strain curve instead of the initial "elastic" modulus should be used in practical analyses as done by Hahn and Tsai (Ref. 2-13). Nevertheless, most composites analyses are performed with the initial elastic modulus from Eq. (2.107).

Another test used to determine the shear modulus and shear strength of a composite is the "cross-beam" test due to Shockey and described in Ref. 2-14. The composite lamina being evaluated is installed as the facing of a sandwich beam whose core elastic modulus is about two orders of magnitude less than that of the lamina. A cross-shaped configuration is subjected to the loads shown schematically in Fig. 2-17. A membrane state of stress results which at 45° to the x axis is supposedly a uniform pure shear stress. However, due to stress concentrations at the corners of the cross, a uniform stress state is approached only in the very center of the cross. Failure initiates in the corners of the cross; thus, the cross-beam test is no longer regarded as an adequate measuring tool for shear strength and shear stiffness.

Yet another shear strength and shear stiffness test is the "rail shear" test as described by Whitney, Stansbarger, and Howell (Ref. 2-15). Basically, two pieces of rail are bolted together along two opposite edges of a lamina as shown schematically in Fig. 2-18. One pair protrudes at the top of the laminate and the other pair at the bottom. The assembly is placed between the loading heads of a universal testing machine and compressed. Thus, shear is induced in the lamina. However, the geometry of such a specimen must be carefully selected to account for end effects such as the free edges at the top and bottom of the lamina. These and other effects may lead to strength evaluations that are lower than physical reality. Nevertheless, the rail shear test is widely used in the aerospace industry because it is simple, inexpensive, and can be used for tests at both higher and lower than room temperature.

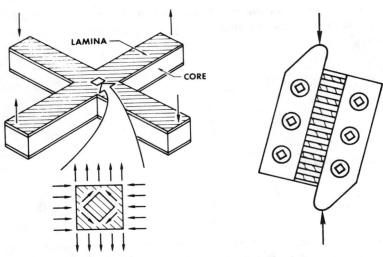

FIG. 2-17. Sandwich cross-beam test.   **FIG. 2-18.** Rail shear test.

2.8.3 Summary of Mechanical Properties

As an illustration of the results of the experiments just described, the mechanical properties for three unidirectionally reinforced composite materials, glass/epoxy, boron/epoxy, and graphite/epoxy, are given in Table 2-3. These values are representative of the strength and elastic stiffness that can be obtained with such materials. Again, recall the essential linearity of the normal stress-normal strain results and the nonlinearity of the shear stress-shear strain results (especially for boron/epoxy and graphite/epoxy). Typical stress-strain curves for these three materials are shown in Appendix C. The specific values will change when the fiber and matrix content of the composite changes. The method of changing the values will be described in Chap. 3 on micromechanics of a lamina. The values in

TABLE 2-3. Typical mechanical properties of some composites

Property	Composite material		
	Glass/epoxy	Boron/epoxy	Graphite/epoxy
E_1	7.8×10^6 psi	30×10^6 psi	30×10^6 psi
E_2	2.6×10^6 psi	3×10^6 psi	$.75 \times 10^6$ psi
ν_{12}	.25	.3	.25
G_{12}	1.3×10^6 psi	1×10^6 psi	$.375 \times 10^6$ psi
X_t	150×10^3 psi	200×10^3 psi	150×10^3 psi
Y_t	4×10^3 psi	12×10^3 psi	6×10^3 psi
S	6×10^3 psi	18×10^3 psi	10×10^3 psi
X_c	150×10^3 psi	400×10^3 psi	100×10^3 psi
Y_c	20×10^3 psi	40×10^3 psi	17×10^3 psi

Table 2-3 will be used in example problems throughout the book, and, as a matter of fact, were already used to obtain Figs. 2-5 and 2-6.

Now that the basic stiffnesses and strengths have been defined for the principal material directions, we can proceed to determine how an orthotropic lamina behaves under biaxial stress states in Sec. 2.9. There, we must combine the information in principal material directions in order to define the stiffness and strength of a lamina at arbitrary orientations under arbitrary biaxial stress states.

2.9 BIAXIAL STRENGTH THEORIES FOR AN ORTHOTROPIC LAMINA

Most experimental determinations of the strength of a material are based on uniaxial stress states. However, the general practical problem involves at least a biaxial if not a triaxial state of stress. Thus, a logical method of using uniaxial strength information in the analysis of multiaxial loading problems is required. Due to a multitude of possible microscopic failure mechanisms, a tensor transformation of strengths is very difficult. Moreover, tensor transformations of strength properties are much more complicated than the tensor transformation of stiffness properties. (The strength tensor is of higher order than the stiffness tensor.) Nevertheless, tensor transformations of strength are performed and used as a phenomenological failure criterion (phenomenological because only the occurrence of failure is predicted, not the mode of failure). A somewhat empirical approach will be adopted — that of comparing actual failure envelopes in stress space with theoretical failure envelopes. The theoretical failure envelopes differ little from the concept of yield surfaces in the theory of plasticity. Both the failure envelopes (or surfaces) and the yield surfaces (or envelopes) represent a termination of linear elastic behavior under a multiaxial stress state. The limits of linear elastic behavior are shown by the symbol x in the stress-strain curves for a failure condition and two types of yielding conditions in Fig. 2-19. Actually, the failure envelopes are not restricted to be the limit of multiaxial linear elastic behavior. The envelopes mask the actual

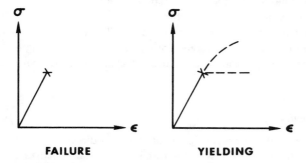

FAILURE   YIELDING

FIG. 2-19. Comparison of failure with yielding.

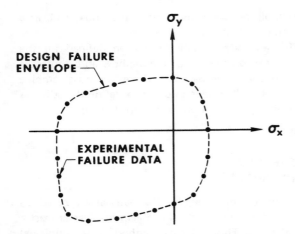

FIG. 2-20. Hypothetical two-dimensional failure data and design curve.

material phenomena that are occurring and merely represent where failure occurs even though other events such as yielding have taken place at lower stress levels. What we are really after is the analytical definition of the failure surface or envelope in stress space. For example, the failure data for a hypothetical material are shown in two dimensions in Fig. 2-20. There, the material has unequal strengths in tension and compression (tension is positive). We must in some manner describe that data by a curve or set of curves (such as shown by the dashed curve in Fig. 2-20) each of which has an equation that is suitable for design use.

Our attention in this section will be restricted to biaxial loading. Some of the biaxial strength theories that have been studied are (1) maximum stress theory, (2) maximum strain theory, (3) Tsai-Hill theory, and (4) Tsai-Wu tensor theory. In all theories, the material, although orthotropic, must be homogeneous. Thus, some of the microscopic failure mechanisms inherently cannot be accounted for. At the same time, the strength theories tend to be smoother than the actual behavior which often exhibits considerable data scatter due to testing technique, manufacturing nonuniformities, etc. The final goal of a theoretical strength envelope that is in agreement with an actual strength envelope would enable the ready designing of structural elements made with composite materials.

2.9.1 Maximum Stress Theory

In the maximum stress theory, the stresses in principal material directions must be less than the respective strengths, otherwise fracture is said to have occurred, that is, for tensile stresses,

$$\sigma_1 < X_t$$
$$\sigma_2 < Y_t \qquad (2.108)$$
$$|\tau_{12}| < S$$

and for compressive stresses

$$\sigma_1 > X_c$$
$$\sigma_2 > Y_c \qquad (2.109)$$

Note that the shear strength is independent of the sign of τ_{12}. If any one of the foregoing inequalities is not satisfied, then the *assumption* is made that the material has failed by the failure mechanism associated with X_t, X_c, Y_t, Y_c, or S, respectively. Note that there is *no interaction* between modes of failure in this criterion — there are actually three subcriteria.

In applications of the maximum stress criterion, the stresses in the body under consideration must be transformed to stresses in the principal material directions. For example, Tsai (Ref. 2-16) considered a unidirectionally reinforced composite subjected to uniaxial load at angle θ to the fibers as shown in Fig. 2-21. The stresses in the principal material directions are obtained by transformation as

$$\sigma_1 = \sigma_x \cos^2\theta$$
$$\sigma_2 = \sigma_x \sin^2\theta$$
$$\tau_{12} = -\sigma_x \sin\theta \cos\theta \qquad (2.110)$$

Then by inversion of Eq. (2.110) and substitution of Eq. (2.108), the maximum uniaxial stress, σ_x, is the smallest of

$$\sigma_x < \frac{X}{\cos^2\theta}$$
$$\sigma_x < \frac{Y}{\sin^2\theta} \qquad (2.111)$$
$$\sigma_x < \frac{S}{\sin\theta \cos\theta}$$

This theory is illustrated in Fig. 2-22 for an E glass/epoxy composite with the properties given in Table 2-3.

The uniaxial strength of the unidirectional composite is plotted in Fig. 2-22 versus the angle θ between the loading direction and the principal material directions. The tension data are denoted by solid circles and the compression data by solid squares. The tension data were obtained by use of dog bone specimens whereas the compression data were obtained by use of specimens with uniform rectangular cross sections. The maximum stress theory is shown as several solid curves, the lowest one of which governs the strength. The theoretical cusps in

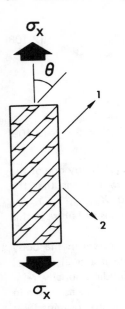

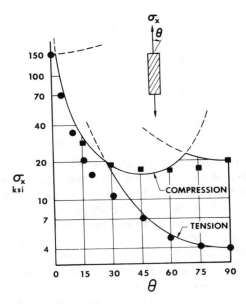

FIG. 2-21. Off-axis
uniaxial loading.

FIG. 2-22. Maximum stress failure theory.
(*After Tsai, Ref. 2-16.*)

the strength variation are not borne out by the experimental data. Moreover, the theoretical strength variation does not adequately represent the experimental strength variation. Thus, another biaxial strength theory must be sought.

2.9.2 Maximum Strain Theory

The maximum strain theory is quite similar to the maximum stress theory. Here, strains are limited rather than stresses. Specifically, the material is said to have failed if one or more of the following inequalities is not satisfied:

$$\epsilon_1 < X_{\epsilon_t}$$
$$\epsilon_2 < Y_{\epsilon_t} \qquad (2.112)$$
$$|\gamma_{12}| < S_\epsilon$$

including for materials with different strength in tension and compression

$$\epsilon_1 > X_{\epsilon_c} \qquad (2.113)$$
$$\epsilon_2 > Y_{\epsilon_c}$$

where

$X_{\epsilon_t}(X_{\epsilon_c})$ = maximum tensile (compressive) normal strain in the 1-direction

$Y_{\epsilon_t}(Y_{\epsilon_c})$ = maximum tensile (compressive) normal strain in the 2-direction

S_ϵ = maximum shear strain in the 1-2 plane

As with the shear strength, the maximum shear strain is unaffected by the sign of the shear stress. The strains in principal material directions, $\epsilon_1, \epsilon_2, \gamma_{12}$, must be found from the strains in body coordinates by transformation before the criterion can be applied.

For a unidirectionally reinforced composite subjected to uniaxial load at angle θ to the fibers (the example problem in the section on maximum stress theory), the allowable stresses can be found from the allowable strains $X_{\epsilon_t}, Y_{\epsilon_t}$, etc. in the following manner.

First, given that the stress-strain relations are

$$\epsilon_1 = \frac{1}{E_1}(\sigma_1 - \nu_{12}\sigma_2)$$

$$\epsilon_2 = \frac{1}{E_2}(\sigma_2 - \nu_{21}\sigma_1)$$

$$\gamma_{12} = \frac{\tau_{12}}{G_{12}}$$

(2.114)

upon substitution of the transformation equations

$$\sigma_1 = \sigma_x \cos^2\theta$$

$$\sigma_2 = \sigma_x \sin^2\theta$$

$$\tau_{12} = -\sigma_x \sin\theta \cos\theta$$

(2.115)

in the stress-strain relations, Eq. (2.114), the strains can be expressed as

$$\epsilon_1 = \frac{1}{E_1}(\cos^2\theta - \nu_{12}\sin^2\theta)\sigma_x$$

$$\epsilon_2 = \frac{1}{E_2}(\sin^2\theta - \nu_{21}\cos^2\theta)\sigma_x$$

$$\gamma_{12} = -\frac{1}{G_{12}}(\sin\theta \cos\theta)\sigma_x$$

(2.116)

Finally, if the usual restriction to linear elastic behavior to failure is made,

$$X_{\epsilon_t} = \frac{X_t}{E_1}$$

$$Y_{\epsilon_t} = \frac{Y_t}{E_2}$$

(2.117)

$$S_\epsilon = \frac{S}{G_{12}}$$

and

$$X_{\epsilon_c} = \frac{X_c}{E_1}$$

(2.118)

$$Y_{\epsilon_c} = \frac{Y_c}{E_2}$$

(which could equally well come from measured values in an experiment), then the maximum strain criterion for this example can be expressed as

$$\sigma_x < \frac{X}{\cos^2\theta - \nu_{12}\sin^2\theta}$$

$$\sigma_x < \frac{Y}{\sin^2\theta - \nu_{21}\cos^2\theta}$$

(2.119)

$$\sigma_x < \frac{S}{\sin\theta \cos\theta}$$

By comparison of the maximum strain criterion, Eq. (2.119), with the maximum stress criterion, Eq. (2.111), it is obvious that the only difference is the inclusion of Poisson's ratio terms in the maximum strain criterion.

As with the maximum stress theory, the maximum strain theory can be plotted against available experimental results for uniaxial loading of an off-angle composite. The discrepancies between experimental results and the theoretical prediction in Fig. 2-23 are similar to, but more pronounced than, those for the maximum stress criterion. Thus, the appropriate failure criterion for this E glass/epoxy composite has not been found.

2.9.3 Tsai-Hill Theory

Hill (Ref. 2-17) proposed a yield criterion for anisotropic materials:

$$(G + H)\sigma_1^2 + (F + H)\sigma_2^2 + (F + G)\sigma_3^2 - 2H\sigma_1\sigma_2 - 2G\sigma_1\sigma_3 - 2F\sigma_2\sigma_3$$
$$+ 2L\tau_{23}^2 + 2M\tau_{13}^2 + 2N\tau_{12}^2 = 1 \quad (2.120)$$

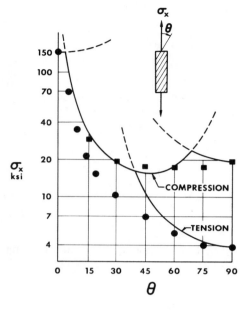

FIG. 2-23. Maximum strain failure theory. (*After Tsai, Ref. 2-16.*)

This anisotropic yield criterion will be used as an anisotropic strength criterion in the spirit of both being limits of linear elastic behavior. Thus, Hill's yield strengths F, G, H, L, M, and N will be regarded as failure strengths. Hill's theory is an extension of von Mises' isotropic yield criterion. The von Mises criterion, in turn, can be related to the amount of energy that is used to distort the body rather than to change its volume. However, distortion cannot be separated from dilatation in orthotropic materials so Eq. (2.120) is not related to distortional energy. Some authors still call the theory of this section a distortional energy failure theory.

The failure strength parameters F, G, H, L, M, and N were related to the usual failure strengths X, Y, and S for a lamina by Tsai (Ref. 2-16). First, if only τ_{12} acts on the body, then, since its maximum value is S;

$$2N = \frac{1}{S^2} \qquad (2.121)$$

Similarly, if only σ_1 acts on the body, then

$$G + H = \frac{1}{X^2} \qquad (2.122)$$

and if only σ_2 acts, then

$$F + H = \frac{1}{Y^2} \qquad (2.123)$$

If the strength in the 3-direction is denoted by Z and only σ_3 acts, then

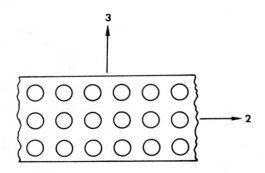

FIG. 2-24. Cross section of a unidirectional lamina with fibers in the 1-direction.

$$F + G = \frac{1}{Z^2} \qquad (2.124)$$

Then, upon combination of Eqs. (2.122), (2.123), and (2.124), the following relations between F, G, H and X, Y, Z result:

$$2H = \frac{1}{X^2} + \frac{1}{Y^2} - \frac{1}{Z^2}$$

$$2G = \frac{1}{X^2} + \frac{1}{Z^2} - \frac{1}{Y^2} \qquad (2.125)$$

$$2F = \frac{1}{Y^2} + \frac{1}{Z^2} - \frac{1}{X^2}$$

For plane stress in the 1-2 plane of a unidirectional lamina with the fibers in the 1-direction, $\sigma_3 = \tau_{13} = \tau_{23} = 0$. However, from the cross section of such a lamina in Fig. 2-24, $Y = Z$ from geometrical symmetry considerations. Thus, Eq. (2.120) leads to

$$\frac{\sigma_1^2}{X^2} - \frac{\sigma_1 \sigma_2}{X^2} + \frac{\sigma_2^2}{Y^2} + \frac{\tau_{12}^2}{S^2} = 1 \qquad (2.126)$$

as the governing failure criterion in terms of the familiar lamina strengths X, Y, and S.

Finally, for the off-axis composite example, substitution of the stress transformation equations,

$$\sigma_1 = \sigma_x \cos^2 \theta$$

$$\sigma_2 = \sigma_x \sin^2 \theta \qquad (2.127)$$

$$\tau_{12} = -\sigma_x \sin \theta \cos \theta$$

in Eq. (2.126) yields the Tsai-Hill failure criterion

$$\frac{\cos^4\theta}{X^2} + \left(\frac{1}{S^2} - \frac{1}{X^2}\right)\cos^2\theta \sin^2\theta + \frac{\sin^4\theta}{Y^2} = \frac{1}{\sigma_x^2} \tag{2.128}$$

which is *one* criterion, not three as in previous criteria.

Results for this criterion are plotted in Fig. 2-25 along with the experimental data for *E* glass/epoxy. The agreement between theory and experiment is quite good. Thus, a suitable failure criterion has apparently been found for *E* glass/epoxy laminae at various orientations in biaxial stress fields.

The Tsai-Hill failure theory appears to be much more applicable to failure prediction for this *E* glass/epoxy composite than either the maximum stress or the maximum strain criterion. Other less obvious advantages of the Tsai-Hill criterion are:

(1) The variation of strength with angle of orientation is smooth rather than possessing cusps.

(2) The strength continuously decreases as θ grows from $0°$ rather than the rise in uniaxial strength that is characteristic of both the maximum stress and the maximum strain criteria.

(3) The agreement between theory and experiment is even better than at first glance because Fig. 2-22, 2-23, and 2-25 are plotted at a logarithmic scale. The maximum stress and strain criteria are incorrect by 100 percent at $30°$!

(4) Considerable *interaction* exists between the failure strengths X, Y, S in the Tsai-Hill criterion, but *none* exists in the other criteria where axial, transverse, and shear failures are presumed to occur independently.

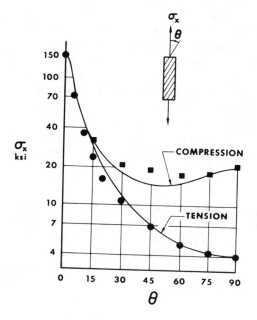

FIG. 2-25. Tsai-Hill theory. (*After Tsai, Ref. 2-16.*)

(5) The maximum stress criterion does not reduce correctly to an isotropic material result (neither does the maximum strain criterion, but obviously they are alike in this regard) from maximum octahedral shear stress theory:

$$X = Y = \sqrt{3}\,S \qquad (2.129)$$

With Eq. (2.129), the maximum stress criterion reduces to

$$\sigma_x < \frac{X}{\cos^2\theta}$$

$$\sigma_x < \frac{X}{\sin^2\theta} \qquad (2.130)$$

$$\sigma_x < \frac{X}{\sqrt{3}\,\sin\theta\,\cos\theta}$$

which is incorrectly dependent on θ. However, we would never have used Eq. (2.110) for an isotropic material. For the Tsai-Hill criterion, substitution of Eq. (2.129) yields

$$\sigma_x < X \qquad (2.131)$$

which is invariant under rotation, θ, as it must be for an isotropic material. This result should be no surprise since the maximum octahedral shear stress is directly related to the distortional energy of isotropic materials.

For this material, E glass/epoxy, the Tsai-Hill failure criterion seems the most accurate of the criteria discussed. However, the applicability of a particular failure criterion depends on whether the material being studied is ductile or brittle. Other composite materials may be better treated with the maximum stress or the maximum strain criteria or even something else.

There are certainly some faults with the Tsai-Hill failure criterion. Among them is the fact that an orthotropic material under a biaxial stress field cannot distort without extending and vice versa. Thus, the amount of distortional energy may not be directly related to failure as it was for many isotropic materials. There may be another component of energy to consider. An improved and more physically attractive, but more complicated, failure criterion is discussed in the next section.

2.9.4 Tsai-Wu Tensor Theory

The preceding biaxial strength theories suffer from various inadequacies in their description of experimental data. One obvious way to improve the correlation between theory and experiment is to increase the number of terms in the prediction equation. This increase in curve fitting ability plus the added feature of representing the various strengths in tensor form was used by Tsai and Wu (Ref. 2-18). In the process, several new strength definitions are required, mainly having to do with interaction between stresses in two directions.

Tsai and Wu postulated that a failure surface in stress space exists in the form

$$F_i \sigma_i + F_{ij}\sigma_i \sigma_j = 1 \qquad i,j = 1, \ldots, 6 \tag{2.132}$$

wherein F_i and F_{ij} are strength tensors of the second and fourth rank, respectively and the usual contracted stress notation is used except that $\sigma_4 = \tau_{23}$, $\sigma_5 = \tau_{31}$, and $\sigma_6 = \tau_{12}$. Equation (2.132) is obviously very complicated; we will restrict our attention to the reduction of Eq. (2.132) for an orthotropic lamina under plane stress conditions:

$$F_1 \sigma_1 + F_2 \sigma_2 + F_6 \sigma_6 + F_{11} \sigma_1^2 + F_{22} \sigma_2^2 + F_{66} \sigma_6^2 + 2F_{12} \sigma_1 \sigma_2 = 1 \tag{2.133}$$

The terms that are linear in the stresses are useful in representing different strengths in tension and compression. The terms that are quadratic in the stresses are the more or less usual terms to represent an ellipsoid in stress space. However, the term involving F_{12} is entirely new to us and is used to represent the interaction between normal stresses in the 1- and 2-directions in a manner quite unlike the shear strength.

Some of the components of the strength tensors are defined in terms of the engineering strengths already discussed. For example, consider a uniaxial load on a specimen in the 1-direction. Under tensile load, the engineering strength is X_t whereas under compressive load, it is X_c (for example, $X_c = -400,000$ psi for boron/epoxy). Thus, under tensile load,

$$F_1 X_t + F_{11} X_t^2 = 1 \tag{2.134}$$

and under compressive load,

$$F_1 X_c + F_{11} X_c^2 = 1 \tag{2.135}$$

Upon simultaneous solution of Eqs. (2.134) and (2.135),

$$F_1 = \frac{1}{X_t} + \frac{1}{X_c} \tag{2.136}$$

$$F_{11} = -\frac{1}{X_t X_c} \tag{2.137}$$

Similarly,

$$F_2 = \frac{1}{Y_t} + \frac{1}{Y_c} \tag{2.138}$$

$$F_{22} = -\frac{1}{Y_t Y_c} \tag{2.139}$$

Similar reasoning, along with our observation in Sec. 2.8 that the shear strength in principal material directions is independent of shear stress sign, leads to

$$F_6 = 0 \tag{2.140}$$

$$F_{66} = \frac{1}{S^2} \tag{2.141}$$

The determination of the fourth rank tensor term F_{12} remains. Basically, F_{12} cannot be determined from any uniaxial test in the principal material directions. Instead, a biaxial test must be used. This fact should not be surprising since F_{12} is the coefficient of σ_1 *and* σ_2 in the failure criterion, Eq. (2.133). Thus, for example, we can impose a state of biaxial tension described by $\sigma_1 = \sigma_2 = \sigma$ and all other stresses are zero. Accordingly, from Eq. (2.133),

$$(F_1 + F_2)\sigma + (F_{11} + F_{22} + 2F_{12})\sigma^2 = 1 \tag{2.142}$$

Now solve for F_{12} after substituting the definitions just derived for F_1, F_2, F_{11}, and F_{22} :

$$F_{12} = \frac{1}{2\sigma^2}\left[1 - \left(\frac{1}{X_t} + \frac{1}{X_c} + \frac{1}{Y_t} + \frac{1}{Y_c}\right)\sigma + \left(\frac{1}{X_t X_c} + \frac{1}{Y_t Y_c}\right)\sigma^2\right] \tag{2.143}$$

The value of F_{12} then depends on the various engineering strengths plus the biaxial tensile failure stress, σ. Tsai and Wu also discuss the use of off-axis uniaxial tests to determine the interaction strengths such as F_{12}.

At this point, recall that all interaction between normal stresses σ_1 and σ_2 in the Tsai-Hill theory is related to the strength in the 1-direction:

$$\frac{\sigma_1^2}{X^2} - \frac{\sigma_1\sigma_2}{X^2} + \frac{\sigma_2^2}{Y^2} + \frac{\tau_{12}^2}{S^2} = 1 \tag{2.144}$$

Thus, Tsai-Wu tensor failure theory is obviously of more general character than Tsai-Hill theory. Specific advantages of Tsai-Wu theory include (1) invariance under rotation or redefinition of coordinates; (2) transformation via known tensor transformation laws; and (3) symmetry properties akin to those of the stiffnesses and compliances. Accordingly, the mathematical operations with this tensor failure theory are well-known and relatively straightforward.

Pipes and Cole (Ref. 2-19) measured the interaction term F_{12} in various off-axis tests for boron/epoxy. They reported significant variation of F_{12} for off-axis tension tests and acceptable variation for off-axis compression tests. However, compression tests are much more difficult to perform than "simple" off-axis tension tests on a flat specimen with a high length-to-width ratio. A compression specimen with a high length-to-width ratio to avoid shear coupling effects is extremely susceptible to buckling. Hence, a tubular specimen (with a rotating end to avoid shear coupling effects) must be used. Although the determination of F_{12} was not precise, Pipes and Cole obtained the excellent agreement between Tsai-Wu tensor theory and the experimental data shown in

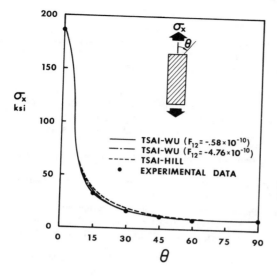

FIG. 2-26. Tsai-Wu tensor theory.
(After Pipes and Cole, Ref. 2-19.)

Fig. 2-26. Changes of F_{12} by a factor of 4 resulted in only slight changes in predicted strength over the range $5° < \theta < 25°$. Also, the difference between Tsai-Wu tensor theory and Tsai-Hill theory was less than 5 percent over the range $5° < \theta < 75°$.

The Tsai-Wu tensor failure theory is a relatively new multiaxial strength theory. As such, it has not been explored or applied to the extent of its potential. The interested reader should be alert to new developments. Other strength theories are described by Sendeckyj (Ref. 2-20).

Problem Set 2.9

Exercise 2.9.1 Identify which subcriterion for failure applies for each segment of the maximum stress and maximum strain failure curves in Figs. 2-22 and 2-23.

Exercise 2.9.2 Derive Eq. (2.125) from Eqs. (2.122), (2.123), and (2.124).

Exercise 2.9.3 Derive Eq. (2.126) from Eqs. (2.125) and (2.120).

Exercise 2.9.4 Derive Eq. (2.128) from Eqs. (2.126) and (2.127).

Exercise 2.9.5 What is the Tsai-Hill failure criterion when the fibers of a unidirectional lamina in the 1-2 plane are aligned in the 2-direction? Denote the laminate strength in the fiber direction by X as usual; thus, the strength in the 1-direction is Y. Compare this criterion with Eq. (2.126).

Exercise 2.9.6 Find the Tsai-Hill failure criterion for pure shear loading at various angles θ to the principal material directions, i.e., the shear analogue of Eq. (2.128).

Exercise 2.9.7 Note for Tsai's E glass/epoxy data that the uniaxial strength at angles between $0°$ and $90°$ is actually less than Y. The correct inference to be made is that E glass/epoxy has a low shear strength. This situation is the strength analogue of the stiffness variation studied in Exercise 2.6.7. Find the relation between S, X, and Y such that such low values of off-axis uniaxial strength occur and also the relation for the case where values of off-axis uniaxial strength higher than X occur.

REFERENCES

2-1 Tsai, Stephen W.: Mechanics of Composite Material, Part II, Theoretical Aspects, *Air Force Materials Laboratory Tech. Rept.* AFML-TR-66-149, November, 1966.

2-2 Lempriere, B. M.: Poisson's Ratio in Orthotropic Materials, *AIAA Journal,* November, 1968, pp. 2226–2227.

2-3 Dickerson, E. O., and B. DiMartino: Off-Axis Strength and Testing of Filamentary Materials for Aircraft Application, in "Advanced Fibrous Reinforced Composites," vol. 10, p. H-23, Society of Aerospace Materials and Process Engineers, 1966.

2-4 Reuter, Robert C., Jr.: Concise Property Transformation Relations for an Anisotropic Lamina, *J. Composite Materials,* April, 1971, pp. 270-272.

2-5 Lekhnitski, S. G.: "Theory of Elasticity of an Anisotropic Elastic Body," Holden-Day, San Francisco, 1963.

2-6 Jones, Robert M.: Stiffness of Orthotropic Materials and Laminated Fiber-Reinforced Composites, *AIAA Journal,* January, 1974, pp. 112-114.

2-7 Tsai, Stephen W., and Nicholas J. Pagano: Invariant Properties of Composite Materials, in S. W. Tsai, J. C. Halpin, and Nicholas J. Pagano (eds.), "Composite Materials Workshop," Technomic Publishing Co., Westport, Conn., 1968, pp. 233-253. Also AFML-TR-67-349, March, 1968.

2-8 Pagano, N. J., and P. C. Chou: The Importance of Signs of Shear Stress and Shear Strain in Composites, *J. Composite Materials,* January, 1969, pp. 166-173.

2-9 Ambartsumyan, S. A.: The Axisymmetric Problem of a Circular Cylindrical Shell Made of Material with Different Stiffness in Tension and Compression, *Izvestiya akademii nauk SSSR Mekhanika,* No. 4, 1965, pp. 77-85; English translation N69-11070, STAR.

2-10 Jones, Robert M.: Buckling of Stiffened Multilayered Circular Cylindrical Shells with Different Orthotropic Moduli in Tension and Compression, *AIAA Journal,* May, 1971, pp. 917-923.

2-11 Pagano, N. J., and J. C. Halpin: Influence of End Constraint in the Testing of Anisotropic Bodies, *J. Composite Materials,* January, 1968, pp. 18-31.

2-12 Whitney, J. M., N. J. Pagano, and R. B. Pipes: Design and Fabrication of Tubular Specimens for Composite Characterization, in "Composite Materials: Testing and Design (Second Conference)," ASTM STP 497, American Society for Testing and Materials, 1972.

2-13 Hahn, Hong T., and Stephen W. Tsai: Nonlinear Elastic Behavior of Unidirectional Composite Laminae, *J. Composite Materials,* January, 1973, pp. 102-118.

2-14 Waddoups, Max E.: Characterization and Design of Composite Materials, in S. W. Tsai, J. C. Halpin, and Nicholas J. Pagano (eds.), "Composite Materials Workshop," Technomic Publishing Co., Westport, Conn., 1968, pp. 254-308.

2-15 Whitney, J. M., D. L. Stansbarger, and H. B. Howell: "Analysis of the Rail Shear Test—Applications and Limitations," *J. Composite Materials,* January, 1971, pp. 24-34.

2-16 Tsai, Stephen W.: Strength Theories of Filamentary Structures, in R. T. Schwartz and H. S. Schwartz (eds.), "Fundamental Aspects of Fiber Reinforced Plastic Composites," Wiley Interscience, New York, 1968, pp. 3-11.

2-17 Hill, R.: "The Mathematical Theory of Plasticity," Oxford University Press, London, 1950.

2-18 Tsai, Stephen W., and Edward M. Wu: A General Theory of Strength for Anisotropic Materials, *J. Composite Materials,* January, 1971, pp. 58-80.

2-19 Pipes, R. Byron, and B. W. Cole: On the Off-Axis Strength Test for Anisotropic Materials, *J. Composite Materials,* April, 1973, pp. 246-256.

2-20 Sendeckyj, G. P.: A Brief Survey of Empirical Multiaxial Strength Criteria for Composites, in "Composite Materials: Testing and Design (Second Conference)," ASTM STP 497, American Society for Testing and Materials, 1972, pp. 41-51.

Chapter 3
MICROMECHANICAL
BEHAVIOR OF A LAMINA

3.1 INTRODUCTION

We have so far talked in Chap. 2 about the "apparent" properties of a lamina. That is, a large enough piece of the lamina has been considered so that the fact that the lamina is made of two or more constituent materials cannot be detected. Thus, almost magically, we were able to say that a boron/epoxy composite with unidirectional boron fibers has a certain stiffness and strength that we have measured. However, this question has not been asked: "How can the stiffness and strength be varied with the amount of boron in the composite?" Just as there must be some rationale for selecting a particular stiffness and/or strength for a particular structural application, there must also be a rationale for determining how best to achieve that stiffness and strength for a composite of two or more materials. That is, how can the percentages of the constituent materials be varied so as to arrive at the desired conglomerate stiffness and strength?

An appropriate division of the efforts just mentioned is brought about by the definition of two areas of composite material behavior, micromechanics and macromechanics:

Micromechanics — The study of composite material behavior wherein the *interaction* of the constituent materials is examined in detail as part of the definition of the behavior of the *heterogeneous* composite material.

Macromechanics — The study of composite material behavior wherein the material is presumed *homogeneous* and the effects of the constituent materials are detected only as averaged apparent properties of the composite.

Thus, the properties of a lamina can be experimentally determined in the "as made" state or can be mathematically derived on the basis of the properties of the constituent materials. That is, we can *predict* lamina properties by the procedures of *micromechanics* and we can *measure* lamina properties by physical means and use the properties in a *macromechanical* analysis of the structure. Knowledge of how to predict properties is essential to constructing composites that must have certain apparent or macroscopic properties. Thus, micromechanics is a natural adjunct to macromechanics when viewed from a *design* rather than an *analysis* environment. Real design power is evidenced when the micromechanical prediction of the properties of a lamina agree with the measured properties. However, recognize that a micromechanical analysis has inherent limitations. For example, a perfect bond between fibers and matrix is a usual analysis restriction that might well not be satisfied by some composites. An imperfect bond would presumably yield a material with properties degraded from those of the micromechanic analysis. Thus, the micromechanical theories must be validated by careful experimental work. With such broad statements as background, let us now turn to the study of some specific micromechanic theories.

There are two basic approaches to the micromechanics of composite materials:

1. Mechanics of materials
2. Elasticity

The mechanics of materials (or strength of materials or resistance of materials) approach embodies the usual concept of vastly simplifying assumptions regarding the hypothesized behavior of the mechanical system. The elasticity approach actually is at least three approaches: (1) bounding principles, (2) exact solutions, and (3) approximate solutions. The foregoing approaches will be discussed in the following sections.

The objective of all of the micromechanics approaches is to determine the elastic moduli or stiffnesses or compliances of a composite material in terms of the elastic moduli of the constituent materials. For example, the elastic moduli of a fiber-reinforced composite must be determined in terms of the properties of the fibers and the matrix and in terms of the relative volumes of fibers and matrix:

$$C_{ij} = C_{ij}(E_f, \nu_f, V_f, E_m, \nu_m, V_m)$$

where

$E_f =$ Young's modulus for an isotropic fiber

$\nu_f =$ Poisson's ratio for an isotropic fiber

$$V_f = \frac{\text{Volume of fibers}}{\text{Total volume of composite}}$$

with analogous definitions applying for the matrix material.

An additional and complementary objective of the micromechanics approaches to composite materials analysis is to determine the strengths of the composite material in terms of the strengths of the constituent materials. For example, the strength of a fiber-reinforced composite must be determined in terms of the strengths of the fibers and the matrix and their relative volumes (relative to the total volume of the composite). In functional form,

$$X_i = X_i(X_{if}, V_f, X_{im}, V_m)$$

where

$X_i = X, Y, S$ = Composite strengths

$X_{if} = X_f, Y_f, S_f$ = Fiber strengths

$$V_f = \frac{\text{Volume of fibers}}{\text{Total volume of composite}}$$

with analogous definitions applying for the matrix material. The foregoing definitions could be modified to account for different strengths under tensile and compressive loading. Also, the definitions could be simplified for isotropic fibers and/or isotropic matrix materials.

Not much work is available regarding micromechanical theories of strength. However, considerable work has been done on micromechanical theories of stiffness. We will concentrate on those aspects of stiffness theory that are most prominent in usage (e.g., the Halpin-Tsai equations) in addition to those that clearly illustrate the thrust of micromechanics. Available strength information will be summarized with the same intent as for stiffness theories.

Irrespective of the micromechanical stiffness approach used, the basic restrictions on the composite material that can be treated are:

The lamina is
    macroscopically homogeneous
    linearly elastic
    macroscopically orthotropic
    initially stress-free
The fibers are
    homogeneous
    linearly elastic
    isotropic
    regularly spaced
    perfectly aligned.
The matrix is
    homogeneous
    linearly elastic
    isotropic.

In addition, no voids can exist in the fibers or matrix or in between them (i.e., the bonds between the fibers and matrix are perfect).

Basic to the discussion of micromechanics is the representative volume element. It is the smallest region or piece of material over which the stresses and strains are macroscopically uniform. Microscopically, however, the stresses and strains are nonuniform owing to the heterogeneity of the material. Thus, scale of the volume element is important. Generally, only a single fiber appears in a representative volume element, but more than one fiber can be required. The fiber spacing in a composite lamina with unidirectional fibers constitutes one

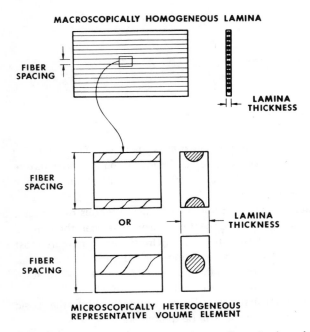

FIG. 3-1. Representative volume element for a lamina with uni-directional fibers.

dimension of the representative volume element. One of the other two dimensions is the lamina thickness or fiber spacing in the thickness direction if the lamina is more than one fiber thick. The third dimension is arbitrary. A typical representative volume element for a lamina with unidirectional fibers is shown in Fig. 3-1.

For a lamina with fibers in two directions, the microscopic representation is more complex than for unidirectional fibers. If the weaving geometry is neglected, two of the dimensions of the representative volume element are the spacings of the respective fibers. Finally, the third dimension is governed by the number of fibers in the thickness. If the actual weave geometry (curved fibers) is to be considered, finite element representation of the representative volume element, as in Fig. 3-2, may be desirable. Triangles and quadrilaterals, including squares, are used in Fig. 3-2 to represent both the fiber and the surrounding matrix. The properties assigned to each element depend on whether it is part of the fiber or the matrix. In finite element analyses, the bond between the fiber and the matrix is presumed perfect.

The results of the micromechanics studies of composites with unidirectional fibers will be presented as plots of an individual mechanical property versus the volume percentage of fibers. A schematic representation of several possible functional relationships between a property and the percentage of fibers is

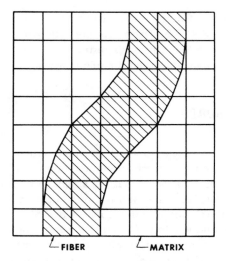

FIBER   MATRIX

FIG. 3-2. Finite element model of a representative volume element for a woven lamina.

shown in Fig. 3-3. In addition, both upper and lower bounds on those functional relationships will be obtained.

The mechanics of materials approach to the micromechanics of material stiffnesses is discussed in Sec. 3.2. There, simple approximations to the engineering constants for an orthotropic material are discussed. In Sec. 3.3, the parallel elasticity approach to the micromechanics of material stiffnesses is addressed. Bounding techniques, exact solutions, the concept of contiguity, and the Halpin-Tsai approximate equations are all discussed. Next, the various approaches to prediction of stiffness are compared in Sec. 3.4 with experimental data for both particulate composites and fiber-reinforced composites. Parallel to

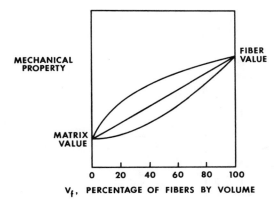

MECHANICAL PROPERTY

FIBER VALUE

MATRIX VALUE

0   20   40   60   80   100

V_f, PERCENTAGE OF FIBERS BY VOLUME

FIG. 3-3. Typical micromechanics results.

the study of the micromechanics of material stiffnesses is the micromechanics of material strengths which is introduced in Sec. 3.5. There, the predictions of tensile and compressive strengths are described. Attention is restricted to the mechanics of materials approach. Summary remarks on all the micromechanics work are made in Sec. 3.6.

3.2 MECHANICS OF MATERIALS APPROACH TO STIFFNESS

The key feature of the mechanics of materials approach is that certain simplifying *assumptions* are made regarding the mechanical behavior of a composite material. The most prominent assumption is that the strains in the fiber direction of a unidirectional fibrous composite are the same in the fibers as in the matrix as shown in Fig. 3-4. Since the strains in both the matrix and fiber are the same, then it is obvious that sections normal to the 1-axis that were plane before being stressed remain plane after stressing. The foregoing is a prominent assumption in the usual mechanics of materials approaches such as in beam, plate, and shell theories. We shall derive, on that basis, the mechanics of materials expressions for the apparent orthotropic moduli of a unidirectionally reinforced fibrous composite material.

3.2.1 Determination of E_1

The first modulus to be determined is that of the composite in the 1-direction, that is, in the fiber direction. From Fig. 3-4,

$$\epsilon_1 = \frac{\Delta_L}{L} \tag{3.1}$$

where ϵ_1 applies for both the fibers and the matrix according to the basic assumption. Then, if both constituent materials behave elastically, the stresses are

$$\sigma_f = E_f \epsilon_1$$
$$\sigma_m = E_m \epsilon_1 \tag{3.2}$$

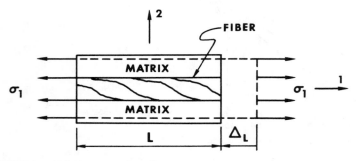

FIG. 3-4. Representative volume element loaded in 1-direction.

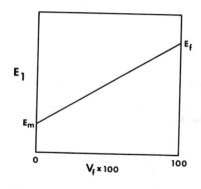

FIG. 3-5. Variation of E_1 with fiber volume fraction.

The average stress σ_1 acts on a cross-sectional area A, σ_f acts on the cross-sectional area of the fibers A_f, and σ_m acts on the cross-sectional area of the matrix A_m. Thus, the resultant force on the element of composite material is

$$P = \sigma_1 A = \sigma_f A_f + \sigma_m A_m \tag{3.3}$$

By substitution of Eq. (3.2) in Eq. (3.3) and recognition that

$$\sigma_1 = E_1 \epsilon_1 \tag{3.4}$$

apparently

$$E_1 = E_f \frac{A_f}{A} + E_m \frac{A_m}{A} \tag{3.5}$$

But the volume fractions of fibers and matrix can be written as

$$V_f = \frac{A_f}{A} \qquad V_m = \frac{A_m}{A} \tag{3.6}$$

Thus,

$$E_1 = E_f V_f + E_m V_m \tag{3.7}$$

which is known as the *rule of mixtures* expression for the apparent Young's modulus in the direction of the fibers. The rule of mixtures is graphically depicted in Fig. 3-5. The rule of mixtures represents a simple linear variation of apparent Young's modulus E_1 from E_m to E_f as V_f goes from 0 to 1.

3.2.2 Determination of E_2

The apparent Young's modulus, E_2, in the direction transverse to the fibers is considered next. In the mechanics of materials approach, the same transverse stress, σ_2, is assumed to be applied to both the fiber and the matrix as in Fig. 3-6. The strains in the fiber and in the matrix are, therefore,

$$\epsilon_f = \frac{\sigma_2}{E_f}$$

$$\epsilon_m = \frac{\sigma_2}{E_m} \tag{3.8}$$

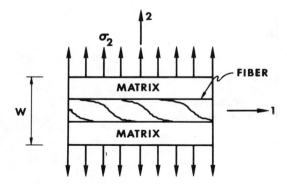

FIG. 3-6. Representative volume element loaded in 2-direction.

The transverse dimension over which, on the average, ϵ_f acts is approximately $V_f W$ whereas ϵ_m acts on $V_m W$. Thus, the total transverse deformation is

$$\epsilon_2 W = V_f W \epsilon_f + V_m W \epsilon_m \tag{3.9}$$

or

$$\epsilon_2 = V_f \epsilon_f + V_m \epsilon_m \tag{3.10}$$

which becomes, upon substitution of Eq. (3.8),

$$\epsilon_2 = V_f \frac{\sigma_2}{E_f} + V_m \frac{\sigma_2}{E_m} \tag{3.11}$$

but

$$\sigma_2 = E_2 \epsilon_2 = E_2 \left(\frac{V_f \sigma_2}{E_f} + \frac{V_m \sigma_2}{E_m} \right) \tag{3.12}$$

whereupon

$$E_2 = \frac{E_f E_m}{V_m E_f + V_f E_m} \tag{3.13}$$

which is the mechanics of materials expression for the apparent Young's modulus in the direction transverse to the fibers. Equation (3.13) can be nondimensionalized as

$$\frac{E_2}{E_m} = \frac{1}{V_m + V_f (E_m / E_f)} \tag{3.14}$$

with values of E_2/E_m being given in Table 3-1 for three values of the matrix to fiber modulus ratio.

Note in Fig. 3-7 that if $V_f = 1$, the modulus predicted is that of the fibers. However, recognize that a perfect bond between fibers is then implied if a tensile σ_2 is applied. No such bond is implied if a compressive σ_2 is applied.

TABLE 3-1.   Values for E_2/E_m for various E_m/E_f and V_f

$\dfrac{E_m}{E_f}$	V_f							
	0	.2	.4	.5	.6	.8	.9	1
1	1	1	1	1	1	1	1	1
1/10	1	1.22	1.56	1.82	2.17	3.57	5.26	10
1/100	1	1.25	1.66	1.98	2.46	4.80	9.17	100

Observe also that more than 50 percent by volume of fibers is required to raise the transverse modulus E_2 to twice the matrix modulus even if $E_f = 10$ E_m! That is, the fibers do not contribute much to the transverse modulus unless the percentage of fibers is very high.

Obviously, the assumptions involved in the foregoing derivation are not entirely consistent. There is a transverse strain mismatch at the boundary between the fiber and the matrix by virtue of Eq. (3.8). Moreover, the transverse stresses in the fiber and in the matrix are not likely to be the same. Rather, a complete match of displacements across the boundary between the fiber and the matrix would constitute a rigorous solution to determine the apparent transverse Young's modulus. Such a solution can be accomplished only by use of the theory of elasticity. The seriousness of such inconsistencies can be measured only by comparison with experimental results.

Another observation on this solution is that if the Poisson's ratios of the fiber and the matrix are not the same (they are likely different), then longitudinal stresses are induced in the fiber and matrix (with a net resultant longitudinal force of zero) with accompanying shearing stresses at the fiber-matrix boundary. Such shearing stresses will naturally arise under some stress states. Thus, this material characteristic cannot be regarded as undesirable or indicative of an inappropriate solution.

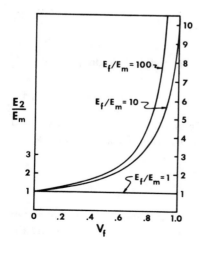

FIG. 3-7. Variation of E_2 with fiber volume fraction.

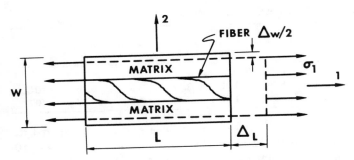

FIG. 3-8. Representative volume element loaded in 1-direction.

3.2.3 Determination of ν_{12}

The so-called major Poisson's ratio, ν_{12}, can be obtained by use of an approach similar to the analysis for E_1. The major Poisson's ratio is defined as

$$\nu_{12} = -\frac{\epsilon_2}{\epsilon_1} \qquad (3.15)$$

for the stress state $\sigma_1 = \sigma$ and all other stresses are zero. The deformations are depicted in Fig. 3-8. The transverse deformation Δ_W is

$$\Delta_W = -W\epsilon_2 = W\nu_{12}\epsilon_1 \qquad (3.16)$$

but is also

$$\Delta_W = \Delta_m W + \Delta_f W \qquad (3.17)$$

In the manner of the analysis for the transverse Young's modulus, E_2, the deformations $\Delta_m W$ and $\Delta_f W$ are approximately

$$\Delta_m W = W V_m \nu_m \epsilon_1 \qquad (3.18)$$

$$\Delta_f W = W V_f \nu_f \epsilon_1$$

Thus, upon combination of Eqs. (3.16), (3.17), and (3.18), division by $\epsilon_1 W$ yields

$$\nu_{12} = V_m \nu_m + V_f \nu_f \qquad (3.19)$$

which is a rule of mixtures for the major Poisson's ratio and is plotted in a manner similar to that for E_1 as in Fig. 3-9. Obviously, if $\nu_m = \nu_f$, then $\nu_{12} = \nu_m = \nu_f = \nu$.

3.2.4 Determination of G_{12}

The in-plane shear modulus of a lamina, G_{12}, is determined in the mechanics of materials approach by assuming that the shearing stresses on the fiber and the

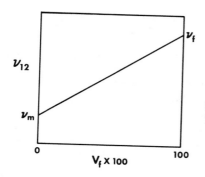

FIG. 3-9. Variation of ν_{12} with fiber volume fraction.

matrix are the same. The loading is shown in Fig. 3-10. By virtue of the basic assumption,

$$\gamma_m = \frac{\tau}{G_m}$$

$$\gamma_f = \frac{\tau}{G_f} \tag{3.20}$$

The nonlinear shear stress–shear strain behavior typical of fiber-reinforced composites is ignored, i.e., the behavior is regarded as linear.

On a microscopic scale, the deformations are shown in Fig. 3-11. The total shearing deformation is defined as

$$\Delta = \gamma W \tag{3.21}$$

and is made up of, approximately,

$$\Delta_m = V_m W \gamma_m$$

$$\Delta_f = V_f W \gamma_f \tag{3.22}$$

Then since $\Delta = \Delta_m + \Delta_f$, division by W yields

$$\gamma = V_m \gamma_m + V_f \gamma_f \tag{3.23}$$

or upon substitution of Eq. (3.20) and realization that

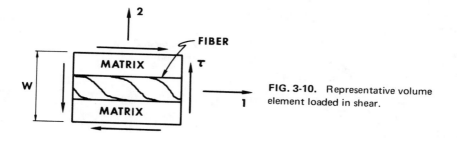

FIG. 3-10. Representative volume element loaded in shear.

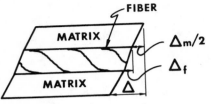

FIG. 3-11. Shear deformation of a representative volume element.

$$\gamma = \frac{\tau}{G_{12}} \tag{3.24}$$

Eq. (3.23) can be written as

$$\frac{\tau}{G_{12}} = V_m \frac{\tau}{G_m} + V_f \frac{\tau}{G_f} \tag{3.25}$$

Finally,

$$G_{12} = \frac{G_m G_f}{V_m G_f + V_f G_m} \tag{3.26}$$

which is the same type of expression as was obtained for the transverse Young's modulus, E_2. As with E_2 the expression for G_{12} can be normalized by a modulus related to the matrix, that is,

$$\frac{G_{12}}{G_m} = \frac{1}{V_m + V_f \dfrac{G_m}{G_f}} \tag{3.27}$$

which is plotted in Fig. 3-12 for several values of G_m/G_f. Note that, as for E_2, the matrix modulus is the dominant term in the expression for G_{12}. Only for a fiber volume of greater than 50 percent of the total volume does G_{12} rise above twice G_m even when $G_f/G_m = 10$!

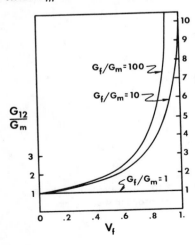

FIG. 3-12. Variation of G_{12} with fiber volume fraction.

3.2.5 Summary Remarks

The foregoing are but examples of the types of mechanics of materials approaches that can be used. Other assumptions of physical behavior lead to different expressions for the four elastic moduli for an orthotropic lamina. For example, Ekvall (Ref. 3-1) obtained a modification of the rule of mixtures expression for E_1 and of the expression for E_2 in which the triaxial stress state in the matrix due to fiber restraint is accounted for:

$$E_1 = V_f E_f + V_m E'_m \tag{3.28}$$

$$E_2 = \frac{E_f E'_m}{V_f E'_m + V_m E_f (1 - \nu_m^2)} \tag{3.29}$$

where

$$E'_m = \frac{E_m}{1 - 2\nu_m^2} \tag{3.30}$$

However, these modifications of the previously derived expressions are not significant for $\nu_m < 1/4$. Other modifications have been made to account for such features as square or rectangular versus round fibers or for stress concentrations due to fibers (Ref. 3-2).

Problem Set 3.2

Exercise 3.2.1 Show that a mechanics of materials type of approach can be used to determine an apparent Young's modulus for the general problem of a composite material with an "inclusion" of arbitrary shape as in Fig. 3-13. Fill in the details to show that if

$$E = \frac{F}{\delta}$$

then the modulus can be written as

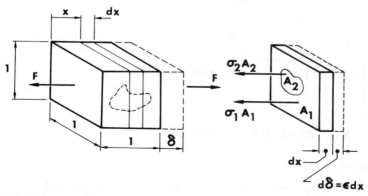

FIG. 3-13. Particulate reinforcement.

$$\frac{1}{E} = \int_0^1 \frac{dx}{E_1 + (E_2 - E_1)A_2(x)}$$

where $A_2(x)$ is the distribution of the inclusion. This result is to be used in Exercises 3.2.2 through 3.2.4.

Exercise 3.2.2 Verify that the general expression for the modulus of a dispersion-stiffened composite material reduces to

$$\frac{E}{E_m} = \frac{E_m + (E_d - E_m) V_d^{2/3}}{E_m + (E_d - E_m) V_d^{2/3} \left(1 - V_d^{1/3}\right)}$$

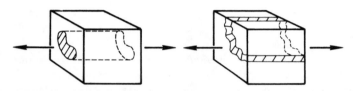

FIG. 3-14. Prismatic reinforcement.

for a cubic particle of modulus E_d in a matrix with modulus E_m. The volume fraction of cubic particles is V_d and that of the matrix is V_m or $1 - V_d$. Hint: the representative volume element is a cube within a cube.

Exercise 3.2.3 Determine the expression for the modulus of a composite material stiffened by particles of any cross section but prismatic along the direction in which the modulus is desired as in Fig. 3-14.

Exercise 3.2.4 Determine the expression for the modulus of a composite material reinforced by a slab of constant thickness in the direction in which the modulus is desired as in Fig. 3-15.

Exercise 3.2.5 What conclusion are you led to if you assume for the determination of E_2 in Sec. 3.2.2 equal strains in the fiber and the matrix instead of equal stresses?

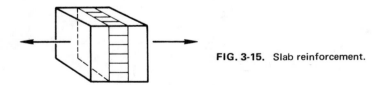

FIG. 3-15. Slab reinforcement.

3.3 ELASTICITY APPROACH TO STIFFNESS

3.3.1 Introduction

The division of micromechanics stiffness evaluation efforts into the mechanics of materials approach and the elasticity approach with its many subapproaches is rather arbitrary. Chamis and Sendeckyj (Ref. 3-3) divide micromechanics stiffness approaches into many more classes: netting analyses,[1] mechanics of materials approaches, self-consistent models, variational techniques using energy bounding principles, exact solutions, statistical approaches, discrete element methods, semiempirical approaches, and microstructure theories. All approaches have the common objective of the prediction of composite stiffnesses. All except the first two approaches use some or all of the principles of elasticity theory to varying degrees so are here classed as elasticity approaches. This simplifying and arbitrary division is useful in this book because the objective here is to merely become acquainted with advanced micromechanics theories after the basic concepts have been introduced by use of mechanics of materials type reasoning. The reader who is interested in micromechanics should supplement this chapter with the excellent critique and extensive bibliography of Chamis and Sendeckyj (Ref. 3-3).

The variational energy principles of classical elasticity theory are used in Sec. 3.3.2 to determine upper and lower bounds on lamina moduli. However, that approach generally leads to bounds that may not be sufficiently close for practical use. In Sec. 3.3.3, all the principles of elasticity theory are invoked to determine the lamina moduli. Because of the resulting complexity of the problem, many advanced analytical techniques and numerical solution procedures are necessary to obtain solutions. However, the assumptions made in such analyses regarding the interaction between the fibers and the matrix are not entirely realistic. An interesting approach to more realistic fiber-matrix interaction, the contiguity approach, is examined in Sec. 3.3.4. The widely used Halpin-Tsai equations are displayed and discussed in Sec. 3.3.5. Finally, summary remarks are presented in Sec. 3.3.6.

3.3.2 Bounding Techniques of Elasticity

Paul (Ref. 3-4) was apparently the first to use the bounding (variational) techniques of linear elasticity to examine the bounds on the moduli of multiphase materials. His work was directed toward analysis of the elastic moduli of alloyed metals rather than toward fibrous composites. Accordingly, the treatment is for an isotropic composite material made of isotropic constituents. The composite is isotropic because the alloyed constituents are uniformly dispersed and have no preferred orientation. The modulus of the basic matrix is E_m and the modulus of the dispersed material is E_d whereas the modulus of the composite material is E. The volume fractions of the constituents are V_m and V_d such that

---

[1] The basic assumption in netting analysis is that the fibers provide all the longitudinal stiffness and the matrix provides all the transverse and shear stiffness as well as the Poisson effect. On the basis of our observations of mechanics of materials results, we recognize the netting analysis assumption to be grossly conservative. Hence, netting analysis will be ignored in this book, and more useful theories will be addressed.

$$V_m + V_d = 1 \tag{3.31}$$

Obviously, any relationship for the composite modulus E must yield $E = E_m$ for $V_m = 1$ and $E = E_d$ for $V_d = 1$.

One of the simplest relationships that satisfies the above restrictions is the rule of mixtures

$$E = E_m V_m + E_d V_d \tag{3.32}$$

wherein the constituents of the composite are presumed to contribute to the composite stiffness in direct proportion to their own stiffnesses and volume fractions. The rule of mixtures will be shown to provide an upper bound on the composite modulus E for the special case in which

$$\nu_m = \nu_d = \nu \tag{3.33}$$

Another simple relationship between the constituent moduli results from the observation that the compliance of the composite, $1/E$, must agree with the compliance of the matrix, $1/E_m$, when $V_m = 1$ and with the compliance of the dispersed material when $V_d = 1$. The resulting rule of mixtures for compliances is

$$\frac{1}{E} = \frac{V_m}{E_m} + \frac{V_d}{E_d} \tag{3.34}$$

which will be shown to yield a lower bound on the composite modulus, E.

In a uniaxial tension test to determine the elastic modulus of the composite material, E, the stress and strain states will be assumed to be macroscopically uniform in consonance with the basic presumption that the composite material is macroscopically isotropic and homogeneous. However, on a microscopic scale, both the stress and strain states will be nonuniform. In the uniaxial tension test,

$$E = \frac{\sigma}{\epsilon} \tag{3.35}$$

where σ is the applied uniaxial stress and ϵ is the resulting axial strain. The resulting strain energy can be written in two forms:

$$U = \frac{1}{2} \frac{\sigma^2}{E} V \tag{3.36}$$

$$U = \frac{1}{2} E \epsilon^2 V \tag{3.37}$$

Lower bound on apparent Young's modulus

The basis for the determination of a lower bound on the apparent Young's modulus is application of the *principle of minimum complementary energy* which can be stated as: Let the *tractions* (forces and moments) be specified over the surface of a body. Let σ_x°, σ_y°, σ_z°, τ_{xy}°, τ_{yz}°, τ_{zx}° be a state of stress that satisfies the stress equations of equilibrium and the specified boundary conditions, i.e., an admissible stress field. Let U° be the strain energy for the stress state σ_x°, σ_y°, σ_z°, τ_{xy}°, τ_{yz}°, τ_{zx}° given by use of the stress-strain relations (a simple rearrangement of the isotropic stress-strain relations in Eq. (2.17) in terms of E and ν)

$$\sigma_x = \frac{\nu E}{(1+\nu)(1-2\nu)} (\epsilon_x + \epsilon_y + \epsilon_z) + \frac{E}{(1+\nu)} \epsilon_x$$

$$\vdots$$

$$\tau_{xy} = G\gamma_{xy} = \frac{E}{2(1+\nu)} \gamma_{xy} \tag{3.38}$$

and the expression for the strain energy

$$U = \frac{1}{2} \int_V (\sigma_x \epsilon_x + \sigma_y \epsilon_y + \sigma_z \epsilon_z + \tau_{xy}\gamma_{xy} + \tau_{yz}\gamma_{yz} + \tau_{zx}\gamma_{zx})dV \tag{3.39}$$

Then, the actual strain energy U in the body due to the specified loads cannot exceed U°, that is,

$$U \leqslant U^\circ \tag{3.40}$$

To find a lower bound on the apparent Young's modulus, E, load the basic uniaxial test specimen with normal stress on the ends. The internal stress field that satisfies this loading and the stress equations of equilibrium is

$$\sigma_x^\circ = \sigma \qquad \sigma_y^\circ = \sigma_z^\circ = \tau_{xy}^\circ = \tau_{yz}^\circ = \tau_{zx}^\circ = 0 \tag{3.41}$$

The strain energy for the stresses in Eq. (3.41) is

$$U^\circ = \frac{1}{2} \int_V \frac{(\sigma_x^\circ)^2}{E} dV = \frac{\sigma^2}{2} \int_V \frac{dV}{E} \tag{3.42}$$

But E is obviously not constant over the volume since the matrix has modulus E_m over volume $V_m V$ and the dispersed material has modulus E_d over volume $V_d V$ where V is the total volume. Thus,

$$\int_V \frac{dV}{E} = \int_{V_m V} \frac{dV}{E_m} + \int_{V_d V} \frac{dV}{E_d} \tag{3.43}$$

or

$$\int_V \frac{dV}{E} = \frac{V_m V}{E_m} + \frac{V_d V}{E_d} \tag{3.44}$$

whereupon

$$U^\circ = \frac{\sigma^2}{2}\left(\frac{V_m}{E_m} + \frac{V_d}{E_d}\right)V \tag{3.45}$$

However, by virtue of the inequality $U \leq U^\circ$ and the definition of U in Eq. (3.36):

$$\frac{1}{2}\frac{\sigma^2}{E}V \leq \frac{\sigma^2}{2}\left(\frac{V_m}{E_m} + \frac{V_d}{E_d}\right)V \tag{3.46}$$

or

$$\frac{1}{E} \leq \frac{V_m}{E_m} + \frac{V_d}{E_d} \tag{3.47}$$

Finally,

$$E \geq \frac{E_m E_d}{V_m E_d + V_d E_m} \tag{3.48}$$

which is a lower bound on the apparent Young's modulus, E, of the composite material in terms of the moduli and volume fractions of the constituent materials. Note that this bound coincides with the value for the modulus transverse to the fibers by the mechanics of materials approach.

Upper bound on apparent Young's modulus

The basis for the determination of an upper bound on the apparent Young's modulus is application of the *principle of minimum potential energy* which can be stated as: Let the *displacements* be specified over the surface of the body except where the corresponding traction is zero. Let $\epsilon_x^*, \epsilon_y^*, \epsilon_z^*, \gamma_{xy}^*, \gamma_{yz}^*, \gamma_{zx}^*$ be any compatible state of strain that satisfies the specified displacement boundary conditions, i.e., an admissible strain field. Let U^* be the strain energy of the strain state ϵ_x^*, etc. by use of the stress-strain relations

$$\sigma_x = \frac{\nu E}{(1+\nu)(1-2\nu)}(\epsilon_x + \epsilon_y + \epsilon_z) + \frac{E}{(1+\nu)}\epsilon_x$$

$$\vdots \tag{3.49}$$

$$\tau_{xy} = G\gamma_{xy} = \frac{E}{2(1+\nu)}\gamma_{xy}$$

and the expression for the strain energy

$$U = \frac{1}{2}\int_V (\sigma_x\epsilon_x + \sigma_y\epsilon_y + \sigma_z\epsilon_z + \tau_{xy}\gamma_{xy} + \tau_{yz}\gamma_{yz} + \tau_{zx}\gamma_{zx})dV \tag{3.50}$$

Then, the actual strain energy U in the body due to the specified displacements cannot exceed U^*, that is,

$$U \leqslant U^* \tag{3.51}$$

To find an upper bound on the apparent Young's modulus, E, subject the basic uniaxial test specimen to an elongation ϵL where ϵ is the average strain and L is the specimen length. The internal strain field that corresponds to the average strain at the boundaries of the specimen is

$$\epsilon_x^* = \epsilon \qquad \epsilon_y^* = \epsilon_z^* = -\nu\epsilon \qquad \gamma_{xy}^* = \gamma_{yz}^* = \gamma_{zx}^* = 0 \tag{3.52}$$

where ν is the apparent Poisson's ratio of the composite. Then, by use of the stress-strain relations, Eq. (3.49), the stresses in the matrix for the given strain field are:

$$\sigma_{x_m}^* = \frac{1 - \nu_m - 2\nu_m\nu}{1 - \nu_m - 2\nu_m^2} E_m \epsilon$$

$$\sigma_{y_m}^* = \sigma_{z_m}^* = \frac{\nu_m - \nu}{1 - \nu_m - 2\nu_m^2} E_m \epsilon \tag{3.53}$$

$$\tau_{xy_m}^* = \tau_{yz_m}^* = \tau_{zx_m}^* = 0$$

and the stresses in the dispersed material are:

$$\sigma_{x_d}^* = \frac{1 - \nu_d - 2\nu_d\nu}{1 - \nu_d - 2\nu_d^2} E_d \epsilon$$

$$\sigma_{y_d}^* = \sigma_{z_d}^* = \frac{\nu_d - \nu}{1 - \nu_d - 2\nu_d^2} E_d \epsilon \tag{3.54}$$

$$\tau_{xy_d}^* = \tau_{yz_d}^* = \tau_{zx_d}^* = 0$$

The strain energy in the composite material is obtained by substituting the strains, Eq. (3.52), and the stresses, Eqs. (3.53) and (3.54), in the strain energy, Eq. (3.50):

$$U^* = \frac{\epsilon^2}{2} \int_{V_d V} \frac{1 - \nu_d - 4\nu_d\nu + 2\nu^2}{1 - \nu_d - 2\nu_d^2} E_d dV$$

$$+ \frac{\epsilon^2}{2} \int_{V_m V} \frac{1 - \nu_m - 4\nu_m\nu + 2\nu^2}{1 - \nu_m - 2\nu_m^2} E_m dV \tag{3.55}$$

or

$$U^* = \frac{\epsilon^2}{2} \left(\frac{1 - \nu_d - 4\nu_d\nu + 2\nu^2}{1 - \nu_d - 2\nu_d^2} E_d V_d \right.$$

$$\left. + \frac{1 - \nu_m - 4\nu_m\nu + 2\nu^2}{1 - \nu_m - 2\nu_m^2} E_m V_m \right) V \quad (3.56)$$

However, by virtue of the inequality $U \leqslant U^*$ and the definition of U in Eq. (3.37),

$$\frac{1}{2} E\epsilon^2 V \leqslant \frac{\epsilon^2}{2} \left(\frac{1 - \nu_d - 4\nu_d\nu + 2\nu^2}{1 - \nu_d - 2\nu_d^2} E_d V_d \right.$$

$$\left. + \frac{1 - \nu_m - 4\nu_m + 2\nu^2}{1 - \nu_m - 2\nu_m^2} E_m V_m \right) V \quad (3.57)$$

whereupon the upper bound on E is, by simple cancellation of terms in Eq. (3.57),

$$E \leqslant \frac{1 - \nu_d - 4\nu_d\nu + 2\nu^2}{1 - \nu_d - 2\nu_d^2} E_d V_d + \frac{1 - \nu_m - 4\nu_m\nu + 2\nu^2}{1 - \nu_m - 2\nu_m^2} E_m V_m \quad (3.58)$$

The value of Poisson's ratio, ν, for the composite material is *unknown* at this stage of the analysis so the upper bound on E is inspecific. In accordance with the principle of minimum potential energy, the expression for the strain energy U^* must be minimized[2] with respect to the unspecified constant ν to specify the bound on E. The minimization procedure consists of determining where

$$\frac{\partial U^*}{\partial \nu} = 0 \quad (3.59)$$

and at the same time

$$\frac{\partial^2 U^*}{\partial \nu^2} > 0 \quad (3.60)$$

First,

$$\frac{\partial U^*}{\partial \nu} = \frac{\epsilon^2 V}{2} \left(\frac{-4\nu_d + 4\nu}{1 - \nu_d - 2\nu_d^2} E_d V_d + \frac{-4\nu_m + 4\nu}{1 - \nu_m - 2\nu_m^2} E_m V_m \right) \quad (3.61)$$

which is zero when

$$\nu = \frac{(1 - \nu_m - 2\nu_m^2)\nu_d E_d V_d + (1 - \nu_d - 2\nu_d^2)\nu_m E_m V_m}{(1 - \nu_m - 2\nu_m^2)E_d V_d + (1 - \nu_d - 2\nu_d^2)E_m V_m} \quad (3.62)$$

[2] Note at this point that the potential energy of external forces is independent of material properties. Thus, its derivatives with respect to ν are zero, and only U^* affects the minimization.

The second derivative of U^* is

$$\frac{\partial^2 U^*}{\partial \nu^2} = \frac{\epsilon^2 V}{2} \left(\frac{4 E_d V_d}{1 - \nu_d - 2\nu_d^2} + \frac{4 E_m V_m}{1 - \nu_m - 2\nu_m^2} \right) \tag{3.63}$$

However, the matrix and dispersed material are isotropic so $\nu_m < \frac{1}{2}$ and $\nu_d < \frac{1}{2}$ (the usual limit on Poisson's ratio for an isotropic material). Thus, upon substitution of these values for ν_m and ν_d, the value of $\partial^2 U^*/\partial \nu^2$ is seen to be always positive (even when $\partial U^*/\partial \nu$ is not zero) since the typical term $(1\text{-}b\text{-}2b^2)$ is always positive when $b < \frac{1}{2}$. Finally, since $\partial^2 U^*/\partial \nu^2$ is always positive, the value of U^* when (3.62) is used, corresponding to a minimum, maximum, or inflection point on the curve for U^* as a function of ν, is proved to be a minimum and, in fact, the absolute minimum.

The value of Poisson's ratio, ν for the composite material has been derived explicitly as Eq. (3.62). Thus, the explicit upper bound on E can be obtained by substituting the expression for ν, Eq. (3.62), in the expression for the upper bound on E in terms of ν, Eq. (3.58). However, the algebra is quite messy so an explicit expression for the upper bound on E is not presented. In practical applications, the value of ν can be calculated from Eq. (3.62) and then substituted in Eq. (3.58) to obtain E. For the special case in which $\nu_m = \nu_d$, the expression for ν, Eq. (3.62), reduces to $\nu = \nu_m = \nu_d$ so the upper bound on E reduces to

$$E \leqslant E_d V_d + E_m V_m \tag{3.64}$$

which is the value of the apparent Young's modulus, E_1, in the fiber direction of a fiber-reinforced composite derived by the mechanics of materials approach. Thus, the expression for E_1 is an upper bound on the actual E_1. In addition, the mechanics of materials solution obviously includes an implicit equality of the Poisson's ratios of the constituent materials.

Paul's work (Ref. 3-4) is primarily applicable to isotropic composites, but it can be interpreted in terms of fibrous composites. For example, Eq. (3.64) is the upper bound on the transverse modulus, E_2, of a fiber-reinforced composite whereas Eq. (3.48) is the lower bound. Obviously, the bounds, as plotted in Fig. 3-16 for a glass/epoxy composite ($E_f = 10.6 \times 10^6$ psi and $E_m = .5 \times 10^6$ psi), are far apart. Bounds on other moduli can be obtained in a similar manner (see Problem Set 3.3).

Hashin (Ref. 3-5) and Hashin and Shtrikman (Ref. 3-6) attempted to tighten Paul's bounds to obtain more useful estimates of moduli for isotropic heterogeneous materials. Their approach was to use a concentric spheres model to treat the heterogeneous material as an elastic sphere inside a concentric spherical portion of elastic matrix material in proportion to the volume content of spherical inclusions in the total volume of the composite material. The included spheres never touch one another in the model although, clearly, as the volume percentage of particles increases, so does the likelihood of particle contact. Moreover, lack of contact may imply perfect particle spacing, an unlikely situation from the practical standpoint.

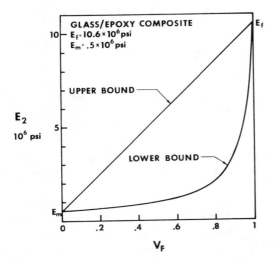

FIG. 3-16. Bounds on E_2 for a glass/epoxy composite.

Hashin and Rosen (Ref. 3-7) extended Hashin's work to fiber-reinforced composite materials. The fibers have a circular cross section and can be hollow or solid. Two cases were treated: (1) identical fibers in a hexagonal array and (2) fibers of various diameters (but same ratio of inside to outside diameter, if hollow) in a random array. The two types of arrays are depicted in Fig. 3-17. In both cases, the basic analysis element is a set of concentric cylinders with their axis in the fiber direction. For the random array, in analogy to Hashin's concentric spheres model, the concentric cylinder model consists of the fiber with matrix material around it in proportion to the volume content of matrix in the total volume. An additional matrix volume term is needed in the hexagonal array case to account for the volume left over when the circles of radius r_m are drawn around each fiber in Fig. 3-17a. The concentric cylinder model itself is displayed in Fig. 3-17a. The Young's modulus in the fiber direction turns out to be, for all practical purposes, the rule of mixtures. The expressions for the transverse Young's modulus for a random array or a hexagonal array of solid or hollow fibers are more complex than the objectives of this book leave room for. Some of the expressions will be plotted later when experimental data are compared with various theoretical predictions. At any rate, the bounds on moduli for fibers in a hexagonal array are rather far apart for large values of the ratio of fiber modulus to matrix modulus in the composite, a typical situation for practical composites. On the other hand, this random array model is not an accurate representation of most practical fiber-reinforced composites, either. If, however, many different sizes of fibers were included so as to fill the matrix voids between the various concentric cylinders as in Fig. 3-17b, the model would presumably be accurate.

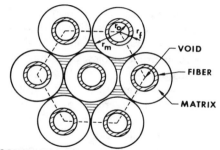

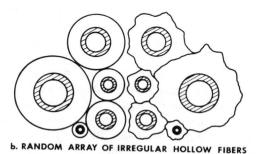

a. HEXAGONAL ARRAY OF REGULAR HOLLOW FIBERS

b. RANDOM ARRAY OF IRREGULAR HOLLOW FIBERS

FIG. 3-17. Hashin and Rosen's fiber reinforcement geometries and composite cylinder model.

3.3.3 Exact Solutions

The problem of determining exact solutions to various cases of elastic inclusions in an elastic matrix is very difficult and well beyond the scope of this book. However, it is appropriate to indicate the types of solutions that are available and to compare them with the mechanics of materials results (in a later section). As in many other elasticity problems, the Saint Venant semi-inverse method is prominent among the available techniques. In brief, the semi-inverse method consists of "dreaming up" or assuming a part of the solution, i.e., some of the components of stress, strain, or displacement, and then seeing if it satisfies the governing differential equations of equilibrium and the boundary conditions. The assumed solution must not be so rigorously specified that the equilibrium and compatibility equations cannot be satisfied. As an example, the assumption that plane sections remain plane is a semi-inverse method approach. In combination with the bounding theorems of elasticity, the semi-inverse method is quite effective.

Problems of inclusions in solids are also treated by exact elasticity approaches such as Muskhelishvili's complex variable mapping techniques (Ref. 3-8). In addition, numerical solution techniques such as finite elements and finite differences have been used extensively.

A strong background in elasticity is required for solution of problems in micromechanics of composite materials. Many of the available papers are quite abstract and of little direct applicability to practical analysis at this stage of development of elasticity approaches to micromechanics. Even the more sophisticated bounding approaches are a bit obscure.

The elasticity approaches depend to a great extent on the specific geometry of the composite material as well as on the characteristics of the fibers and the matrix. The fibers can be hollow or solid, but are usually circular in cross section, although rectangular cross section fibers are not uncommon. In addition, fibers are usually isotropic, but can have more complex material behavior, e.g., graphite fibers are transversely isotropic.

The fibers can exist in many types of arrays. Several typical arrays are shown in Figs. 3-18 through 3-21. There, the representative volume element for each array is shown along with a simplified representative volume element that is just as representative by virtue of symmetry, but does not include a whole fiber. Note in Fig. 3-20 that if the rows of the staggered array with round fibers are offset by one-half the fiber spacing, the representative volume element is the same as for the square array, but with principal loading directions rotated by 45°. Also, the staggered array of rectangular cross-section fibers in Fig. 3-21 is sometimes called a diamond array. Herrman and Pister (Ref. 3-9) were apparently the first to use the representative volume element and recognize its inherent symmetry.

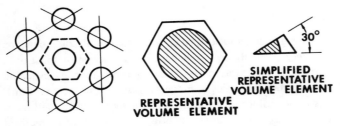

FIG. 3-18. Hexagonal array and representative volume elements.

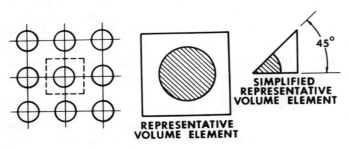

FIG. 3-19. Square array and representative volume elements.

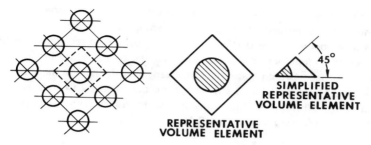

FIG. 3-20. Staggered square array of round fibers and representative volume elements.

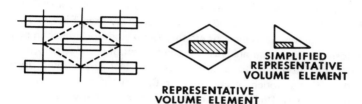

FIG. 3-21. Staggered square array of rectangular fibers and representative volume elements.

Adams and Tsai (Ref. 3-10) studied random arrays of two types: (1) square random arrays and (2) hexagonal random arrays. Both arrays have repeating elements so are not truly random. However, results of the hexagonal random array analysis agree better with experiments than do results of the square random array analysis. This observation is more satisfying than a previous result that (nonrandom) square array analyses agreed better with experiments than the more physically realistic hexagonal array analyses.

A variation on the exact solutions is the so-called self-consistent model. The self-consistent model that is explained in simplest engineering terms is due to Whitney and Riley (Ref. 3-11). Their model has a single hollow fiber embedded in a concentric cylinder of matrix material as in Fig. 3-22. That is, only one inclusion is considered. The volume fraction of the inclusion in the composite cylinder is the same as that of the entire body of fibers in the composite

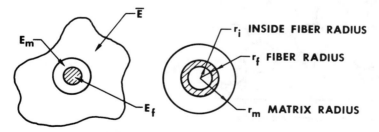

FIG. 3-22. Self-consistent composite cylinder model.

material. Such an assumption is not entirely valid since the matrix material may tend to coat the fibers imperfectly and hence leave voids. Note that there is no association of this model with any particular array of fibers. Also recognize the similarity between this model and the concentric cylinder model of Hashin and Rosen (Ref. 3-7). Other more complex self-consistent models include those by Hill (Ref. 3-12) and Hermans (Ref. 3-13) which are discussed by Chamis and Sendeckyj (Ref. 3-3). Whitney extended his model to transversely isotropic fibers (Ref. 3-14) and to twisted fibers (Ref. 3-15).

3.3.4 Elasticity Solutions with Contiguity

In the fabrication of filamentary composites, the fibers are often somewhat randomly placed rather than being packed in a regular array (see Fig. 3-23). (This random nature is much more typical of graphite/epoxy composites than of boron/epoxy composites.) Thus, the analyses for the moduli of composites with regular arrays must be modified to account for the fact that fibers are contiguous, i.e., that fibers touch each other rather than being entirely surrounded by matrix material. But, the fibers do not touch in many instances. Rather, some are contiguous and some are not. From an analytical point of view, a linear combination of (1) a solution in which all fibers are isolated from one another and (2) a solution in which all fibers contact each other provides the correct modulus. If C denotes degree of contiguity, then $C = 0$ corresponds to no contiguity (isolated fibers) and $C = 1$ corresponds to perfect contiguity (all fibers in contact) as in Fig. 3-24. Naturally, with high volume fractions of fibers, C would be expected to approach $C = 1$. This approach is an example of what Chamis and Sendeckyj (Ref. 3-3) call a semiempirical method. It could also be classified as a bounding technique.

For the elasticity approach in which the contiguity is considered, Tsai (Ref. 3-16) obtains for the modulus transverse to the fibers

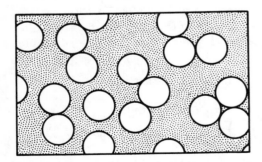

FIG. 3-23. Schematic diagram of actual fiber arrangement.

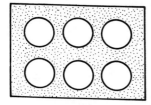
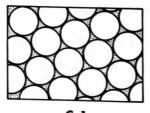

C=0

**ISOLATED FIBERS
RESIN CONTIGUOUS**

C=1

**ISOLATED MATRIX
FIBERS CONTIGUOUS**

FIG. 3-24. Extremes of fiber contiguity.

$$E_2 = 2[1 - v_f + (v_f - v_m)V_m] \left[(1-C) \frac{K_f(2K_m + G_m) - G_m(K_f - K_m)V_m}{(2K_m + G_m) + 2(K_f - K_m)V_m} \right.$$

$$\left. + C \frac{K_f(2K_m + G_f) + G_f(K_m - K_f)V_m}{(2K_m + G_f) - 2(K_m - K_f)V_m} \right] \quad (3.65)$$

where

$$K_f = \frac{E_f}{2(1 - v_f)}$$

$$K_m = \frac{E_m}{2(1 - v_m)}$$

$$G_f = \frac{E_f}{2(1 + v_f)} \quad\quad (3.66)$$

$$G_m = \frac{E_m}{2(1 + v_m)}$$

and C lies between 0 and 1. From a practical point of view, C would be determined by comparison of curves of E_2 versus V_f (or V_m) for various values of C. The value of C for the prediction that best agrees with experiment is then the appropriate design value for the given material. Since $C = 0$ corresponds to the case where each fiber is isolated and $C = 1$ corresponds to the much less likely case where all fibers are in contact, low values of C should be expected.

Tsai also obtains

$$v_{12} = (1-C) \frac{K_f v_f(2K_m + G_m)V_f + K_m v_m(2K_f + G_m)V_m}{K_f(2K_m + G_m) - G_m(K_f - K_m)V_m}$$

$$+ C \frac{K_m v_m(2K_f + G_f)V_m + K_f v_f(2K_m + G_f)V_f}{K_f(2K_m + G_m) + G_f(K_m - K_f)V_m} \quad (3.67)$$

wherein the definitions of Eq. (3.66) apply and C takes on the same value as in Eq. (3.65).

Finally, for the shear modulus, Tsai obtains

$$G_{12} = (1 - C)G_m \frac{2G_f - (G_f - G_m)V_m}{2G_m + (G_f - G_m)V_m}$$
$$+ CG_f \frac{(G_f + G_m) - (G_f - G_m)V_m}{(G_f + G_m) + (G_f - G_m)V_m} \quad (3.68)$$

where again the definitions of Eq. (3.66) apply and C has the same value as in Eqs. (3.65) and (3.67).

For the modulus in the direction of the fibers, Tsai modified the rule of mixtures to account for imperfections in fiber alignment:

$$E_1 = k(V_f E_f + V_m E_m) \quad (3.69)$$

The fiber misalignment factor, k, ordinarily varies from .9 to 1 so Eq. (3.69) does not represent a very significant departure from the rule of mixtures. Of course, k is an experimentally determined constant and is highly dependent on the manufacturing process.

Tsai (Ref. 3-16) performed some interesting parametric studies for the values of E_1, E_2, ν_{12} and G_{12} for glass filament–epoxy resin composites. The baseline composite has the properties $E_f = 10.6 \times 10^6$ psi, $\nu_f = .22$, $E_m = .5 \times 10^6$ psi, and $\nu_m = .35$. By use of Eqs. (3.69), (3.65), (3.67), and (3.68), E_1, E_2, ν_{12} and G_{12} are plotted in Figs. 3-25 through 3-27 for the baseline composite. In addition, the influence of fiber modulus is assessed by using $E_f = 16 \times 10^6$ psi and $E_f = 6 \times 10^6$ psi in the governing equations and is shown graphically in Fig. 3-25. Similarly, the influence of the matrix modulus is shown in Fig. 3-26 and that of the matrix and fiber Poisson's ratios in Fig. 3-27. In all figures, the fiber misalignment factor, k, is unity and the filament contiguity factor, C, is .2. Both values were found to be reasonable by comparison with experimental data (see Sec. 3.4, Comparison of Approaches to Stiffness).

The influence of the fiber modulus is felt most by the composite modulus in the fiber direction, E_1, as evidenced by Fig. 3-25. The composite modulus transverse to the fiber direction, E_2, is most strongly influenced by the matrix modulus as seen in Fig. 3-26 where E_m takes on values of 1.2×10^6 psi and $.2 \times 10^6$ psi in addition to the baseline value. The composite shearing modulus, G_{12}, appears to be more strongly influenced by the matrix modulus than by the fiber modulus when Figs. 3-25 and 3-26 are compared. From Fig. 3-27, it is clear that the Poisson's ratios of the fiber and the matrix have little effect on the composite moduli for practical values of the Poisson's ratios. In fact, they have no effect on the composite modulus in the direction of the fibers, E_1, since they do not appear in the expression for that modulus. No study similar to that shown in Figs. 3-25 through 3-27 was performed for the composite major Poisson's ratio, ν_{12}.

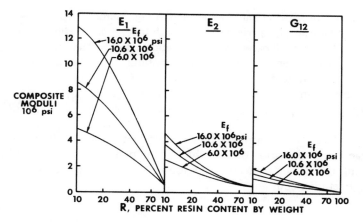

FIG. 3-25. Contribution of E_f to E_1, E_2, and G_{12}. (*After Tsai, Ref. 3-16.*)

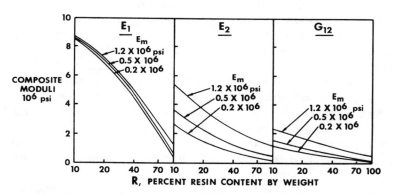

FIG. 3-26. Contribution of E_m to E_1, E_2, and G_{12}. (*After Tsai, Ref. 3-16.*)

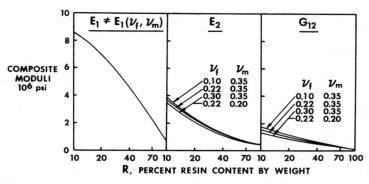

FIG. 3-27. Contribution of ν_f and ν_m to E_1, E_2, and G_{12}. (*After Tsai, Ref. 3-16.*)

The contiguity factor, C, is actually a so-called "fudge factor" used to make sense out of the comparison of experimental data with theoretical predictions. It is useful only when the data fall between the theoretical bounds. The concept of a contiguity factor, i.e., some expression of the continuity of one phase of a composite material relative to another, is more easily seen to affect the tensile properties of a lamina than the compressive properties. There may be some interesting relation of contiguity to granuie and fiber stiffnesses in tension and compression.

The contiguity factor can apparently take on different values for the same fiber volume fraction. The relation of C to the packing pattern is not obvious. Moreover, it is not clear why a higher modulus is obtained for a higher C even if the fiber volume fraction is constant.

3.3.5 The Halpin-Tsai Equations

All of the preceding micromechanics results are given by complicated equations and/or curves. The equations are usually somewhat awkward to use. The curves are generally restricted to a relatively small portion of the potential design regime. Thus, a need clearly exists for simpler results to be used in the design of composite materials.

Halpin and Tsai (Ref. 3-17) developed an interpolation procedure that is an approximate representation of more complicated micromechanics results. The beauty of the procedure is twofold. First, it is simple so it can readily be used in the design process. Second, it enables the generalization of usually limited, although more exact, micromechanics results. Moreover, the procedure is apparently quite accurate if the fiber volume fraction (V_f) does not approach one.

The essence of the procedure is that Halpin and Tsai (Ref. 3-17) showed that Hermans' solution (Ref. 3-13) generalizing Hill's self-consistent model (Ref. 3-12) can be reduced to the approximate form:

$$E_1 \cong E_f V_f + E_m V_m \tag{3.70}$$

$$\nu_{12} = \nu_f V_f + \nu_m V_m \tag{3.71}$$

and

$$\frac{M}{M_m} = \frac{1 + \xi \eta V_f}{1 - \eta V_f} \tag{3.72}$$

where

$$\eta = \frac{(M_f/M_m) - 1}{(M_f/M_m) + \xi} \tag{3.73}$$

in which

M = composite modulus E_2, G_{12}, or ν_{23}

M_f = corresponding fiber modulus E_f, G_f, or ν_f

M_m = corresponding matrix modulus E_m, G_m, or ν_m

and ξ is a measure of fiber reinforcement of the composite that depends on the fiber geometry, packing geometry, and loading conditions. The values of ξ are obtained by comparing Eqs. (3-72) and (3-73) with exact elasticity solutions and assessing a value of, or function for, ξ by curve fitting techniques.

Note that the expressions for E_1 and ν_{12} are the generally accepted rule of mixtures results. The Halpin-Tsai equations are equally applicable to fiber, ribbon, or particulate composites. For example, Halpin and Thomas (Ref. 3-18) successfully applied Eqs. (3-72) and (3-73) to analysis of the stiffness of glass ribbon-reinforced composites.

The only difficulty in using the Halpin-Tsai equations seems to be in the determination of a suitable value for ξ. Halpin and Tsai obtained excellent agreement with Adams and Doner's results (Refs. 3-19 and 3-20) for circular fibers in a square array when $\xi = 2$ for calculation of E_2 and $\xi = 1$ for calculation of G_{12} at a fiber volume fraction of .55 (see Figs. 3-28 and 3-29). For the same values of ξ, excellent agreement was also obtained with Foye's results (Refs. 3-21 and 3-22) for fibers with square cross sections in a diamond array when the fiber volume fraction ranged up through .9 as in Figs. 3-30 and 3-31. When Foye's rectangular cross-section fibers were addressed, Halpin and Tsai found that correlation with their equations required the value of ξ for transverse modulus calculations to be

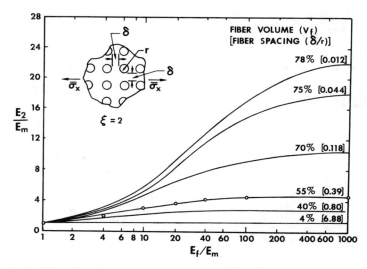

FIG. 3-28. Comparison of Halpin-Tsai calculations (circles) with Adams and Doner's calculations for E_2 of circular fibers in a square array. (*After Halpin and Tsai, Ref. 3-17.*)

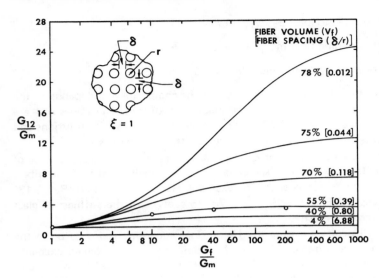

FIG. 3-29. Comparison of Halpin-Tsai calculations (circles) with Adams and Doner's calculations for G_{12} of circular fibers in a square array. (*After Halpin and Tsai, Ref. 3-17.*)

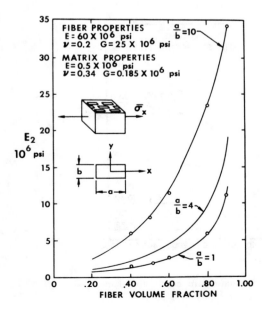

FIG. 3-30. Comparison of Halpin-Tsai calculations (circles) with Foye's calculations for E_2 of rectangular cross-section fibers in a diamond array. (*After Halpin and Tsai, Ref. 3-17.*)

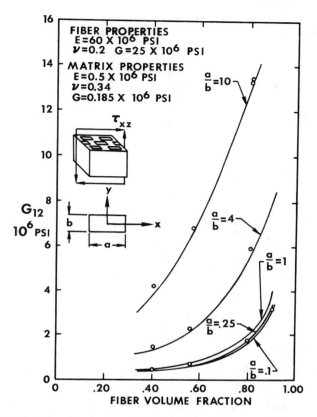

FIG. 3-31. Comparison of Halpin-Tsai calculations (circles) with Foye's calculations for G_{12} of rectangular cross-section fibers in a diamond array. (*After Halpin and Tsai, Ref. 3-17.*)

$$\xi_{E_2} = 2 \frac{a}{b} \tag{3.74}$$

where *a/b* is the rectangular cross-section aspect ratio. Also, the value of ξ for shear modulus calculations had to be

$$\log \xi_{G_{12}} = 1.73 \log \frac{a}{b} \tag{3.75}$$

to obtain the agreement with Foye's results shown in Figs. 3-30 and 3-31.

Predictions of the Halpin-Tsai equations for glass/epoxy and boron/epoxy composites are shown in Fig. 3-32 and 3-33. There, Foye's solutions for square arrays and hexagonal arrays are plotted in addition to Hermans' solution (to which the Halpin-Tsai equations are, of course, related). Note that the Halpin-Tsai predictions with $\xi = 2$ generally fall below the square array results but

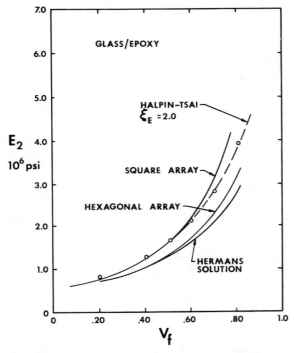

FIG. 3-32. Comparison of E_2 calculations for a glass/epoxy composite. (*After Halpin and Tsai, Ref. 3-17.*)

above the hexagonal array results for $V_f > .65$. Below that fiber volume fraction, the Halpin-Tsai results are quite close to Foye's square array results.

However, Hewitt and de Malherbe (Ref. 3-23) point out that the Halpin-Tsai equations yield an underestimate of the shear modulus G_{12} of composites with circular fibers in a square array for fiber volume fractions greater than .5. Specifically, the underestimate is 30 percent at $V_f = .75$ for $G_f/G_m = 20$, a realistic value for both glass/epoxy and graphite/epoxy composites. They suggested that, instead of Halpin and Tsai's value of one for ξ, the value determined from

$$\xi = 1 + 40V_f^{10} \tag{3.76}$$

correlates better in the Halpin-Tsai equations with Adams and Donner's numerical solution as shown in Fig. 3-34. Such a relation for ξ, like any other, is empirically determined. More refined estimates of ξ could be found, but care must be taken not to fall into the pit of deriving an expression that exceeds both the necessary accuracy requirements and defeats the original intent of a simple design tool that is easy to use.

The mere existence of different predicted stiffnesses for different arrays leads to an important physical observation: Variations in composite material

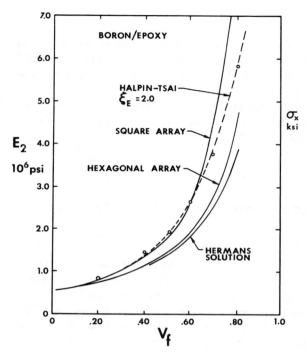

FIG. 3-33. Comparison of E_2 calculations for a boron/epoxy composite. (*After Halpin and Tsai, Ref. 3-17.*)

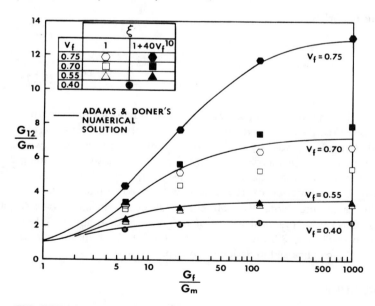

FIG. 3-34. Comparison of modified Halpin-Tsai calculations with Adams and Donner's calculations for G_{12} of circular fibers in a square array. (*After Hewitt and de Malherbe, Ref. 3-23.*)

manufacturing will always yield variations in array geometry and hence in composite moduli. Thus, we cannot hope to predict composite moduli precisely, *nor is there any need to*! Approximations such as the Halpin-Tsai equations should satisfy all practical requirements.

Some physical insight into the Halpin-Tsai equations can be gained by examining their behavior for the ranges of values of ξ and η. First, although it is not obvious, ξ can range from 0 to ∞. When $\xi = 0$,

$$\frac{1}{M} = \frac{V_f}{M_f} + \frac{V_m}{M_m} \tag{3.77}$$

which is the series-connected model generally associated with a lower bound of a composite modulus. When $\xi = \infty$,

$$M = V_f M_f + V_m M_m \tag{3.78}$$

which is the parallel-connected model, known as the rule of mixtures, generally associated with an upper bound of a composite modulus. Thus, ξ is a measure of the reinforcement of the composite by the fibers. For small values of ξ, the fibers are not very effective, whereas for large values of ξ, the fibers are extremely effective in increasing the composite stiffness above the matrix stiffness. Next, the limiting values of η can be shown to be: For rigid inclusions,

$$\eta = 1 \tag{3.79}$$

for homogeneous materials,

$$\eta = 0 \tag{3.80}$$

and for voids,

$$\eta = -\frac{1}{\xi} \tag{3.81}$$

The term ηV_f in Eq. (3.72) can be interpreted as a reduced fiber volume fraction. The word "reduced" is used because $\eta \leqslant 1$. Moreover, it is apparent from Eq. (3.73) that η is affected by the constituent material properties as well as by the reinforcement geometry factor ξ. To further assist in gaining appreciation of the Halpin-Tsai equations, the basic equation, Eq. (3.72), is plotted in Fig. 3-35 as a function of ηV_f. Curves with intermediate values of ξ can be quickly generated. Note that all curves approach infinity as ηV_f approaches one. Obviously, practical values of ηV_f are less than about .6, but most curves are shown in Fig. 3-35 for values up to about .9. Such master curves for various values of ξ can be used in design of composite materials.

3.3.6 Summary Remarks

There is much controversy associated with micromechanical analyses. Much of the controversy has to do with which approximations should be used. The

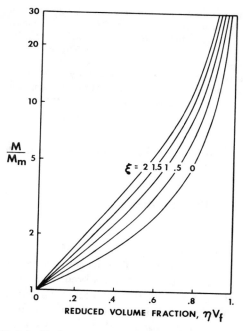

FIG. 3-35. Master M/M_m curves for various ξ.

Halpin-Tsai equations seem to be a flag around which many factions can gather; thus, special attention should be paid to them. The reader is alerted to examine the current literature (especially the *Journal of Composite Materials*) for additional results.

One of the important conclusions of some of the microstructure work (see Chamis and Sendeckyj, Ref. 3-3) is that macroscopic homogeneity may *not* exist for composites. That is, microstructural considerations may be required. The reader is also alerted to developments in that field.

Problem Set 3.3

Exercise 3.3.1 Derive Eqs. (3.53) and (3.54), and use them to derive Eq. (3.55).

Exercise 3.3.2 Consider a dispersion-stiffened composite material. Determine the influence on the upper bound for the apparent Young's modulus of different Poisson's ratios in the matrix and in the dispersed material. Consider the following three combinations of material properties of the constituent materials:

Case	E_m	ν_m	E_d	ν_d
1	5×10^6 psi	0	50×10^6 psi	.3
2	5×10^6 psi	.3	50×10^6 psi	.3
3	5×10^6 psi	.3	50×10^6 psi	0

The values of E for values of $V_d = 0$, .2, .4, .6, .8, and 1.0 (and any values in between necessary to plot representative curves) should be tabulated and plotted as E versus V_d so that both specific numerical and visual differences can be examined.

Exercise 3.3.3 Use the bounding techniques of elasticity to determine upper and lower bounds on the shear modulus, G, of a dispersion-stiffened composite material. Express the results in terms of the shear moduli of the constituents (G_m for the matrix and G_d for the dispersed particles) and their respective volume fractions (V_m and V_d). The representative volume element of the composite material should be subjected to a macroscopically uniform shear stress τ which results in a macroscopically uniform shear strain γ.

Exercise 3.3.4 Determine the bounds on E for a dispersion-stiffened composite material of more than two constituents, i.e., more than one type of particle is dispersed in a matrix material.

Exercise 3.3.5 Derive Eq. (3.77).

Exercise 3.3.6 Derive Eq. (3.78).

Exercise 3.3.7 Show that the limiting values of η are given correctly by Eqs. (3.79) through (3.81).

3.4 COMPARISON OF APPROACHES TO STIFFNESS

3.4.1 Particulate Composites

The mechanics of materials approach to the estimation of stiffness of a composite material has been shown to be an upper bound on the actual stiffness. Paul (Ref. 3-4) compared the upper and lower bound stiffness predictions with experimental data (Refs. 3-24 and 3-25) for an alloy of tungsten carbide in cobalt. Tungsten carbide (WC) has a Young's modulus of 102×10^6 psi and a Poisson's ratio of .22. Cobalt (Co) has a Young's modulus of 30×10^6 psi and a Poisson's ratio of .3.

The constituent material properties are substituted in Eqs. (3.62) and (3.58) to obtain the upper bound on E of the composite and in Eq. (3.48) to obtain the lower bound on E. In addition, the mechanics of materials approach studied in Exercises 3.2.1 through 3.2.4 is also compared with the experimental data. Specifically, the result for the modulus of a composite material that is stiffened by dispersion of cube-shaped particles is used, that is,

$$\frac{E}{E_m} = \frac{E_m + (E_d - E_m)V_d^{2/3}}{E_m + (E_d - E_m)V_d^{2/3}\left(1 - V_d^{1/3}\right)} \tag{3.82}$$

The predictions of the various approaches are plotted along with the experimental data in Fig. 3-36.

Note in Fig. 3-36 that the upper bound on Young's modulus, E, is indistinguishable from a straight line. Thus, the effect of Poisson's ratio, ν, in Eq. (3.62) on the result of Eq. (3.58) is negligible for the Poisson's ratios and Young's moduli of cobalt and tungsten carbide. For practical purposes, then, the upper bound is given by the simple mechanics of materials expression known as the rule of mixtures:

$$E = E_m V_m + E_d V_d \tag{3.83}$$

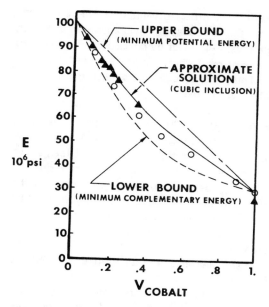

FIG. 3-36. Comparison of predicted modulus bounds with experimental data for tungsten carbide in cobalt. (*After Paul, Ref. 3-4.*)

in which $\nu_m = \nu_d$. The experimental data for E fall between the upper and lower bounds for E as they must for the bounds to be true bounds. Moreover, the approximate mechanics of materials prediction Eq. (3.82), appears to agree very well with the experimental data indicated by the triangles. The data indicated by open circles do not correlate particularly well with the approximate prediction of Eq. (3.82). However, no information is available regarding the shape of the dispersed particles. In addition, the constituent materials may not be precisely the same in both sets of experimental data. Hashin and Shtrikman (Ref. 3-6) obtained much closer bounds on the same experimental data.

3.4.2 Fiber-Reinforced Composites

Tsai (Ref. 3-16) conducted experiments to measure the various moduli of glass filament–epoxy resin composites. The glass fibers had a Young's modulus of 10.6 $\times$ 10^6 psi and a Poisson's ratio of .22. The epoxy resin had a Young's modulus of .5 $\times$ 10^6 psi and a Poisson's ratio of .35.

For various volume fractions of fibers, the experimental results were compared with the theoretical results in Eqs. (3.69), (3.65), (3.67), and (3.68) for E_1, E_2, ν_{12} and G_{12}, respectively. Since the theoretical results depend on k, the fiber misalignment factor, and C, the contiguity factor, theoretical curves can be drawn for a wide range of values of k and C. The objective of the comparison of theoretical and experimental results is to demonstrate both

qualitative and quantitative agreement in order to validate a theoretical prediction. If the theoretical results have the same shape as the experimental results,
then the agreement is termed qualitative. Further, if by consistent adjustment of
the parameters k and C the two sets of results agree in value as well, then the
agreement is termed quantitative. Thus, the concepts of a fiber misalignment
factor and a contiguity factor will be investigated.

The experimental and theoretical results for E_1 are shown in Fig. 3-37 for a
resin content by weight ranging from 10 to 100 percent. Since E_1 is not a
function of C, only k was varied — two values were chosen: $k = 1$ and $k = .9$.
Some experimental results in Fig. 3-37 lie above the curve for $k = 1$ (i.e., above
the upper bound!); some results lie below $k = .9$. However most results lie
between $k = .9$ and $k = 1$ with $k = .9$ being a conservative estimate of the
behavior. The actual specimens were handmade so the resin content may not be
precise and fiber misalignment is not unexpected. Thus, the results above the
upper bound are not unusual nor is the basic fact of variation in E_1.

The theoretical and experimental results for E_2 are shown in Fig. 3-37 as a
function of resin content by weight. Theoretical results from Eq. (3.65) are
shown for $C = 0$, .2, .4, and 1. Note that the experimental data are bounded by
the curves for $C = 0$ and $C = .4$. The theoretical curve labeled "glass-resin
connected in series" is a lower, lower bound than the $C = 0$ curve and is an
overly conservative estimate of the stiffness.

The existence of a contiguity factor, C, has been reasonably well demonstrated by the results of Fig. 3-37. However, a more critical examination of the
concept of a contiguity factor is in order. For this purpose, some special
experiments were devised in which steel-epoxy composites were used. In order
to obtain a composite with $C = 0$ (no contiguity of fibers, i.e., no fibers touch),
steel rods were inserted in holes in an epoxy bar. For a composite with $C = 1$
(perfect contiguity of fibers, i.e., all fibers touch), epoxy resin was placed in
holes in a steel bar. In both cases, there were 54 holes transverse to the

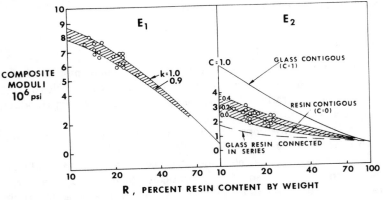

FIG. 3-37. E_1 and E_2 vs. resin content. (*After Tsai, Ref. 3-16.*)

FIG. 3-38. Steel/epoxy specimens. (*After Tsai, Ref. 3-16.*)

longitudinal axis of the bar. Thus, when the bars are pulled in their longitudinal direction, the modulus E_2 can be measured. The bars are shown in Fig. 3-38. The steel is always regarded as the fiber so $E_f = 30 \times 10^6$ psi and $\nu_f = .3$. The epoxy material has values $E_m = .45 \times 10^6$, $.60 \times 10^6$, and $.50 \times 10^6$ psi for three successive cases in addition to $\nu_m = .35$. The results for E_2 in the three cases are summarized in Fig. 3-39. Obviously the experimental data agree very well with the results from Eq. (3.65) for the cases $C = 0$ and $C = 1$. The data for

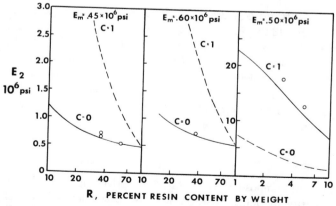

FIG. 3-39. E_2 of steel/epoxy composites with $C = 0$ and $C = 1$. (*After Tsai, Ref. 3-16.*)

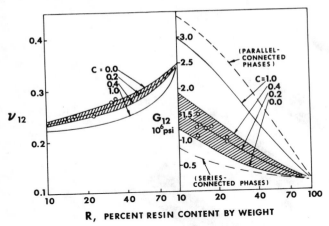

FIG. 3-40. ν_{12} and G_{12} of a glass/epoxy composite. (*After Tsai, Ref. 3-16.*)

$C = 1$ are fairly close to the theoretical results, but recognize that such small percentages of matrix are concerned that the comparison is difficult. On the other hand, the data for $C = 0$ agree to an extent that is a little surprising. Thus, the physical significance of the contiguity factor has been established.

The experimental results for ν_{12} of a glass/epoxy composite are shown along with the theoretical prediction from Eq. (3.67) as a function of resin content by weight in Fig. 3-40. Theoretical results are shown for contiguity factors of $C = 0, .2, .4,$ and 1. From the figure, apparently $C = 0$ is the upper limit of the data whereas $C = .4$ is the lower limit. Thus, the concept of a contiguity factor is further reinforced.

The experimental results for G_{12} are shown also in Fig. 3-40, along with theoretical results from Eq. (3.68) for $C = 0, .2, .4,$ and 1. As with the previous moduli, the experimental data are bounded by curves for $C = 0$ and $C = .4$. The upper (parallel-connected phases) and lower (series-connected phases) bounds due to Paul (see Sec. 3.3) are shown to demonstrate the accuracy of the bounds in the present case where E_f is much greater than E_m. The lower bound results of Hashin and Rosen (Ref. 3-7) correspond to $C = 0$, but their upper bound is below some of the experimental data in Fig. 3-40.

3.4.3 Summary Remarks

For particulate-reinforced composites, Paul (Ref. 3-4) derived upper and lower bounds on the composite modulus. His approximate mechanics of materials solution agrees fairly well with experimental data for tungsten carbide particles in cobalt.

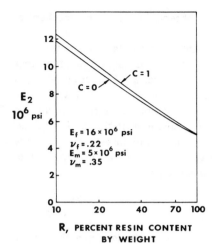

FIG. 3-41. E_2 of a fictitious glass/epoxy composite. (*After Tsai, Ref. 3-16.*)

For fiber-reinforced composites, Tsai (Ref. 3-16) gives expressions for E_1, E_2, ν_{12} and G_{12} that are in good agreement with experimental data for a glass fiber-reinforced resin composite. A contiguity factor, C, is the key to the agreement. Thus, the constituent material properties have the following effect on the properties of the composite:

(1) E_f makes a significant contribution to E_1.

(2) E_m makes a significant contribution to E_2 and G_{12}.

(3) ν_f and ν_m have little effect on E_2 and G_{12} and have no effect on E_1.

The contiguity factor is very important for glass fiber-reinforced composites wherein $E_f/E_m = 20$. However, for composites wherein E_f/E_m is close to unity, contiguity is probably not important. This latter conclusion is deduced from the results for $C = 0$ and $C = 1$ in Fig. 3-41 where a fictitious glass-epoxy composite is considered. There, a fictitious matrix with $E_m = 5 \times 10^6$ is combined with a high modulus glass fiber ($E_f = 16 \times 10^6$ psi) to give $E_f/E_m = 3.2$. Note in Fig. 3-41 that E_2, as calculated from Eq. (3.65), changes very little between $C = 0$ and $C = 1$.

3.5 MECHANICS OF MATERIALS APPROACH TO STRENGTH

3.5.1 Introduction

Prediction of the strength of fiber-reinforced composites has not achieved the near-esoteric levels of the stiffness predictions studied in the preceding sections. Nevertheless, there are many interesting physical models for the strength characteristics of a matrix reinforced by fibers. Most of the models represent a very high degree of integration of physical observation with the mechanical description of a phenomenon.

Two major topics will be addressed in this section: tensile strength and compressive strength of a unidirectionally reinforced lamina in the fiber direction. The tensile strength will be examined in Sec. 3.5.2 by use of a model with fibers that all have the same strength in addition to a model in which the fibers have a statistical strength distribution. The compressive strength will be examined in Sec. 3.5.3 by use of a model for buckling of fibers in a matrix. These two topics have occupied the attention of many fine investigators over the past ten years or so. However, to date, little work has been done on other topics of obvious importance such as prediction of shear strength.

3.5.2 Tensile Strength in the Fiber Direction

A unidirectional fiber-reinforced composite deforms as the load increases in the following four stages, more or less, depending on the relative brittleness or ductility of the fibers and the matrix:

1. Both fibers and matrix deform elastically.
2. The fibers continue to deform elastically, but the matrix deforms plastically.
3. Both the fibers and the matrix deform plastically.
4. The fibers fracture followed by fracture of the composite.

Of course, for brittle fibers, stage 3 may not be realized. Similarly, a brittle matrix may not achieve either stage 2 or 3. Whether fracture of the composite occurs as a fiber failure or a matrix failure depends on the relative ductility of the fibers and matrix.

Fibers of equal strength

Consider the case of fibers that all have the same strength and are relatively brittle in comparison to the matrix as studied by Kelly and Davies (Ref. 3-26). If the composite has more than a certain minimum volume fraction of fibers, V, the ultimate strength is achieved when the fibers are strained to correspond to their maximum (ultimate) stress. That is, in terms of strains,

$$\epsilon_{c_{max}} = \epsilon_{f_{max}} \tag{3.84}$$

Since the fibers are more brittle than the matrix, they cannot elongate as much as the matrix. Thus, the fibers are the weak link, from the strain viewpoint, in the strength chain.

The schematic stress-strain curves for the fibers and the matrix shown in Fig. 3-42 are useful in interpreting the reasoning to obtain the composite strength. Thus, if the fiber strain is presumed equal to the matrix strain in the direction of the fibers, then the ultimate strength of the composite is

$$\sigma_{c_{max}} = \sigma_{f_{max}} V_f + (\sigma_m)_{\epsilon_{f_{max}}} (1 - V_f) \tag{3.85}$$

where

$$\sigma_{f_{max}} = \text{maximum fiber tensile stress}$$

$$(\sigma_m)_{\epsilon_{f_{max}}} = \text{matrix stress at a matrix strain equal to the maximum tensile strain in the fibers.}$$

Obviously, if fiber reinforcement is to effect a greater strength than can be obtained by the matrix alone, then

$$\sigma_{c_{max}} > \sigma_{m_{max}} \tag{3.86}$$

Then, Eqs. (3.85) and (3.86) can be solved for the critical V_f that must be exceeded in order to obtain fiber strengthening:

$$V_{critical} = \frac{\sigma_{m_{max}} - (\sigma_m)_{\epsilon_{f_{max}}}}{\sigma_{f_{max}} - (\sigma_m)_{\epsilon_{f_{max}}}} \tag{3.87}$$

For smaller values of V_f, the behavior of the composite may not follow Eq. (3.85) because there may not be enough fibers to control the matrix elongation. Thus, the fibers would be subjected to high strains with only small loads and would fracture. If all fibers break at the same strain (an occurrence that is unlikely from a statistical standpoint), then the composite will fracture unless the matrix can take the entire load imposed on the composite, that is,

$$\sigma_{c_{max}} < \sigma_{m_{max}} V_m \tag{3.88}$$

Finally, the entire composite fails after fracture of the fibers if

$$\sigma_{c_{max}} = \sigma_{f_{max}} V_f + (\sigma_m)_{\epsilon_{f_{max}}} (1 - V_f) \geqslant \sigma_{m_{max}} (1 - V_f) \tag{3.89}$$

from which a minimum V_f for validity of Eq. (3.89) can be obtained as

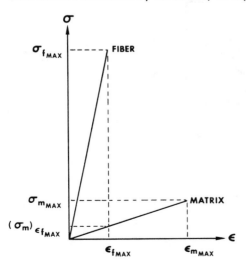

FIG. 3-42. Schematic stress-strain curves for fibers and matrix.

$$V_{minimum} = \frac{\sigma_{m_{max}} - (\sigma_m)_{\epsilon_{f_{max}}}}{\sigma_{f_{max}} + \sigma_{m_{max}} - (\sigma_m)_{\epsilon_{f_{max}}}} \qquad (3.90)$$

The preceding expressions, Eqs. (3.85) through (3.90), are more easily understood when they are plotted as in Fig. 3-43. There, the composite strength (i.e., the maximum composite stress) is plotted as a function of the fiber volume fraction. When V_f is less than $V_{minimum}$, the composite strength is controlled by the matrix deformation and is actually less than the matrix strength. When V_f is greater than $V_{minimum}$, but less than $V_{critical}$, the composite strength is controlled by the fiber deformation, but the composite strength is still less than the inherent matrix strength. Only when V_f exceeds $V_{critical}$ does the composite gain strength from having fiber reinforcement. Then, the composite strength is controlled by the fiber deformations since V_f is greater than $V_{minimum}$. Note that the shape of Fig. 3-43 will vary as $V_{critical}$ varies. Also, from Eq. (3.87), $V_{critical}$ is small when

$$\sigma_{m_{max}} \cong (\sigma_m)_{\epsilon_{f_{max}}} \qquad (3.91)$$

as is the case for glass fibers reinforcing a resin matrix. In the latter case, the composite strength is always fiber controlled since $V_{critical}$ always exceeds $V_{minimum}$.

The preceding analysis is premised on having continuous fibers of equal strength all of which fracture at the same longitudinal position. However, fibers under tension do not all have the same fracture strength nor do they fracture in the same place. Rather, because of the variable nature of fiber surface imperfections, the fibers have different fracture strengths. A statistical analysis is then necessary to rationally define the strength of a composite.

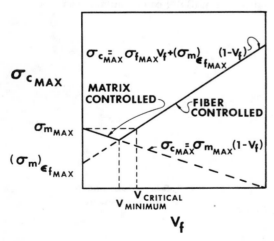

FIG. 3-43. Composite strength vs. fiber volume fraction. (*After Kelly and Davies, Ref. 3-26.*)

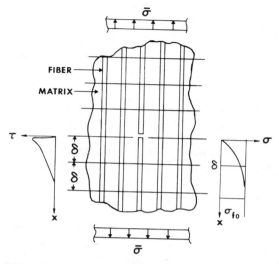

FIG. 3-44. Rosen's tensile failure model. (*After Dow and Rosen, Ref. 3-28.*)

Fibers with a statistical strength distribution

Rosen (Ref. 3-27) analyzed the strength of composites reinforced by fibers with a statistical strength distribution by use of the model shown in Fig. 3-44. There, the representative volume element is meant to include several fibers that are not broken and one fiber that is broken. Obviously, the representative volume element either changes size during loading and subsequent fiber fracture or else the number of fiber fractures in a fixed size volume element increases. The broken fiber has presumably been subjected to a stress high enough to initiate fracture at a surface imperfection. The broken fiber causes redistribution of stresses around the fracture. Stress must then pass from one end of the broken fiber past the break to the other end. The mechanism for accomplishing this stress transfer is by developing high shear stresses in the matrix over a short distance from the fiber break as shown in Fig. 3-44. The longitudinal fiber stress is thereby increased from zero at the break to the stress level, σ_{fo}, of any other fiber in the composite.

Failure of the composite can then occur in two ways. First, the matrix shear stress around the fiber could exceed the allowable matrix shear stress. More precisely, the bond between the fiber and the matrix might be broken owing to high shear stress in the aforementioned mechanism for transfer of stress between broken fibers. Second, the fiber fracture could actually propagate across the matrix through other fibers and hence cause overall fracture of the composite. If a good bond is achieved between the fiber and the matrix and if the matrix fracture toughness is high, then the fiber fractures can continue until the statistical accumulation is sufficient to cause gross composite fracture.

By use of statistical analysis, Rosen (Ref. 3-28) obtained

$$\sigma_{c_{max}} = \sigma_{ref} V_f \left(\frac{1 - V_f^{1/2}}{V_f^{1/2}} \right)^{-1/(2\beta)} \tag{3.92}$$

where σ_{ref} is a reference stress level that is a function of the fiber and matrix properties and β is a statistical parameter in the Weibull distribution of fiber strength.

Rosen's results are plotted in Fig. 3-45 for $\beta = 7.7$, a representative value for commercial E-glass filaments. Also plotted in Fig. 3-45 is the rule of mixtures expression

$$\sigma_{c_{max}} = \sigma_{ref} V_f \tag{3.93}$$

in which the tensile strength of the matrix has been ignored since it is much less than the fiber tensile strength. Thus, σ_{ref} must be interpreted as essentially the fiber tensile strength, but with some statistical implications. Note in Fig. 3-45 that Rosen's results from Eq. (3.92) do not go to one at $V_f = 1$. This behavior occurs because the fiber packing has a maximum density as a hexagonal array of uniform diameter fibers for which $V_f = .904$. Note that Rosen's results are close to the rule of mixtures expression. However, the two expressions are based on such widely differing approaches that it is falacious to infer from mere agreement with each other and with experimental data that the correct physical theory has been found.

Several interesting conclusions can be drawn from observation of Fig. 3-45. Basically, the fracture strength of the composite exceeds that of an individual fiber since Rosen's results lie above the rule of mixtures expression. Moreover,

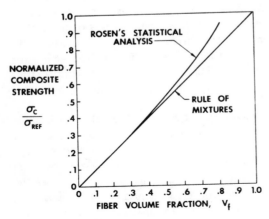

FIG. 3-45. Composite strength vs. fiber volume fraction. (*After Dow and Rosen, Ref. 3-28.*)

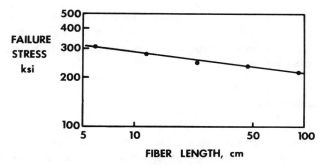

FIG. 3-46. Fiber strength vs. fiber length. (*After Dow and Rosen, Ref. 3-28.*)

the energy absorption capacity of the composite also exceeds that of the fibers. These characteristics follow from observation of Fig. 3-46 and 3-47 where, first, fiber strength (*E*-glass fibers) is shown to be inversely proportional to fiber length and, second, the number of fiber fractures is seen to increase as the ultimate load for the composite is approached. Note in Fig. 3-47 that fibers actually fracture at half the ultimate load and that the number of fractures rapidly accumulates until the overall composite fractures. Figure 3-46 on fiber strength versus length can be rationalized by likening a fiber with surface imperfections to a chain; the longer the chain (fiber), the higher the probability of a weak link (surface imperfection).

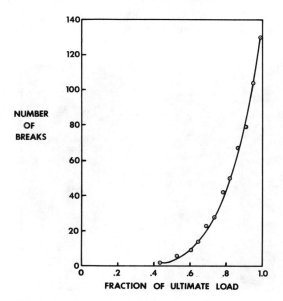

FIG. 3-47. Number of fiber fractures vs. percentage of ultimate composite strength. (*After Rosen, Dow, and Hashin, Ref. 3-29.*)

Definitive studies of composite strength in tension from a micromechanics viewpoint simply do not exist. Obviously, much work remains in this area before composite materials can be accurately designed, i.e., constituents chosen and proportioned to resist a specified tensile stress.

3.5.3 Compressive Strength in the Fiber Direction

When fiber-reinforced composites are compressed, the mode of failure appears to be fiber buckling (Ref. 3-28). One indication of such failures is the periodic nature of the photoelastic stress pattern for the E-glass fibers of three different diameters in an epoxy matrix shown in Fig. 3-48. If fiber buckling were to occur in the matrix, then a column on an elastic foundation model would appear to be reasonable. For such a model, the buckle wavelength can be shown to be directly proportional to the fiber diameter. This theoretical result is verified by the experimental data shown in Fig. 3-49 where a best-fit linear relation is represented as a $45°$ line on the log-log plot. Moreover, the overall hypothesis that fiber buckling is responsible for compressive failure is further strengthened.

In addition to being caused by mechanical compressive loads, fiber buckling can be caused by shrinkage stresses developed during curing of the composite. The shrinkage stresses result from the matrix having a higher thermal coefficient of expansion than the fibers. As a matter of fact, the photoelastic stress patterns in Fig. 3-48 were due to shrinkage during curing.

Two modes of fiber buckling are possible. First, the fibers can buckle out of phase relative to one another (symmetric about a line halfway between the

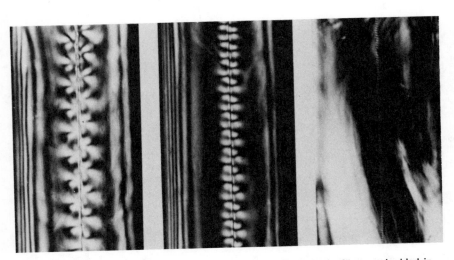

FIG. 3-48. Photoelastic stress patterns for three individual E-glass fibers embedded in an epoxy matrix. (*Courtesy of B. Walter Rosen, Materials Sciences Corporation.*)

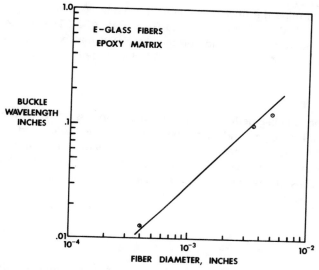

FIG. 3-49. Experimental results for fibers buckle wavelength vs. fiber diameter. (After Dow and Rosen, Ref. 3-28.)

fibers) to give the "transverse" or "extensional" buckling mode in Fig. 3-50a. The mode is so named because the matrix alternately deforms in extension and compression transverse to the fibers. The second mode, the "shear" mode, is so named because the matrix is subjected to shearing deformation since the fibers buckle in phase with one another (antisymmetrically with respect to the line halfway between the fibers) as shown in Fig. 3-50b.

In the model for both buckling modes, the fibers are regarded as plates h thick separated by matrix $2c$ wide. Thus, the problem is made two-dimensional since the dimension out of the Figure is disregarded. Each fiber is subjected to compressive load P and is of length L. The fibers are also regarded as being much stiffer than the matrix (that is, $G_f \gg G_m$) so the fiber shearing deformations are neglected.

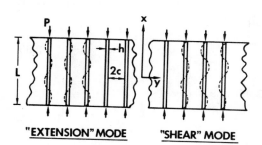

FIG. 3-50. Transverse or extensional mode and shear mode of fiber buckling.

The analysis to determine the fiber buckling load in each mode is based on the energy method described by Timoshenko and Gere (Ref. 3-30). The change in strain energy for the fiber, ΔU_f, and for the associated matrix, ΔU_m, is equated to the work done by the fiber force, ΔW, during deformation to a buckled state, that is,

$$\Delta U_f + \Delta U_m = \Delta W \tag{3.94}$$

In the energy method, buckle deflection configurations are assumed for the various buckle modes. The corresponding buckling loads are then calculated by use of Eq. (3.94). An important feature of the energy method is that the calculated buckling loads are an upper bound to the actual buckling load for the problem considered. Thus, if the buckling displacement of an individual fiber in the y-direction (transverse to the fibers in Fig. 3-50) is represented by the series

$$v = \sum_{n=1}^{\infty} a_n \sin \frac{n\pi x}{L} \tag{3.95}$$

a buckling load will be obtained that is higher than the actual buckling load. If Eq. (3.95) is used in energy expressions for transverse buckling and for shear buckling of the fiber-reinforced composite, then the lowest of the two buckling loads governs the fiber buckling of the composite. A buckling mode having deformations intermediate to the transverse and shear modes (i.e., fiber deformations that are neither in phase nor perfectly out of phase) would be expected to have a higher buckling load than either of the two simple modes.

Transverse mode

For the transverse buckling mode in Fig. 3-50, the strain in the y-direction (transverse to the fibers) is presumed to be independent of y, that is,

$$\epsilon_y = \frac{2v}{2c} \tag{3.96}$$

whereupon

$$\sigma_y = E_m \frac{v}{c} \tag{3.97}$$

The change in strain energy is presumed to be dominated by the energy due to transverse stresses. Thus, for the matrix

$$\Delta U_m = \frac{1}{2} \int_V \sigma_y \epsilon_y \, dV \tag{3.98}$$

which, upon substitution of Eqs. (3.95)-(3.97), becomes

$$\Delta U_m = \frac{E_m L}{2c} \sum_n a_n^2 \tag{3.99}$$

For the fibers, in the manner of Timoshenko and Gere (Ref. 3-30),

$$\Delta U_f = \frac{\pi^4 E_f h^3}{48 L^3} \sum_n n^4 a_n^2 \tag{3.100}$$

Finally, the work done by external forces is

$$\Delta W = \frac{P \pi^2}{4L} \sum_n n^2 a_n^2 \tag{3.101}$$

in which, for this two-dimensional problem, the fiber load per unit width perpendicular to the plane of Fig. 3-50 is

$$P = \sigma_f h \tag{3.102}$$

i.e., the fiber axial stress times the fiber thickness. Upon substitution of the foregoing energy expressions in the buckling criterion, Eq. (3.94), the fiber buckling load is

$$P = \frac{\pi^2 E_f h^3}{12 L^2} \left(\frac{\sum\limits_n n^4 a_n^2 + \dfrac{24 L^4 E_m}{\pi^4 c h^3 E_f} \sum\limits_n a_n^2}{\sum\limits_n n^2 a_n^2} \right) \tag{3.103}$$

Now presume that P achieves a minimum for a particular sine wave, say the mth wave. Thus,

$$\sigma_{f_{cr}} = \frac{\pi^2 E_f h^2}{12 L^2} \left[m^2 + \frac{24 L^4 E_m}{\pi^4 c h^3 E_f} \left(\frac{1}{m^2} \right) \right] \tag{3.104}$$

From the aforementioned photoelasticity investigations (Ref. 3-28), m is obviously a large number. Thus, $\sigma_{f_{cr}}$ can be treated as a continuous function of m and the minimum $\sigma_{f_{cr}}$ is obtained from

$$\frac{\partial \sigma_{f_{cr}}}{\partial m} = 0 \tag{3.105}$$

subject to the condition that

$$\frac{\partial^2 \sigma_{f_{cr}}}{\partial m^2} > 0 \tag{3.106}$$

If m were small, the minimum $\sigma_{f_{cr}}$ must be found for *discrete* (integer) values of m. The following reasoning is offered in support of the preceding contention. Consider a hypothetical plot of $\sigma_{f_{cr}}$ versus m where $\sigma_{f_{cr}}$ only has values at integer m in Fig. 3-51. The buckling load for the indicated lowest minimum at $m = 2.7$ does not physically exist and deviates substantially in value from the physical minimum at $m = 2$ because the mode number is

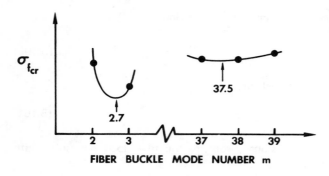

FIBER BUCKLE MODE NUMBER m

FIG. 3-51. Hypothetical relative minima of $\sigma_{f_{cr}}$.

squared in the buckling expression. However, the buckling load for the second minimum at $m = 37.5$, although it does not physically exist, is a reasonably close approximation to the actual minimum since the percentage difference between the buckling load for $m = 37.5$ and that for $m = 37$ or 38 is negligible. The minimum of $\sigma_{f_{cr}}$ as a continuous function of m is

$$\sigma_{f_{cr}} = 2 \sqrt{\frac{V_f E_m E_f}{3(1 - V_f)}} \tag{3.107}$$

as can easily be verified (see Exercise 3.5.7). In the preceding derivation, recognize that

$$V_f = \frac{h}{h + 2c} \tag{3.108}$$

The critical stress in the composite is then

$$\sigma_{c_{max}} = V_f \sigma_{f_{cr}} = 2V_f \sqrt{\frac{V_f E_m E_f}{3(1 - V_f)}} \tag{3.109}$$

wherein the matrix is assumed to be essentially unstressed in comparison to the fibers. The strain at buckling can be calculated from Eq. (3.107) and the uniaxial stress-strain relation as

$$\epsilon_{f_{cr}} = 2 \sqrt{\frac{V_f}{3(1 - V_f)}} \left(\frac{E_m}{E_f}\right)^{1/2} \tag{3.110}$$

Then, if the matrix is assumed to have the same strain in the fiber direction as the fiber,

$$\sigma_m = E_m \epsilon_{f_{cr}} \tag{3.111}$$

whereupon the maximum composite stress is

$$\sigma_{c_{max}} = V_f \sigma_{f_{cr}} + V_m \sigma_m \tag{3.112}$$

or

$$\sigma_{c_{max}} = \left[V_f + (1 - V_f) \frac{E_m}{E_f} \right] \sigma_{f_{cr}} \tag{3.113}$$

Finally,

$$\sigma_{c_{max}} = 2 \left[V_f + (1 - V_f) \frac{E_m}{E_f} \right] \sqrt{\frac{V_f E_m E_f}{3(1 - V_f)}} \tag{3.114}$$

The difference between Eqs. (3.109) and (3.114) is slight for high ratios of E_f to E_m as in practical fiber-reinforced composites.

Shear mode

For the shear instability mode in Fig. 3-50, the fiber displacements are equal and in phase with one another. The shear strains are presumed to be a function of the fiber direction coordinate alone. The matrix is sheared according to

$$\gamma_{xy} = \frac{\partial v}{\partial x} + \frac{\partial u}{\partial y} \tag{3.115}$$

where v is the displacement in the y-direction and u is the displacement in the x-direction. Then, since the transverse displacement is independent of the transverse coordinate y,

$$\left. \frac{dv}{dx} \right|_{matrix} = \left. \frac{dv}{dx} \right|_{fiber} \tag{3.116}$$

Since the shear strain is independent of y,

$$\frac{\partial u}{\partial y} = \frac{1}{2c} [u(c) - u(-c)] \tag{3.117}$$

as can be verified by examination of Fig. 3-52. Next, since the shear deformation of the fiber is ignored,

$$u(c) = \frac{h}{2} \left. \frac{dv}{dx} \right|_{fiber} \tag{3.118}$$

But from substitution of Eq. (3.118) in Eq. (3.117),

$$\frac{\partial u}{\partial y} = \frac{h}{2c} \left. \frac{dv}{dx} \right|_{fiber} \tag{3.119}$$

Now substitute Eq. (3.119) and Eq. (3.116) in Eq. (3.115) to get

$$\gamma_{xy} = \left(1 + \frac{h}{2c} \right) \left. \frac{dv}{dx} \right|_{fiber} \tag{3.120}$$

Recall that

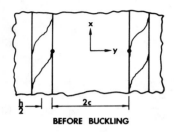

BEFORE BUCKLING

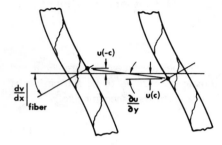

AFTER BUCKLING

FIG. 3-52. Fiber deformations during shear mode buckling.

$$\tau_{xy} = G_m \gamma_{xy} \tag{3.121}$$

The change in strain energy of the matrix is merely that due to shear:

$$\Delta U_m = \frac{1}{2} \int_V \tau_{xy} \gamma_{xy} \, dV \tag{3.122}$$

Substitute the deflection function, Eq. (3.95), the shear strain expression, Eq. (3.120), and the stress-strain relation, Eq. (3.121), in Eq. (3.122) to get

$$\Delta U_m = G_m c \left(1 + \frac{h}{2c}\right)^2 \frac{\pi^2}{2L} \sum_n n^2 a_n^2 \tag{3.123}$$

The change in strain energy of the fiber is still given by Eq. (3.100), and the work done is still that in Eq. (3.101). Thus, on application of the buckling criterion, Eq. (3.94),

$$\sigma_{f_{cr}} = \frac{G_m}{V_f(1 - V_f)} + \frac{\pi^2 E_f}{12} \left(\frac{mh}{L}\right)^2 \tag{3.124}$$

Since the buckle wave length is L/m, the second term in Eq. (3.124) is small when the wave length is large relative to the fiber diameter, h. Thus, the fiber buckling stress is approximately

$$\sigma_{f_{cr}} = \frac{G_m}{V_f(1 - V_f)} \tag{3.125}$$

The maximum composite stress is then

$$\sigma_{c_{max}} = \frac{G_m}{1 - V_f}$$

(3.126)

and the critical strain is

$$\epsilon_{cr} = \frac{1}{V_f(1 - V_f)}\left(\frac{G_m}{E_f}\right)$$

(3.127)

The maximum composite stress expressions, Eqs. (3.109) and (3.126), are plotted in Fig. 3-53 for a glass/epoxy composite. Note that the shear mode has the lowest strength for the composite over a wide range of fiber volume fractions. However, the transverse or extensional mode does govern the composite strength for low fiber volume fractions. For fiber volume fractions of between .6 and .7, the predicted compressive strength is between 450 and 600 ksi. These strength levels have not been obtained for glass-reinforced epoxy composites. If such a composite were to have a strength on the order of 500 ksi, the strain would have to exceed 5 percent. Under these conditions, the matrix would deform plastically. Thus, the predicted strength should be below the curve labeled "elastic shear mode" in Fig. 3-53. As an approximation to the inelastic behavior, Dow and Rosen (Ref. 3-28) replaced the matrix shear

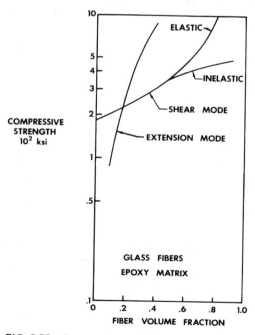

FIG. 3-53. Compressive strength of glass/epoxy composites. (*After Dow and Rosen, Ref. 3-28.*)

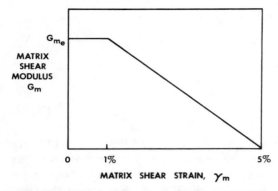

FIG. 3-54. Variation of matrix shear modulus with shear strain.

modulus in Eq. (3.126) by a shear modulus that varies linearly from the elastic value at 1 percent strain to a zero value at 5 percent strain as in Fig. 3-54. The resulting strength curve is labeled "inelastic shear mode" in Fig. 3-53. The predicted compressive strengths then appear more reasonable for glass/epoxy composites.

Dow and Rosen's results are plotted in another form, composite strain at buckling versus fiber volume fraction, in Fig. 3-55. These results are Eq. (3.127) for two values of the ratio of fiber Young's modulus to matrix shear modulus (E_f/G_m) at a matrix Poisson's ratio of .25. As in the previous form of Dow and

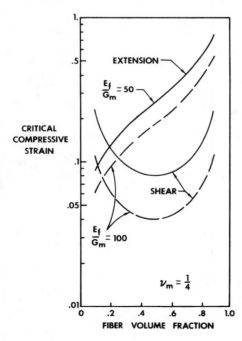

FIG. 3-55. Maximum compressive strain for fiber-reinforced composites. (*After Dow and Rosen, Ref. 3-28.*)

Rosen's results, the shear mode governs the composite behavior for a wide range of fiber volume fractions. Moreover, note that a factor of 2 change in the ratio E_f/G_m causes a factor of 2 change in the maximum composite compressive strain. Thus, the importance of the matrix shear modulus reduction due to inelastic deformation is quite evident.

Schuerch (Ref. 3-31) examined boron/metal composites parametrically with Rosen's equations and found them to require plastic buckling analysis. Moreover, so do S-glass/epoxy composites, but boron/epoxy composites apparently buckle elastically according to Schuerch.

Lager and June (Ref. 3-32) compared Dow and Rosen's theoretical predictions with experimental results for boron/epoxy composites with two different matrix materials. The theory appears to correlate well with the data if the matrix moduli in Eqs. (3.109) and (3.126) are multiplied by .63, that is,

$$\left(\sigma_{c\max}\right)_{\text{transverse}} = 2V_f \sqrt{\frac{V_f(.63E_m)E_f}{3(1 - V_f)}} \tag{3.128}$$

$$\left(\sigma_{c\max}\right)_{\text{shear}} = \frac{.63G_m}{1 - V_f} \tag{3.129}$$

The fibers were laid up in a near-perfect square array as verified by inspection of magnified pictures of machined cross sections. The resulting comparison of theory is given by Eqs. (3.128) and (3.129) and the data are shown in Fig. 3-56. Note that the initial (elastic) modulus was used with the reduction factor of .63. The "influence" coefficient of .63 is apparently due to the matrix not becoming plastic to the same degree in all directions. (That is, Lager and June

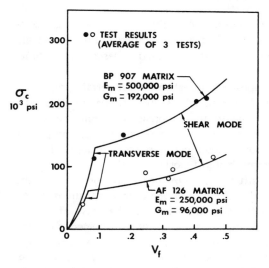

FIG. 3-56. Compressive strength of boron/epoxy composites. (*After Lager and June, Ref. 3-32.*)

disagree with Schuerch's contention that boron/epoxy composites buckle elastically). The influence coefficient is believed to be a strong function of the matrix modulus. For example, if reinforcing cloth such as fiberglass cloth is added to the matrix, the influence coefficient increases to .97.

Sadowsky, Pu, and Hussain (Ref. 3-33) treated the fiber as a column in an elastic matrix. Thus, one limit of their model's behavior is the Euler column buckling load.

For more information on the strength of fiber-reinforced composites, the reader is urged to consult the excellent survey articles by Corten (Ref. 3-34) and Rosen (Refs. 3-35 and 3-36).

Problem Set 3.5

Exercise 3.5.1 Derive Eq. (3.87).
Exercise 3.5.2 Derive Eq. (3.90).
Exercise 3.5.3 Derive Eq. (3.99).
Exercise 3.5.4 Derive Eq. (3.100).
Exercise 3.5.5 Derive Eq. (3.101).
Exercise 3.5.6 Derive Eq. (3.103).
Exercise 3.5.7 Derive Eq. (3.107) and verify that it is a minimum.
Exercise 3.5.8 Derive Eq. (3.123).
Exercise 3.5.9 Derive Eq. (3.124).

3.6 SUMMARY REMARKS ON MICROMECHANICS

In all the micromechanics results presented in this book, we have seen an attempt to *predict* the mechanical properties of the composite material based on the mechanical properties of its constituent materials. In nearly all cases of fiber-reinforced composites, there is considerable difference between expectation and realization. Thus, we must ask what is the usefulness of micromechanical analysis. Basically, there are two answers; one related to designing a material and one related to designing a structure.

First, if we are designing a composite material to achieve certain properties, we must have a design rationale. That rationale is micromechanics. However, obviously adjustments must be made to the rationale to obtain agreement between the predicted and the actual properties for given constituent properties and volume percentages. That is, something must be done to make up for the quantitative shortcomings of micromechanics theories.

Second, if we are designing a structure, we might ideally wish to have the freedom to design the material for the structure as well as the structure itself. In such a case, we would have need for micromechanics in the sense of answer number one. However, we would much more likely be obliged to standardize the material (e.g., a particular boron/epoxy tape) and concentrate on how to use the standard material to best advantage. Specifically, how to orient laminae of known (*measured,* not predicted) properties to achieve design goals would be the thrust of our efforts. Thus, the second possible answer to the question of the usefulness of micromechanics is that in many cases there is virtually *no* need for micromechanics. That is, the structural designer will probably rely almost exclusively on the results of mechanical tests for his material property data. He cannot risk using unsubstantiated micromechanics predictions that are often considerably in error.

Irrespective of the interest of the reader, he should be exposed to both micromechanics and macromechanics in order to function effectively in either material design or structural design. The main thrust of this book is in line with structural design and analysis requirements. Thus, the point of our addressing micromechanics is to better understand how composites function.

REFERENCES

3-1 Ekvall, J. C.: Structural Behavior of Monofilament Composites, *Proc. AIAA 6th Structures and Materials Conf.,* AIAA, New York, April, 1965.

3-2 Ekvall, J. C.: Elastic Properties Of Orthotropic Monofilament Composites, *ASME Paper* 61-AV-56, Aviation Conference, Los Angeles, Calif., March 12-16, 1961.

3-3 Chamis, C. C. and G. P. Sendeckyj: Critique on Theories Predicting Thermoelastic Properties of Fibrous Composites, *J. Composite Materials,* July, 1968, pp. 332-358.

3-4 Paul, B.: Prediction of Elastic Constants of Multiphase Materials, *Trans. Metallurgical Society of AIME,* February, 1960, pp. 36-41.

3-5 Hashin, Zvi: The Elastic Moduli of Heterogeneous Materials, *J. Appl. Mech.,* March, 1962, pp. 143-150.

3-6 Hashin, Zvi, and S. Shtrikman: A Variational Approach to the Theory of the Elastic Behaviour of Multiphase Materials, *J. Mech. Phys. Solids,* March-April 1963, pp. 127-140.

3-7 Hashin, Zvi, and B. Walter Rosen: The Elastic Moduli of Fiber-Reinforced Materials, *J. Appl. Mech.* June, 1964, pp. 223-232. Errata, March, 1965, p. 219.

3-8 Muskhelishvili, N. I.: "Some Basic Problems of the Mathematical Theory of Elasticity," P. Noordhoff, Groningen, The Netherlands, 1953.

3-9 Herrmann, L. R., and K. S. Pister: Composite Properties of Filament-Resin Systems, *ASME Paper* 63-WA-239, presented at the ASME Winter Annual Meeting, Philadelphia, Pa., November 17-22, 1963.

3-10 Adams, Donald F., and Stephen W. Tsai: The Influence of Random Filament Packing on the Elastic Properties of Composite Materials, *J. Composite Materials,* July, 1969, pp. 368-381.

3-11 Whitney, J. M., and M. B. Riley: Elastic Properties of Fiber Reinforced Composite Materials, *AIAA Journal,* September, 1966, pp. 1537-1542.

3-12 Hill, R.: Theory of Mechanical Properties of Fibre-Strengthened Materials — III. Self-Consistent Model, *J. Mech. Phys. Solids,* August, 1965, pp. 189-198.

3-13 Hermans, J. J.: The Elastic Properties of Fiber Reinforced Materials when the Fibers are Aligned, *Proc. Kon. Ned. Akad. Wetensch.,* Amsterdam, Ser B. Vol. 70, No. 1, pp. 1-9, 1967.

3-14 Whitney, J. M.: Elastic Moduli of Unidirectional Composites with Anisotropic Filaments, *J. Composite Materials,* April, 1967, pp. 188-193.

3-15 Whitney, James M.: Geometrical Effects of Filament Twist on the Modulus and Strength of Graphite Fiber-Reinforced Composites, *Textile Res. J.,* September, 1966, pp. 765-770.

3-16 Tsai, Stephen W.: Structural Behavior of Composite Materials, *NASA* CR-71, July, 1964.

3-17 Halpin, J. C., and S. W. Tsai: Effects of Environmental Factors on Composite Materials, AFML-TR 67-423, June, 1969.

3-18 Halpin, J. C., and R. L. Thomas: Ribbon Reinforcement of Composites, *J. Composite Materials,* October, 1968, pp. 488-497.

3-19 Adams, D. F., and D. R. Doner: Transverse Normal Loading of a Unidirectional Composite, *J. Composite Materials,* April, 1967, pp. 152-164.

3-20 Adams, D. F., and D. R. Doner: Longitudinal Shear Loading of a Unidirectional Composite, *J. Composite Materials,* January, 1967, pp. 4-17.

3-21 Foye, R. L.: An Evaluation of Various Engineering Estimates of the Transverse Properties of Unidirectional Composites, *Proc. 10th Nat. Symp. Soc. Aerosp. Mater. Process Eng.,* San Diego, Calif., Nov. 9-11, 1966, pp. G-31 to G-42.

3-22 Foye, R. L.: Structural Composites, *Quarterly Progress Reports Nos. 1 and 2,* AFML Contract No. AF 33(615)-5150, 1966.

3-23 Hewitt, R. L., and M. C. de Malherbe: An Approximation for the Longitudinal Shear Modulus of Continuous Fibre Composites, *J. Composite Materials,* April, 1970, pp. 280-282.

3-24 Nishimatsu, C., and J. Gurland: Experimental Survey of the Deformation of a Hard-Ductile Two-Phase Alloy System, WC-Co, *Brown Univ. Div. Eng. Tech. Rept. No. 2,* September, 1958.

3-25 Kieffer, R., and P. Schwartzkopf: "Hartstoffe und Hartmetalle," Springer, Vienna, 1953.

3-26 Kelly, A., and G. J. Davies: The Principles of the Fibre Reinforcement of Metals, *Met. Rev.,* 1965, pp. 1-77.

3-27 Rosen, B. Walter: Tensile Failure of Fibrous Composites, *AIAA Journal,* November, 1964, pp. 1985-1991.

3-28 Dow, Norris F., and B. Walter Rosen: Evaluations of Filament-Reinforced Composites for Aerospace Structural Applications, *NASA* CR-207, April, 1965.

3-29 Rosen, B. Walter, Norris F. Dow, and Zvi Hashin: Mechanical Properties of Fibrous Composites, *NASA* CR-31, April, 1964.

3-30 Timoshenko, S. P., and J. M. Gere: "Theory of Elastic Stability," 2d ed., McGraw-Hill Book Company, New York, 1961.

3-31, Schuerch, H.: Compressive Strength of Boron-Metal Composites, *NASA* CR-202, April, 1965.

3-32 Lager, John R., and Reid R. June: Compressive Strength of Boron-Epoxy Composites, *J. Composite Materials,* January, 1969, pp. 48-56.

3-33 Sadowsky, M. A., S. L. Pu, and M. A. Hussain: Buckling of Microfibers, *J. Appl. Mech.,* December, 1967, pp. 1011-1016.

3-34 Corten, Herbert T., Micromechanics and Fracture Behavior of Composites, chap. 2 in Lawrence J. Broutman and Richard H. Krock (eds.), "Modern Composite Materials," Addison-Wesley, Reading, Mass., 1967.

3-35 Rosen, B. Walter: Mechanics of Composite Strengthening, chap. 3 in "Fiber Composite Materials," American Society for Metals, Metals Park, Ohio, 1965.

3-36 Rosen, B. Walter: Strength of Uniaxial Fibrous Composites, in F. W. Wendt, H. Liebowitz, and N. Perrone (eds.), "Mechanics of Composite Materials," *Proc. 5th Symp. Naval Structural Mechanics,* 1967, Pergamon, New York, 1970, pp. 621-651.

Chapter 4
MACROMECHANICAL
BEHAVIOR OF A LAMINATE

4.1 INTRODUCTION

A laminate is two or more laminae bonded together to act as an integral structural element (see, for example, Fig. 1-10). The laminae principal material directions are oriented to produce a structural element capable of resisting load in several directions. The stiffness of such a composite material configuration is obtained from the properties of the constituent laminae by procedures derived in this chapter. The procedures enable the analysis of laminates that have individual laminae with principal material directions oriented at arbitrary angles to the chosen or natural axes of the laminate. As a consequence of the arbitrary orientations, the laminate may not have definable principal directions.

Classical lamination theory is derived in Sec. 4.2. Special stiffnesses of practical interest are classified and examined in Sec. 4.3. The theoretical stiffnesses obtained by classical lamination theory are compared with experimental results in Sec. 4.4. In Sec. 4.5, the strength of various laminates is addressed. The stresses between the laminae of a laminate are examined in Sec. 4.6 and found to be a probable cause of delamination of some laminates. The broad problem of laminate design is discussed briefly in Sec. 4.7.

4.2 CLASSICAL LAMINATION THEORY

Classical lamination theory embodies a collection of stress and deformation hypotheses that are described in this section. By use of this theory, we can consistently proceed from the basic building block, the lamina, to the end result, a structural laminate.

Actually, because of the stress and deformation hypotheses that are an inseparable part of classical lamination theory, a more correct name would be classical thin lamination theory, or even classical laminated plate theory. We will use the common simplification classical lamination theory, but recognize that it is a convenient oversimplification of the rigorous nomenclature. In the composites literature, classical lamination theory is often abbreviated as CLT.

First, the stress-strain behavior of an individual lamina is reviewed in Sec. 4.2.1, and expressed in equation form for the kth lamina of a laminate. Then, the variations of stress and strain through the thickness of the laminate are determined in Sec. 4.2.2. Finally, the laminate stiffnesses, including the stiffnesses that are used to relate coupling between bending and extension, are derived in Sec. 4.2.3. The derivations in this section are quite similar to the classical work by Pister and Dong (Ref. 4-1) and Reissner and Stavsky (Ref. 4-2).

4.2.1 Lamina Stress-Strain Behavior

The stress-strain relations in principal material coordinates for a lamina of an orthotropic material under plane stress are

$$\begin{Bmatrix} \sigma_1 \\ \sigma_2 \\ \tau_{12} \end{Bmatrix} = \begin{bmatrix} Q_{11} & Q_{12} & 0 \\ Q_{12} & Q_{22} & 0 \\ 0 & 0 & Q_{66} \end{bmatrix} \begin{Bmatrix} \epsilon_1 \\ \epsilon_2 \\ \gamma_{12} \end{Bmatrix} \tag{4.1}$$

The reduced stiffnesses, Q_{ij}, are defined in terms of the engineering constants in Eq. (2.61). In any other coordinate system in the plane of the lamina, the stresses are

$$\begin{Bmatrix} \sigma_x \\ \sigma_y \\ \tau_{xy} \end{Bmatrix} = \begin{bmatrix} \bar{Q}_{11} & \bar{Q}_{12} & \bar{Q}_{16} \\ \bar{Q}_{12} & \bar{Q}_{22} & \bar{Q}_{26} \\ \bar{Q}_{16} & \bar{Q}_{26} & \bar{Q}_{66} \end{bmatrix} \begin{Bmatrix} \epsilon_x \\ \epsilon_y \\ \gamma_{xy} \end{Bmatrix} \tag{4.2}$$

where the transformed reduced stiffnesses, $\bar{Q}_{ij}$, are given in terms of the reduced stiffnesses, Q_{ij}, in Eq. (2.80).

The stress-strain relations in arbitrary coordinates, Eq. (4.2), are useful in the definition of the laminate stiffnesses because of the arbitrary orientation of the constituent laminae. Both Eqs. (4.1) and (4.2) can be thought of as stress-strain relations for the kth layer of a multilayered laminate. Thus, Eq. (4.2) can be written as

$$\{\sigma\}_k = [\bar{Q}]_k \{\epsilon\}_k \tag{4.3}$$

We will proceed in the next section to define the strain and stress variations through the thickness of a laminate. The resultant forces and moments on a laminate will then be obtained in Sec. 4.2.3 by integrating Eq. (4.3) through the laminate thickness subject to the stress and strain variations determined in Sec. 4.2.2.

4.2.2 Strain and Stress Variation in a Laminate

Knowledge of the variation of stress and strain through the laminate thickness is essential to the definition of the extensional and bending stiffnesses of a laminate. The laminate is presumed to consist of perfectly bonded laminae. Moreover, the bonds are presumed to be infinitesimally thin as well as non–shear-deformable. That is, the displacements are continuous across lamina boundaries so that no lamina can slip relative to another. Thus, the laminate acts as a single layer with very *special* properties, but nevertheless acts as a single layer of material.

Accordingly, if the laminate is thin, a line originally straight and perpendicular to the middle surface of the laminate is assumed to remain straight and perpendicular to the middle surface when the laminate is extended and bent. Requiring the normal to the middle surface to remain straight and normal under deformation is equivalent to ignoring the shearing strains in planes perpendicular to the middle surface, that is, $\gamma_{xz} = \gamma_{yz} = 0$ where z is the direction of the normal to the middle surface in Fig. 4-1. In addition, the normals are presumed to have constant length so that the strain perpendicular to the middle surface is ignored as well, that is, $\epsilon_z = 0$. The foregoing collection of assumptions of the behavior of the single layer that represents the laminate constitutes the familiar Kirchhoff hypothesis for plates and the Kirchhoff–Love hypothesis for shells. Note that

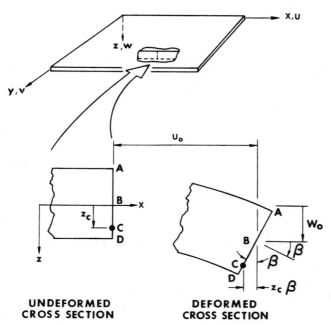

UNDEFORMED CROSS SECTION   **DEFORMED CROSS SECTION**

FIG. 4-1. Geometry of deformation in the *xz* plane.

no restriction has been made to flat laminates; the laminates can, in fact, be curved or shell-like.

The implications of the Kirchhoff or the Kirchhoff-Love hypothesis on the laminate displacements u, v, and w in the x-, y-, and z-directions are derived by use of the laminate cross section in the xz plane shown in Fig. 4-1. The displacement in the x-direction of point B from the undeformed to the deformed middle surface is u_0. Since line ABCD remains straight under deformation of the laminate,

$$u_c = u_0 - z_c \beta \tag{4.4}$$

But since, under deformation, line ABCD further remains perpendicular to the middle surface, β is the slope of the laminate middle surface in the x-direction, that is,

$$\beta = \frac{\partial w_0}{\partial x} \tag{4.5}$$

Then, the displacement, u, at any point z through the laminate thickness is

$$u = u_0 - z \frac{\partial w_0}{\partial x} \tag{4.6}$$

By similar reasoning, the displacement, v, in the y-direction is

$$v = v_0 - z \frac{\partial w_0}{\partial y} \tag{4.7}$$

The laminate strains have been reduced to ϵ_x, ϵ_y, and γ_{xy} by virtue of the Kirchhoff-Love hypothesis. That is, $\epsilon_z = \gamma_{xz} = \gamma_{yz} = 0$. For small strains (linear elasticity), the remaining strains are defined in terms of displacements as

$$\epsilon_x = \frac{\partial u}{\partial x}$$

$$\epsilon_y = \frac{\partial v}{\partial y} \tag{4.8}$$

$$\gamma_{xy} = \frac{\partial u}{\partial y} + \frac{\partial v}{\partial x}$$

Thus, for the derived displacements u and v in Eqs. (4.6) and (4.7), the strains are

$$\epsilon_x = \frac{\partial u_0}{\partial x} - z \frac{\partial^2 w_0}{\partial x^2}$$

$$\epsilon_y = \frac{\partial v_0}{\partial y} - z \frac{\partial^2 w_0}{\partial y^2} \tag{4.9}$$

$$\gamma_{xy} = \frac{\partial u_0}{\partial y} + \frac{\partial v_0}{\partial x} - 2z \frac{\partial^2 w_0}{\partial x \partial y} \qquad \begin{array}{l} (4.9) \\ (\text{cont'd.}) \end{array}$$

or

$$\begin{Bmatrix} \epsilon_x \\ \epsilon_y \\ \gamma_{xy} \end{Bmatrix} = \begin{Bmatrix} \epsilon_x^0 \\ \epsilon_y^0 \\ \gamma_{xy}^0 \end{Bmatrix} + z \begin{Bmatrix} \kappa_x \\ \kappa_y \\ \kappa_{xy} \end{Bmatrix} \qquad (4.10)$$

where the middle surface strains are

$$\begin{Bmatrix} \epsilon_x^0 \\ \epsilon_y^0 \\ \gamma_{xy}^0 \end{Bmatrix} = \begin{Bmatrix} \dfrac{\partial u_0}{\partial x} \\[2mm] \dfrac{\partial v_0}{\partial y} \\[2mm] \dfrac{\partial u_0}{\partial y} + \dfrac{\partial v_0}{\partial x} \end{Bmatrix} \qquad (4.11)$$

and the middle surface curvatures are

$$\begin{Bmatrix} \kappa_x \\ \kappa_y \\ \kappa_{xy} \end{Bmatrix} = - \begin{Bmatrix} \dfrac{\partial^2 w_0}{\partial x^2} \\[2mm] \dfrac{\partial^2 w_0}{\partial y^2} \\[2mm] 2\dfrac{\partial^2 w_0}{\partial x \partial y} \end{Bmatrix} \qquad (4.12)$$

(The last term in Eq. (4.12) is the twist curvature of the middle surface). Thus, the Kirchhoff or the Kirchhoff-Love hypothesis has been readily verified to imply a linear variation of strain through the laminate thickness. Because of the strain-displacement relations in Eq. (4.8), the foregoing strain analysis is valid only for plates. For shells, the ϵ_y term in Eq. (4.8) must be supplemented by w_0/r where r is the radius of a circular cylindrical shell; other shells have more complicated strain-displacement relations.

By substitution of the strain variation through the thickness, Eq. (4.10), in the stress-strain relations, Eq. (4.3), the stresses in the kth layer can be expressed in terms of the laminate middle surface strains and curvatures as

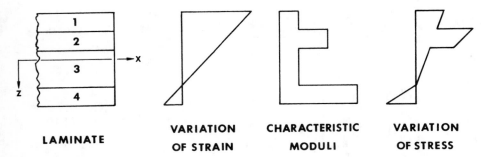

| LAMINATE | VARIATION OF STRAIN | CHARACTERISTIC MODULI | VARIATION OF STRESS |

FIG. 4-2. Hypothetical variation of strain and stress through the laminate thickness.

$$\left\{\begin{matrix} \sigma_x \\ \sigma_y \\ \tau_{xy} \end{matrix}\right\}_k = \begin{bmatrix} \bar{Q}_{11} & \bar{Q}_{12} & \bar{Q}_{16} \\ \bar{Q}_{12} & \bar{Q}_{22} & \bar{Q}_{26} \\ \bar{Q}_{16} & \bar{Q}_{26} & \bar{Q}_{66} \end{bmatrix}_k \left\{ \left\{\begin{matrix} \epsilon_x^0 \\ \epsilon_y^0 \\ \gamma_{xy}^0 \end{matrix}\right\} + z \left\{\begin{matrix} \kappa_x \\ \kappa_y \\ \kappa_{xy} \end{matrix}\right\} \right\} \qquad (4.13)$$

Since the $\bar{Q}_{ij}$ can be different for each layer of the laminate, the stress variation through the laminate thickness is not necessarily linear, even though the strain variation is linear. Instead, typical strain and stress variations are shown in Fig. 4-2.

4.2.3 Resultant Laminate Forces and Moments

The resultant forces and moments acting on a laminate are obtained by integration of the stresses in each layer or lamina through the laminate thickness, for example,

$$N_x = \int_{-t/2}^{t/2} \sigma_x \, dz$$

$$M_x = \int_{-t/2}^{t/2} \sigma_x z \, dz \qquad (4.14)$$

Actually, N_x is a force per unit length (width) of the cross section of the laminate as shown in Fig. 4-3. Similarly, M_x is a moment per unit length as shown in Fig. 4-4. The entire collection of force and moment resultants for an N-layered laminate is depicted in Figs. 4-3 and 4-4 and is defined as

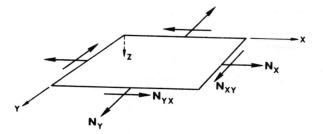

FIG. 4-3. In-plane forces on a flat laminate.

$$\begin{Bmatrix} N_x \\ N_y \\ N_{xy} \end{Bmatrix} = \int_{-t/2}^{t/2} \begin{Bmatrix} \sigma_x \\ \sigma_y \\ \tau_{xy} \end{Bmatrix} dz = \sum_{k=1}^{N} \int_{z_{k-1}}^{z_k} \begin{Bmatrix} \sigma_x \\ \sigma_y \\ \tau_{xy} \end{Bmatrix}_k dz \qquad (4.15)$$

and

$$\begin{Bmatrix} M_x \\ M_y \\ M_{xy} \end{Bmatrix} = \int_{-t/2}^{t/2} \begin{Bmatrix} \sigma_x \\ \sigma_y \\ \tau_{xy} \end{Bmatrix} z\,dz = \sum_{k=1}^{N} \int_{z_{k-1}}^{z_k} \begin{Bmatrix} \sigma_x \\ \sigma_y \\ \tau_{xy} \end{Bmatrix}_k z\,dz \qquad (4.16)$$

where z_k and z_{k-1} are defined in Fig. 4-5. Note that $z_0 = -t/2$. These force and moment resultants do not depend on z after integration, but are functions of x and y, the coordinates in the plane of the laminate middle surface.

The integration indicated in Eqs. (4.15) and (4.16) can be rearranged to take advantage of the fact that the stiffness matrix for a lamina is constant within the lamina.[1] Thus, the stiffness matrix goes outside the integration over each layer, but is within the summation of force and moment resultants for each layer. When the lamina stress-strain relations, Eq. (4.13), are substituted,

[1] Unless the lamina has temperature-dependent properties and a temperature gradient across the lamina exists.

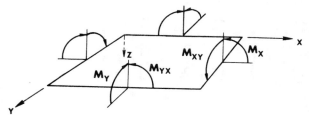

FIG. 4-4. Moments on a flat laminate.

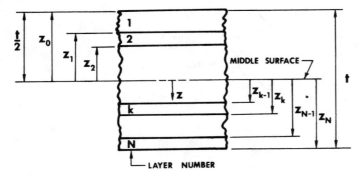

FIG. 4-5. Geometry of an N-layered laminate.

$$\begin{Bmatrix} N_x \\ N_y \\ N_{xy} \end{Bmatrix} = \sum_{k=1}^{N} \begin{bmatrix} \bar{Q}_{11} & \bar{Q}_{12} & \bar{Q}_{16} \\ \bar{Q}_{12} & \bar{Q}_{22} & \bar{Q}_{26} \\ \bar{Q}_{16} & \bar{Q}_{26} & \bar{Q}_{66} \end{bmatrix}_k \left\{ \int_{z_{k-1}}^{z_k} \begin{Bmatrix} \epsilon_x^0 \\ \epsilon_y^0 \\ \gamma_{xy}^0 \end{Bmatrix} dz + \int_{z_{k-1}}^{z_k} \begin{Bmatrix} \kappa_x \\ \kappa_y \\ \kappa_{xy} \end{Bmatrix} z\,dz \right\}$$

(4.17)

$$\begin{Bmatrix} M_x \\ M_y \\ M_{xy} \end{Bmatrix} = \sum_{k=1}^{N} \begin{bmatrix} \bar{Q}_{11} & \bar{Q}_{12} & \bar{Q}_{16} \\ \bar{Q}_{12} & \bar{Q}_{22} & \bar{Q}_{26} \\ \bar{Q}_{16} & \bar{Q}_{26} & \bar{Q}_{66} \end{bmatrix}_k \left\{ \int_{z_{k-1}}^{z_k} \begin{Bmatrix} \epsilon_x^0 \\ \epsilon_y^0 \\ \gamma_{xy}^0 \end{Bmatrix} z\,dz + \int_{z_{k-1}}^{z_k} \begin{Bmatrix} \kappa_x \\ \kappa_y \\ \kappa_{xy} \end{Bmatrix} z^2\,dz \right\}$$

(4.18)

However, we should now recall that ϵ_x^0, ϵ_y^0, γ_{xy}^0, κ_x, κ_y, and κ_{xy} are not functions of z but are middle surface values so can be removed from under the summation signs. Thus, Eqs. (4.17) and (4.18) can be written as

$$\begin{Bmatrix} N_x \\ N_y \\ N_{xy} \end{Bmatrix} = \begin{bmatrix} A_{11} & A_{12} & A_{16} \\ A_{12} & A_{22} & A_{26} \\ A_{16} & A_{26} & A_{66} \end{bmatrix} \begin{Bmatrix} \epsilon_x^0 \\ \epsilon_y^0 \\ \gamma_{xy}^0 \end{Bmatrix} + \begin{bmatrix} B_{11} & B_{12} & B_{16} \\ B_{12} & B_{22} & B_{26} \\ B_{16} & B_{26} & B_{66} \end{bmatrix} \begin{Bmatrix} \kappa_x \\ \kappa_y \\ \kappa_{xy} \end{Bmatrix}$$

(4.19)

$$\begin{Bmatrix} M_x \\ M_y \\ M_{xy} \end{Bmatrix} = \begin{bmatrix} B_{11} & B_{12} & B_{16} \\ B_{12} & B_{22} & B_{26} \\ B_{16} & B_{26} & B_{66} \end{bmatrix} \begin{Bmatrix} \epsilon_x^0 \\ \epsilon_y^0 \\ \gamma_{xy}^0 \end{Bmatrix} + \begin{bmatrix} D_{11} & D_{12} & D_{16} \\ D_{12} & D_{22} & D_{26} \\ D_{16} & D_{26} & D_{66} \end{bmatrix} \begin{Bmatrix} \kappa_x \\ \kappa_y \\ \kappa_{xy} \end{Bmatrix}$$

(4.20)

where

$$A_{ij} = \sum_{k=1}^{N} (\bar{Q}_{ij})_k (z_k - z_{k-1})$$

(4.21)

$$B_{ij} = \frac{1}{2} \sum_{k=1}^{N} (\bar{Q}_{ij})_k (z_k^2 - z_{k-1}^2)$$

(4.21)
(cont'd.)

$$D_{ij} = \frac{1}{3} \sum_{k=1}^{N} (\bar{Q}_{ij})_k (z_k^3 - z_{k-1}^3)$$

In Eqs. (4.19), (4.20), and (4.21), the A_{ij} are called extensional stiffnesses, the B_{ij} are called coupling stiffnesses, and the D_{ij} are called bending stiffnesses. The presence of the B_{ij} implies coupling between bending and extension of a laminate. Thus, it is impossible to pull on a laminate that has B_{ij} terms without at the same time bending and/or twisting the laminate. That is, an extensional force results in not only extensional deformations, but twisting and/or bending of the laminate. Also, such a laminate cannot be subjected to moment without at the same time suffering extension of the middle surface. The first observation is borne out by the experiment depicted in Fig. 4-6. There, a two-layered, nylon-reinforced laminate is subjected to the force resultant N_x and, because of the manner of support, $N_y = N_{xy} = M_x = M_{xy} = 0$. When the principal material directions of the laminae are oriented at $+\alpha$ and $-\alpha$ to the laminate x axis, we can show that

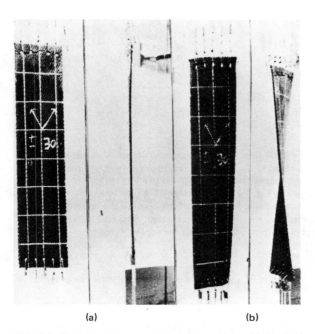

(a)   (b)

FIG. 4-6. Twisting of a two-layered unsymmetrical laminate under tensile load. (*After Ashton, Halpin, and Petit, Ref. 4-3.*)

$$N_x = A_{11} \epsilon_x^0 + A_{12} \epsilon_y^0 + B_{16} \kappa_{xy} \qquad (4.22)$$

Thus, the force resultant N_x produces twisting of the laminate as evidenced by the κ_{xy} term in addition to the usual extensional strains ϵ_x^0 and ϵ_y^0.

Problem Set 4.2

Exercise 4.2.1 Verify for a single layer of isotropic material with material properties E and ν and thickness t that the extensional and bending stiffnesses are

$$A_{11} = A_{22} = \frac{Et}{1 - \nu^2}$$

$$D_{11} = D_{22} = \frac{Et^3}{12(1 - \nu^2)}$$

which are commonly called B and D, respectively. What are the coupling stiffnesses?

Exercise 4.2.2 Derive the summation expressions for extensional, coupling, and bending stiffnesses for laminates with constant properties in each lamina; that is, derive Eq. (4.21) from Eqs. (4.17) and (4.18).

Exercise 4.2.3 Show that the stiffnesses in Eq. (4.21) can be written as

$$A_{ij} = \sum_{k=1}^{N} (\bar{Q}_{ij})_k t_k$$

$$B_{ij} = \sum_{k=1}^{N} (\bar{Q}_{ij})_k t_k \bar{z}_k$$

$$D_{ij} = \sum_{k=1}^{N} (\bar{Q}_{ij})_k \left(t_k \bar{z}_k^2 + \frac{t_k^3}{12} \right)$$

wherein t_k is the thickness and $\bar{z}_k$ is the distance to the centroid of the kth layer. What is the physical meaning of the coefficients of $(\bar{Q}_{ij})_k$ in each of the foregoing expressions?

Exercise 4.2.4 Determine the extensional, coupling, and bending stiffnesses of an equal thickness bimetallic strip as shown in Fig. 1-3 (a beam made of two different isotropic materials with E_1, ν_1, E_2, and ν_2). Use the middle surface of the beam as the reference surface.

Exercise 4.2.5 Demonstrate that the force per unit length on a two-layered laminate with laminae of equal thickness oriented at $+\alpha$ and $-\alpha$ to the applied force is

$$N_x = A_{11} \epsilon_x^0 + A_{12} \epsilon_y^0 + B_{16} \kappa_{xy}$$

What are A_{11}, A_{12}, and B_{16} in terms of the transformed reduced stiffnesses, $\bar{Q}_{ij}$, of a lamina and the lamina thickness, t?

4.3 SPECIAL CASES OF LAMINATE STIFFNESSES

This section is devoted to those special cases of laminates for which the stiffnesses take on certain simplified values as opposed to the general form in Eq.

(4.21). Some of the cases are almost trivial, others are more specialized, but all are contributions to the understanding of the concept of laminate stiffnesses. This section is a useful catalog of the special cases in approximately increasing order of complexity. Many of the cases result from the common practice of constructing laminates from laminae that have the same material properties and thickness, but have different orientations of their principal material directions relative to one another and relative to the laminate axes. Other more general cases are examined as well. The most general case of the stiffnesses is represented by Eq. (4.21) along with the force and moment resultants in Eqs. (4.19) and (4.20).

Stiffnesses for single-layered configurations are treated first to provide a baseline for subsequent discussion. Such stiffnesses should be recognizable in terms of concepts previously encountered by the reader in his study of plates and shells. Next, laminates that are symmetric about their middle surface are discussed and classified. Then, laminates with laminae that are antisymmetrically disposed about their middle surface are described. Finally, laminates with complete lack of middle surface symmetry are discussed.

4.3.1 Single-layered Configurations

The special single-layered configurations treated in this section are isotropic, specially orthotropic, generally orthotropic, and anisotropic. The generally orthotropic configuration cannot, of course, be distinguished from an anisotropic layer from the analysis point of view, but does have only the four independent material properties of an orthotropic material.

Single isotropic layer
For a single isotropic layer with material properties, E and ν, and thickness, t, the laminate stiffnesses of Eq. (4.21) reduce to

$$A_{11} = \frac{Et}{1 - \nu^2} = A \qquad\qquad D_{11} = \frac{Et^3}{12(1 - \nu^2)} = D$$

$$A_{12} = \nu A \qquad\qquad D_{12} = \nu D$$

$$A_{22} = A \qquad\qquad D_{22} = D$$

$$A_{16} = 0 \qquad\quad B_{ij} = 0 \qquad D_{16} = 0 \qquad\qquad (4.23)$$

$$A_{26} = 0 \qquad\qquad D_{26} = 0$$

$$A_{66} = \frac{Et}{2(1 + \nu)} = \frac{1 - \nu}{2} A \qquad D_{66} = \frac{Et^3}{24(1 + \nu)} = \frac{1 - \nu}{2} D$$

whereupon the resultant forces are dependent only on the in-surface strains of the laminate middle surface, and the resultant moments are dependent only on the curvatures of the middle surface:

$$\begin{Bmatrix} N_x \\ N_y \\ N_{xy} \end{Bmatrix} = \begin{bmatrix} A & \nu A & 0 \\ \nu A & A & 0 \\ 0 & 0 & \dfrac{1-\nu}{2} A \end{bmatrix} \begin{Bmatrix} \epsilon_x^0 \\ \epsilon_y^0 \\ \gamma_{xy}^0 \end{Bmatrix} \tag{4.24}$$

$$\begin{Bmatrix} M_x \\ M_y \\ M_{xy} \end{Bmatrix} = \begin{bmatrix} D & \nu D & 0 \\ \nu D & D & 0 \\ 0 & 0 & \dfrac{1-\nu}{2} D \end{bmatrix} \begin{Bmatrix} \kappa_x \\ \kappa_y \\ \kappa_{xy} \end{Bmatrix} \tag{4.25}$$

Thus, there is no coupling between bending and extension of a single isotropic layer. Also note that

$$D = \frac{At^2}{12} \tag{4.26}$$

Single specially orthotropic layer

For a single specially orthotropic layer of thickness, t, and lamina stiffnesses, Q_{ij}, given by Eq. (2.61), the laminate stiffnesses are

$$A_{11} = Q_{11}t \qquad\qquad\qquad D_{11} = \frac{Q_{11}t^3}{12}$$

$$A_{12} = Q_{12}t \qquad\qquad\qquad D_{12} = \frac{Q_{12}t^3}{12}$$

$$A_{22} = Q_{22}t \qquad B_{ij} = 0 \qquad D_{22} = \frac{Q_{22}t^3}{12} \tag{4.27}$$

$$A_{16} = 0 \qquad\qquad\qquad\qquad D_{16} = 0$$

$$A_{26} = 0 \qquad\qquad\qquad\qquad D_{26} = 0$$

$$A_{66} = Q_{66}t \qquad\qquad\qquad D_{66} = \frac{Q_{66}t^3}{12}$$

whereupon, as with a single isotropic layer, the resultant forces depend only on the in-surface strains, and the resultant moments depend only on the curvatures:

$$\begin{Bmatrix} N_x \\ N_y \\ N_{xy} \end{Bmatrix} = \begin{bmatrix} A_{11} & A_{12} & 0 \\ A_{12} & A_{22} & 0 \\ 0 & 0 & A_{66} \end{bmatrix} \begin{Bmatrix} \epsilon_x^0 \\ \epsilon_y^0 \\ \gamma_{xy}^0 \end{Bmatrix} \tag{4.28}$$

$$\begin{Bmatrix} M_x \\ M_y \\ M_{xy} \end{Bmatrix} = \begin{bmatrix} D_{11} & D_{12} & 0 \\ D_{12} & D_{22} & 0 \\ 0 & 0 & D_{66} \end{bmatrix} \begin{Bmatrix} \kappa_x \\ \kappa_y \\ \kappa_{xy} \end{Bmatrix} \qquad (4.29)$$

Single generally orthotropic layer

For a single generally orthotropic layer of thickness, t, and lamina stiffnesses, $\bar{Q}_{ij}$, given by Eq. (2.80), the laminate stiffnesses are

$$A_{ij} = \bar{Q}_{ij}t \qquad B_{ij} = 0 \qquad D_{ij} = \frac{\bar{Q}_{ij}t^3}{12} \qquad (4.30)$$

Again, there is no coupling between bending and extension so the force and moment resultants are:

$$\begin{Bmatrix} N_x \\ N_y \\ N_{xy} \end{Bmatrix} = \begin{bmatrix} A_{11} & A_{12} & A_{16} \\ A_{12} & A_{22} & A_{26} \\ A_{16} & A_{26} & A_{66} \end{bmatrix} \begin{Bmatrix} \epsilon_x^0 \\ \epsilon_y^0 \\ \gamma_{xy}^0 \end{Bmatrix} \qquad (4.31)$$

$$\begin{Bmatrix} M_x \\ M_y \\ M_{xy} \end{Bmatrix} = \begin{bmatrix} D_{11} & D_{12} & D_{16} \\ D_{12} & D_{22} & D_{26} \\ D_{16} & D_{26} & D_{66} \end{bmatrix} \begin{Bmatrix} \kappa_x \\ \kappa_y \\ \kappa_{xy} \end{Bmatrix} \qquad (4.32)$$

Note, in contrast to both an isotropic layer and a specially orthotropic layer, that extensional forces depend on shearing strain as well as on extensional strain. Also, the resultant shearing force, N_{xy}, depends on the extensional strains, ϵ_x^0 and ϵ_y^0, as well as on the shear strain, γ_{xy}^0. Similarly, the moment resultants all depend on both the curvatures, κ_x and κ_y, and on the twist, κ_{xy}.

Single anisotropic layer

The only difference in appearance between a single generally orthotropic layer and an anisotropic layer is that the latter has lamina stiffnesses, Q_{ij}, given implicitly by Eq. (2.84) whereas the generally orthotropic layer has stiffnesses, $\bar{Q}_{ij}$, given by Eq. (2.80). The laminate stiffnesses are

$$A_{ij} = Q_{ij}t \qquad B_{ij} = 0 \qquad D_{ij} = \frac{Q_{ij}t^3}{12} \qquad (4.33)$$

and the force and moment resultants are given by Eqs. (4.31) and (4.32).

4.3.2 Symmetric Laminates

For laminates that are symmetric in *both* geometry and material properties about the middle surface, the general stiffness equations, Eq. (4.21), simplify considerably. In particular, because of the symmetry of the $(\overline{Q}_{ij})_k$ and the thicknesses t_k, all the coupling stiffnesses, that is, the B_{ij}, can be shown to be zero. The elimination of coupling between bending and extension has two important practical ramifications. First, such laminates are usually much easier to analyze than laminates with coupling. Second, symmetric laminates do not have a tendency to twist from the inevitable thermally induced contractions that occur during cooling following the curing process. Consequently, symmetric laminates are commonly used unless special circumstances require an unsymmetric laminate. For example, part of the function of a laminate may be to serve as a heat shield, but the heat comes from only one side; thus, an unsymmetric laminate is likely to be used.

The force and moment resultants for a symmetric laminate are

$$
\left\{ \begin{array}{c} N_x \\ N_y \\ N_{xy} \end{array} \right\} = \begin{bmatrix} A_{11} & A_{12} & A_{16} \\ A_{12} & A_{22} & A_{26} \\ A_{16} & A_{26} & A_{66} \end{bmatrix} \left\{ \begin{array}{c} \epsilon_x^0 \\ \epsilon_y^0 \\ \gamma_{xy}^0 \end{array} \right\} \tag{4.34}
$$

$$
\left\{ \begin{array}{c} M_x \\ M_y \\ M_{xy} \end{array} \right\} = \begin{bmatrix} D_{11} & D_{12} & D_{16} \\ D_{12} & D_{22} & D_{26} \\ D_{16} & D_{26} & D_{66} \end{bmatrix} \left\{ \begin{array}{c} \kappa_x \\ \kappa_y \\ \kappa_{xy} \end{array} \right\} \tag{4.35}
$$

Special cases of symmetric laminates will be described in the following subsections. In each case, the A_{ij} and D_{ij} in Eqs. (4.34) and (4.35) take on different values and some will even vanish.

Symmetric laminates with multiple isotropic layers
If multiple isotropic layers of various thicknesses are arranged symmetrically about a middle surface from both a geometric and a material property standpoint, the resulting laminate does not exhibit coupling between bending and extension. A simple example of a symmetric laminate with three isotropic layers is shown in Fig. 4-7. A more complicated example of a symmetric laminate with six isotropic layers of different elastic properties and thicknesses is given in Table 4-1. Note that layers 3 and 4 in Table 4-1 could be regarded as a single layer of thickness $6t$ without changing the stiffness characterisitics.

The extensional and bending stiffnesses for the general case are calculated from Eq. (4.21) wherein for the kth layer

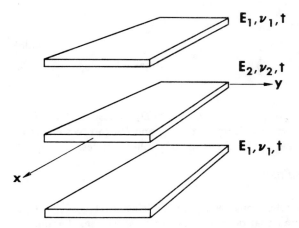

E_1, ν_1, t

E_2, ν_2, t

$\longrightarrow y$

E_1, ν_1, t

x

FIG. 4-7. Exploded (unbonded) view of a three-layered symmetric laminate with isotropic layers.

$$(\bar{Q}_{11})_k = (\bar{Q}_{22})_k = \frac{E_k}{1 - \nu_k^2} \qquad (\bar{Q}_{16})_k = (\bar{Q}_{26})_k = 0$$

$$(\bar{Q}_{12})_k = \frac{\nu_k E_k}{1 - \nu_k^2} \qquad\qquad (\bar{Q}_{66})_k = \frac{E_k}{2(1 + \nu_k)}$$

(4.36)

The force and moment resultants take the form

$$\begin{Bmatrix} N_x \\ N_y \\ N_{xy} \end{Bmatrix} = \begin{bmatrix} A_{11} & A_{12} & 0 \\ A_{12} & A_{11} & 0 \\ 0 & 0 & A_{66} \end{bmatrix} \begin{Bmatrix} \epsilon_x^0 \\ \epsilon_y^0 \\ \gamma_{xy}^0 \end{Bmatrix}$$

(4.37)

TABLE 4-1. Symmetric laminate with six multiple isotropic layers.

Layer	Material properties	Thickness
1	E_1, ν_1	t
2	E_2, ν_2	$2t$
3	E_3, ν_3	$3t$
4	E_3, ν_3	$3t$
5	E_2, ν_2	$2t$
6	E_1, ν_1	t

$$\begin{Bmatrix} M_x \\ M_y \\ M_{xy} \end{Bmatrix} = \begin{bmatrix} D_{11} & D_{12} & 0 \\ D_{12} & D_{22} & 0 \\ 0 & 0 & D_{66} \end{bmatrix} \begin{Bmatrix} \kappa_x \\ \kappa_y \\ \kappa_{xy} \end{Bmatrix} \qquad (4.38)$$

wherein, for isotropic layers, $A_{11} = A_{22}$ and $D_{11} = D_{22}$ because of the first condition of Eq. (4.36). The specific form of the A_{ij} and D_{ij} can be somewhat involved as can easily be verified by the use of some simple examples.

Symmetric laminates with multiple specially orthotropic layers

Because of the analytical complications involving the stiffnesses A_{16}, A_{26}, D_{16}, and D_{26}, a laminate is desired that does not have these stiffnesses. Laminates can be made with orthotropic layers that have principal material directions aligned with the laminate axes. If the thicknesses, locations, and material properties of the laminae are symmetric about the middle surface of the laminate, there is no coupling between bending and extension. A general example is shown in Table 4-2. The extensional and bending stiffnesses are calculated from Eq. (4.21) wherein for the kth layer

$$(\bar{Q}_{11})_k = \frac{E_1^k}{1 - \nu_{12}^k \nu_{21}^k} \qquad (\bar{Q}_{16})_k = 0$$

$$(\bar{Q}_{12})_k = \frac{\nu_{21}^k E_1^k}{1 - \nu_{12}^k \nu_{21}^k} \qquad (\bar{Q}_{26})_k = 0 \qquad (4.39)$$

$$(\bar{Q}_{22})_k = \frac{E_2^k}{1 - \nu_{12}^k \nu_{21}^k} \qquad (\bar{Q}_{66})_k = G_{12}^k$$

TABLE 4-2.   Symmetric laminate with five specially orthotropic layers.

Layer	Material properties				Orientation	Thickness
	Q_{11}	Q_{12}	Q_{22}	Q_{66}		
1	F_1	F_2	F_3	F_4	0°	t
2	G_1	G_2	G_3	G_4	90°	$2t$
3	H_1	H_2	H_3	H_4	90°	$4t$
4	G_1	G_2	G_3	G_4	90°	$2t$
5	F_1	F_2	F_3	F_4	0°	t

Because $(\bar{Q}_{16})_k$ and $(\bar{Q}_{26})_k$ are zero, the stiffnesses A_{16}, A_{26}, D_{16}, and D_{26} vanish. Also, the stiffnesses B_{ij} are zero because of symmetry. This type of laminate could therefore be called a specially orthotropic laminate in analogy to a specially orthotropic lamina. The force and moment resultants take the form of Eqs. (4.37) and (4.38), respectively.

A very common special case of symmetric laminates with multiple specially orthotropic layers occurs when the laminae are all of the same thickness and material properties, but have their major principal material directions alternating at $0°$ and $90°$ to the laminate axes, for example, $0°/90°/0°$. Such laminates are called *regular symmetric cross-ply laminates*. A simple example of a regular symmetric cross-ply laminate with three layers of equal thickness and properties is shown in Fig. 4-8. The fiber directions of each lamina are schematically indicated by the use of light lines in Fig. 4-8. The laminate must have an odd number of layers to satisfy the symmetry requirement by which coupling between bending and extension is eliminated. Cross-ply laminates with an even number of layers are obviously not symmetric and will be discussed in Sec. 4.3.3. The less common case of cross-ply laminates that have odd-numbered layers with equal thicknesses and even-numbered layers with thicknesses equal to each other but not to that of the odd-numbered layers will be discussed in Sec. 4.4, Comparison of Theoretical and Experimental Laminate Stiffnesses. A common example of such a laminate is ordinary plywood.

The logic to establish the various stiffnesses will be traced to illustrate the general procedures. First, consider the extensional stiffnesses

$$A_{ij} = \sum_{k=1}^{N} (\bar{Q}_{ij})_k (z_k - z_{k-1}) \tag{4.40}$$

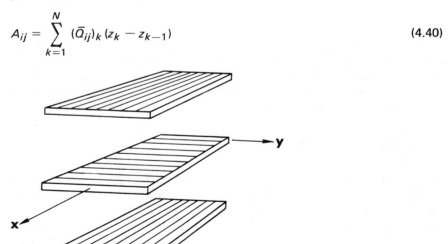

FIG. 4-8. Exploded (unbonded) view of a three-layered regular symmetric cross-ply laminate.

The A_{ij} are the sum of the product of the individual laminae $\bar{Q}_{ij}$ and the laminae thicknesses. Thus, the only ways to obtain a zero individual A_{ij} are for all $\bar{Q}_{ij}$ to be zero or for some $\bar{Q}_{ij}$ to be negative and some positive so that their products with their respective thicknesses sum to zero. From the expressions for the transformed lamina stiffnesses, $\bar{Q}_{ij}$, in Eq. (2.80), apparently $\bar{Q}_{11}$, $\bar{Q}_{12}$, $\bar{Q}_{22}$, and $\bar{Q}_{66}$ are positive definite since all trigonometric functions appear to even powers. Thus, A_{11}, A_{12}, A_{22}, and A_{66} are positive definite since the thicknesses are, of course, always positive. However, $\bar{Q}_{16}$ and $\bar{Q}_{26}$ are zero for lamina orientations of $0°$ and $90°$ to the laminate axes. Thus, A_{16} and A_{26} are zero for laminates of orthotropic laminae oriented at either $0°$ or $90°$ to the laminate axes.

Second, consider the coupling stiffnesses

$$B_{ij} = \frac{1}{2} \sum_{k=1}^{N} (\bar{Q}_{ij})_k (z_k^2 - z_{k-1}^2) \qquad (4.41)$$

If the cross-ply laminate is symmetric about the middle surface, then the B_{ij} all vanish as can easily be shown.

Finally, consider the bending stiffnesses

$$D_{ij} = \frac{1}{3} \sum_{k=1}^{N} (\bar{Q}_{ij})_k (z_k^3 - z_{k-1}^3) \qquad (4.42)$$

The D_{ij} are sum of the product of the individual laminae $\bar{Q}_{ij}$ and the term $(z_k^3 - z_{k-1}^3)$. Since $\bar{Q}_{11}$, $\bar{Q}_{12}$, $\bar{Q}_{22}$, and $\bar{Q}_{66}$ are positive definite and the geometric term is positive definite, then D_{11}, D_{12}, D_{22}, and D_{66} are positive definite. Also, $\bar{Q}_{16}$ and $\bar{Q}_{26}$ are zero for lamina principal material property orientations of $0°$ and $90°$ to the laminate coordinate axes. Thus, D_{16} and D_{26} are zero.

Symmetric laminates with multiple generally orthotropic layers

A laminate of multiple generally orthotropic layers that are symmetrically disposed about the middle surface exhibits no coupling between bending and extension; that is, the B_{ij} are zero. Therefore, the force and moment resultants are represented by Eqs. (4.34) and (4.35), respectively, There, all the A_{ij} and D_{ij} are required because of coupling between normal forces and shearing strain, shearing force and normal strains, normal moments and twist, and twisting moment and normal curvatures. Such coupling is evidenced by the A_{16}, A_{26}, D_{16}, and D_{26} stiffnesses.

A special subclass of this class of symmetric laminates is the *regular symmetric angle-ply laminate*. Such laminates have orthotropic laminae of equal

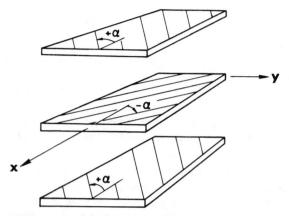

FIG. 4-9. Exploded (unbonded) view of a three-layered regular symmetric angle-ply laminate.

thicknesses. The adjacent laminae have opposite signs of the angle of orientation of the principal material properties with respect to the laminate axes, for example, $+\alpha/-\alpha/+\alpha$. Thus, for symmetry, there must be an odd number of layers. A simple example of a three-layered regular symmetric angle-ply laminate is shown in Fig. 4-9. A more complicated example of a symmetric laminate with generally orthotropic layers is given in Table 4-3.

The aforementioned coupling that involves A_{16}, A_{26}, D_{16}, and D_{26} takes on a special form for symmetric angle-ply laminates. Those stiffnesses can be shown to be largest when $N = 3$ (the lowest N for which this class of laminates exists) and decrease in proportion to $1/N$ as N increases. Actually, in the expressions for the extensional and bending stiffnesses A_{16} and D_{16}

$$A_{16} = \sum_{k=1}^{N} (\bar{Q}_{16})_k (z_k - z_{k-1}) \tag{4.43}$$

TABLE 4-3. Symmetric laminate with five generally orthotropic layers

Layer	Material properties				Orientation	Thickness
	Q_{11}	Q_{12}	Q_{22}	Q_{66}		
1	F_1	F_2	F_3	F_4	$+30°$	t
2	G_1	G_2	G_3	G_4	$-60°$	$3t$
3	H_1	H_2	H_3	H_4	$+15°$	$5t$
4	G_1	G_2	G_3	G_4	$-60°$	$3t$
5	F_1	F_2	F_3	F_4	$+30°$	t

$$D_{16} = \frac{1}{3} \sum_{k=1}^{N} (\bar{Q}_{16})_k (z_k^3 - z_{k-1}^3) \tag{4.44}$$

obviously, A_{16} and D_{16} are sums of terms of alternating signs since

$$(\bar{Q}_{16})_{+\alpha} = -(\bar{Q}_{16})_{-\alpha} \tag{4.45}$$

Thus, for many layered symmetric angle-ply laminates, the values of A_{16}, A_{26}, D_{16}, and D_{26} can be quite small when compared to the other A_{ij} and D_{ij}, respectively.

When the always present advantage of zero B_{ij} because of symmetry is considered in addition to the low A_{16}, A_{26}, D_{16}, and D_{26}, many layered symmetric angle-ply laminates can offer significant, practically advantageous simplifications over some more general laminates. In addition, symmetric angle-ply laminates offer more shear stiffness than do the simpler cross-ply laminates so are used more often. However, knowledge of the effect of A_{16}, A_{26}, D_{16}, and D_{26} on the individual class of problems being considered by an analyst or designer is essential because even a small A_{16} or D_{16} might cause significantly different results from cases in which those stiffnesses are exactly zero. Only in the situation where A_{16}, A_{26}, D_{16}, and D_{26} are exactly zero can they be ignored without further thought or analysis.

Symmetric laminates with multiple anisotropic layers

The general case of a laminate with multiple anisotropic layers symmetrically disposed about the middle surface does not have any stiffness simplifications other than the elimination of the B_{ij} by virtue of symmetry. The A_{16}, A_{26}, D_{16}, and D_{26} stiffnesses all exist and do not necessarily go to zero as the number of layers is increased. That is, the A_{16} stiffness, for example, is derived from the Q_{ij} matrix in Eq. (2.84) for an anisotropic lamina which, of course, has more independent material properties than an orthotropic lamina. Thus, many of the stiffness simplifications possible for other laminates cannot be achieved for this class.

4.3.3 Antisymmetric Laminates

Symmetry of a laminate about the middle surface is often desirable to avoid coupling between bending and extension. However, many physical applications of laminated composites require nonsymmetric laminates to achieve design requirements. For example, coupling is a necessary feature to make jet turbine fan blades with pretwist. As a further example, if the shear stiffness of a laminate made of laminae with unidirectional fibers must be increased, one way to achieve this requirement is to position layers at some angle to the laminate axes. To stay within weight and cost requirements, an even number of such layers may be necessary at orientations that alternate from layer to layer, e.g., $+\alpha/-\alpha/+\alpha/-\alpha$. Therefore, symmetry about the middle surface is destroyed and the behavioral characteristics of the laminate can be substantially changed from the symmetric case. Although the example laminate is not symmetric, it is antisymmetric about the middle surface, and certain stiffness simplifications are possible.

The general class of antisymmetric laminates must have an even number of layers if adjacent laminae have alternating signs of the principal material property directions with respect to the laminate axes. In addition, each pair of laminae must have the same thickness.

The stiffnesses of an antisymmetric laminate of anisotropic laminae do not simplify from those presented in Eqs. (4.19) and (4.20). However, as a consequence of antisymmetry of material properties of generally orthotropic laminae,[2] but symmetry of their thicknesses, the extensional coupling stiffness A_{16}

$$A_{16} = \sum_{k=1}^{N} (\bar{Q}_{16})_k (z_k - z_{k-1})$$ (4.46)

is easily seen to be zero since

$$(\bar{Q}_{16})_{+\alpha} = -(\bar{Q}_{16})_{-\alpha}$$ (4.47)

and layers symmetric about the middle surface have equal thickness and hence the same value of the geometric term multiplying $(\bar{Q}_{16})_k$. Similarly, A_{26} is zero as is the bending twist coupling stiffness D_{16},

$$D_{16} = \frac{1}{3} \sum_{k=1}^{N} (\bar{Q}_{16})_k (z_k^3 - z_{k-1}^3)$$ (4.48)

since again Eq. (4.47) holds and the geometric term multiplying $(\bar{Q}_{16})_k$ is the same for two layers symmetric about the middle surface. The preceding reasoning applies also for D_{26}.

[2] Because of the coupling between bending and extension, the terminology generally orthotropic and specially orthotropic have meaning only with reference to an individual layer and not to a laminate.

The coupling stiffnesses, B_{ij}, vary for different classes of antisymmetric laminates of generally orthotropic laminae, and, in fact, no general representation exists other than in the following force and moment resultants

$$\begin{Bmatrix} N_x \\ N_y \\ N_{xy} \end{Bmatrix} = \begin{bmatrix} A_{11} & A_{12} & 0 \\ A_{12} & A_{22} & 0 \\ 0 & 0 & A_{66} \end{bmatrix} \begin{Bmatrix} \epsilon_x^0 \\ \epsilon_y^0 \\ \gamma_{xy}^0 \end{Bmatrix} + \begin{bmatrix} B_{11} & B_{12} & B_{16} \\ B_{12} & B_{22} & B_{26} \\ B_{16} & B_{26} & B_{66} \end{bmatrix} \begin{Bmatrix} \kappa_x \\ \kappa_y \\ \kappa_{xy} \end{Bmatrix} \tag{4.49}$$

$$\begin{Bmatrix} M_x \\ M_y \\ M_{xy} \end{Bmatrix} = \begin{bmatrix} B_{11} & B_{12} & B_{16} \\ B_{12} & B_{22} & B_{26} \\ B_{16} & B_{26} & B_{66} \end{bmatrix} \begin{Bmatrix} \epsilon_x^0 \\ \epsilon_y^0 \\ \gamma_{xy}^0 \end{Bmatrix} + \begin{bmatrix} D_{11} & D_{12} & 0 \\ D_{12} & D_{22} & 0 \\ 0 & 0 & D_{66} \end{bmatrix} \begin{Bmatrix} \kappa_x \\ \kappa_y \\ \kappa_{xy} \end{Bmatrix} \tag{4.50}$$

The purpose of the remainder of this section is to discuss two important classes of antisymmetric laminates, the antisymmetric cross-ply laminate and the antisymmetric angle-ply laminate.

Antisymmetric cross-ply laminates

An antisymmetric cross-ply laminate consists of an even number of orthotropic laminae laid on each other with principal material directions alternating at $0°$ and $90°$ to the laminate axes as in the simple example of Fig. 4-10. A more complicated example is given in Table 4-4. Such laminates do not have A_{16}, A_{26}, D_{16}, and D_{26}, but do have coupling between bending and extension. We will demonstrate later that the coupling is such that the force and moment resultants are

$$\begin{Bmatrix} N_x \\ N_y \\ N_{xy} \end{Bmatrix} = \begin{bmatrix} A_{11} & A_{12} & 0 \\ A_{12} & A_{22} & 0 \\ 0 & 0 & A_{66} \end{bmatrix} \begin{Bmatrix} \epsilon_x^0 \\ \epsilon_y^0 \\ \gamma_{xy}^0 \end{Bmatrix} + \begin{bmatrix} B_{11} & 0 & 0 \\ 0 & -B_{11} & 0 \\ 0 & 0 & 0 \end{bmatrix} \begin{Bmatrix} \kappa_x \\ \kappa_y \\ \kappa_{xy} \end{Bmatrix} \tag{4.51}$$

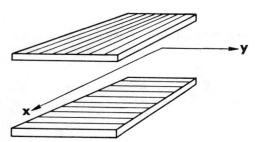

FIG. 4-10. Exploded (unbonded) view of a two-layered regular antisymmetric cross-ply laminate.

TABLE 4-4. Antisymmetric laminate with six specially orthotropic layers

Layer	Material properties				Orientation	Thickness
	Q_{11}	Q_{12}	Q_{22}	Q_{66}		
1	F_1	F_2	F_3	F_4	0°	t
2	G_1	G_2	G_3	G_4	90°	$3t$
3	H_1	H_2	H_3	H_4	90°	$2t$
4	H_1	H_2	H_3	H_4	0°	$2t$
5	G_1	G_2	G_3	G_4	0°	$3t$
6	F_1	F_2	F_3	F_4	90°	t

$$\begin{Bmatrix} M_x \\ M_y \\ M_{xy} \end{Bmatrix} = \begin{bmatrix} B_{11} & 0 & 0 \\ 0 & -B_{11} & 0 \\ 0 & 0 & 0 \end{bmatrix} \begin{Bmatrix} \epsilon_x^0 \\ \epsilon_y^0 \\ \gamma_{xy}^0 \end{Bmatrix} + \begin{bmatrix} D_{11} & D_{12} & 0 \\ D_{12} & D_{22} & 0 \\ 0 & 0 & D_{66} \end{bmatrix} \begin{Bmatrix} \kappa_x \\ \kappa_y \\ \kappa_{xy} \end{Bmatrix} \qquad (4.52)$$

A *regular antisymmetric cross-ply laminate* is defined to have laminae all of equal thickness and is common because of simplicity of fabrication. As the number of layers increases, the coupling stiffness B_{11} can be shown to approach zero.

Antisymmetric angle-ply laminates

An antisymmetric angle-ply laminate has laminae oriented at $+\alpha$ degrees to the laminate coordinate axes on one side of the middle surface and corresponding equal thickness laminae oriented at $-\alpha$ degrees on the other side. A simple example of an antisymmetric angle-ply laminate is shown in Fig. 4-11. A more complicated example is given in Table 4-5.

A *regular antisymmetric angle-ply laminate* has laminae all of the same thickness for ease of fabrication. This class of laminates can be further restricted to have a single value of α as opposed to several orientations as in Table 4-5.

The force and moment resultants for an antisymmetric angle-ply laminate are

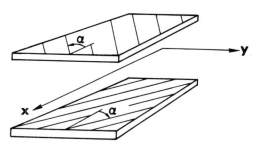

FIG. 4-11. Exploded (unbonded) view of a two-layered regular antisymmetric angle-ply laminate.

TABLE 4-5. Six-layered antisymmetric angle-ply laminate

Layer	Material properties				Orientation	Thickness
	Q_{11}	Q_{12}	Q_{22}	Q_{66}		
1	F_1	F_2	F_3	F_4	$-45°$	t
2	G_1	G_2	G_3	G_4	$+30°$	$2t$
3	H_1	H_2	H_3	H_4	$90°$	$3t$
4	H_1	H_2	H_3	H_4	$0°$	$3t$
5	G_1	G_2	G_3	G_4	$-30°$	$2t$
6	F_1	F_2	F_3	F_4	$+45°$	t

$$\begin{Bmatrix} N_x \\ N_y \\ N_{xy} \end{Bmatrix} = \begin{bmatrix} A_{11} & A_{12} & 0 \\ A_{12} & A_{22} & 0 \\ 0 & 0 & A_{66} \end{bmatrix} \begin{Bmatrix} \epsilon_x^0 \\ \epsilon_y^0 \\ \gamma_{xy}^0 \end{Bmatrix} + \begin{bmatrix} 0 & 0 & B_{16} \\ 0 & 0 & B_{26} \\ B_{16} & B_{26} & 0 \end{bmatrix} \begin{Bmatrix} \kappa_x \\ \kappa_y \\ \kappa_{xy} \end{Bmatrix} \tag{4.53}$$

$$\begin{Bmatrix} M_x \\ M_y \\ M_{xy} \end{Bmatrix} = \begin{bmatrix} 0 & 0 & B_{16} \\ 0 & 0 & B_{26} \\ B_{16} & B_{26} & 0 \end{bmatrix} \begin{Bmatrix} \epsilon_x^0 \\ \epsilon_y^0 \\ \gamma_{xy}^0 \end{Bmatrix} + \begin{bmatrix} D_{11} & D_{12} & 0 \\ D_{12} & D_{22} & 0 \\ 0 & 0 & D_{66} \end{bmatrix} \begin{Bmatrix} \kappa_x \\ \kappa_y \\ \kappa_{xy} \end{Bmatrix} \tag{4.54}$$

The coupling stiffnesses B_{16} and B_{26} can be shown to go to zero as the number of layers in the laminate increases for a fixed laminate thickness.

4.3.4 Nonsymmetric Laminates

For the general case of multiple isotropic layers of thickness t_k and material properties E_k and ν_k, the extensional, coupling, and bending stiffnesses are given by Eq. (4.21) wherein

$$(\bar{Q}_{11})_k = (\bar{Q}_{22})_k = \frac{E_k}{1 - \nu_k^2} \qquad (\bar{Q}_{16})_k = (\bar{Q}_{26})_k = 0$$

$$(\bar{Q}_{12})_k = \frac{\nu_k E_k}{1 - \nu_k^2} \qquad (\bar{Q}_{66})_k = \frac{E_k}{2(1 + \nu_k)} \tag{4.55}$$

No special reduction of the stiffnesses is possible when t_k is arbitrary. That is, coupling between bending and extension can be obtained by unsymmetric arrangement about the middle surface of isotropic layers with different material properties and possibly (but not necessarily) different thicknesses. Thus, coupling between bending and extension is *not* a manifestation of material orthotropy but rather of laminate heterogeneity; that is, a combination of *both* geometric and material properties. The force and moment resultants are

$$\left\{ \begin{matrix} N_x \\ N_y \\ N_{xy} \end{matrix} \right\} = \begin{bmatrix} A_{11} & A_{12} & 0 \\ A_{12} & A_{11} & 0 \\ 0 & 0 & A_{66} \end{bmatrix} \left\{ \begin{matrix} \epsilon_x^0 \\ \epsilon_y^0 \\ \gamma_{xy}^0 \end{matrix} \right\} + \begin{bmatrix} B_{11} & B_{12} & 0 \\ B_{12} & B_{11} & 0 \\ 0 & 0 & B_{66} \end{bmatrix} \left\{ \begin{matrix} \kappa_x \\ \kappa_y \\ \kappa_{xy} \end{matrix} \right\} \tag{4.56}$$

$$\left\{ \begin{matrix} M_x \\ M_y \\ M_{xy} \end{matrix} \right\} = \begin{bmatrix} B_{11} & B_{12} & 0 \\ B_{12} & B_{11} & 0 \\ 0 & 0 & B_{66} \end{bmatrix} \left\{ \begin{matrix} \epsilon_x^0 \\ \epsilon_y^0 \\ \gamma_{xy}^0 \end{matrix} \right\} + \begin{bmatrix} D_{11} & D_{12} & 0 \\ D_{12} & D_{11} & 0 \\ 0 & 0 & D_{66} \end{bmatrix} \left\{ \begin{matrix} \kappa_x \\ \kappa_y \\ \kappa_{xy} \end{matrix} \right\} \tag{4.57}$$

Nonsymmetric laminates with multiple specially orthotropic layers can be shown to have the force and moment resultants in Eqs. (4.56) and (4.57) but with different A_{22}, B_{22}, and D_{22} from A_{11}, B_{11}, and D_{11}, respectively. That is, there are no shear coupling terms, and therefore the solution of problems with this kind of lamination is about as easy as with isotropic layers.

Nonsymmetric laminates with multiple generally orthotropic layers or with multiple anisotropic layers have force and moment resultants no simpler than Eqs. (4.19) and (4.20). All stiffnesses are present. Hence, configurations with either of those two laminae are much more difficult to analyze than configurations with either multiple isotropic layers or multiple specially orthotropic layers.

4.3.5 Summary Remarks

Single layer "laminates" (of course, such configurations are *not* laminates, but laminate stiffnesses must reduce to individual layer stiffnesses) with a reference surface at the middle surface do not exhibit coupling between bending and extension. With any other reference surface, there is indeed such coupling.

Multilayered laminates, in general, develop coupling between bending and extension. The coupling is influenced by the geometrical as well as by the material property characterisitics of laminates. There are, however, combinations of the geometrical and material property characteristics for which there is no coupling between bending and extension. Those special cases have been reviewed in this section along with other special cases. All the special cases find important applications and should be well understood. Note from the collection of special cases that the elastic symmetry of the laminae (whether isotropic, orthotropic, etc.) is not necessarily maintained in the laminate. The symmetry can be increased, decreased, or remain the same. Moreover, the symmetries of the three stiffness matrices, A, B, and D, need not be the same.

The basic concept of coupling between bending and extension must be understood because there are many applications of composite materials where neglect of coupling can be catastrophic. This coupling is the key to the correct analysis of eccentrically stiffened plates and shells. For example, Card and Jones (Ref. 4-4) showed that if longitudinal stiffeners are placed on the outside of an axially loaded circular cylindrical shell, the buckling load is twice the value when the same stiffeners are on the inside of the shell. Previously, the coupling between the stiffener and the shell had been ignored!

The manner of describing a laminate by use of individual layer thicknesses, principal material property orientations, and overall stacking sequence could be quite involved. However, fortunately, all pertinent parameters are represented in a simple, concise fashion by use of the following stacking sequence terminology. For regular (equal thickness layers) laminates, a listing of the layers and their orientations suffices, for example, $[0°/90°/45°]$. Note that only the principal material direction orientations need be given. Many different laminates could be made with the same layers, for example, $[90°/0°/45°]$. For irregular (layers do not have the same thickness) laminates, a notation of layer thicknesses must be appended to the previous notation, for example, $[0°@t/90°@2t/45°@3t]$. Finally, for symmetric laminates, the simplest representation of, for example, $[0°/90°/45°/45°/90°/0°]$ is $[0°/90°/45°]$ symmetric. This notation will be used throughout the remainder of the book.

Problem Set 4.3

Exercise 4.3.1 Prove that the coupling stiffnesses, B_{ij}, are zero for laminates that are symmetric about the middle surface.

Exercise 4.3.2 Consider two laminae with principal material directions at $+\alpha$ and $-\alpha$ with respect to a reference axis. Prove that for orthotropic materials

$$(\bar{Q}_{16})_{+\alpha} = -(\bar{Q}_{16})_{-\alpha}$$

Discuss whether this relation is valid for anisotropic materials. The transformation equations for anisotropic materials are given in Sec. 2.7.

Exercise 4.3.3 The term quasi-isotropic is used to describe laminates that have essentially isotropic extensional stiffnesses (the same in all directions). The simplest example of a quasi-isotropic laminate is a three-ply laminate with a $-60°/0°/60°$ stacking sequence. The next simplest example is a four-ply laminate with a $0°/-45°/45°/90°$ stacking sequence. As the number of layers increases, the angle between the adjacent laminae decreases. Although these laminates are called quasi-isotropic, they do not behave like isotropic homogeneous materials. Discuss why not and describe how they do behave. Why is a two-ply laminate with a $0°/90°$ stacking sequence and equal thickness layers not a quasi-isotropic laminate? Determine whether the extensional stiffnesses are the same irrespective of the laminate axes for the two-ply and three-ply cases.

Exercise 4.3.4 Show that B_{16} and B_{26} for an antisymmetric angle-ply laminate approach zero as the number of layers increases if the total laminate thickness is held constant. What happens if equal thickness layers are added so the total laminate thickness increases, too?

4.4 COMPARISON OF THEORETICAL AND EXPERIMENTAL LAMINATE STIFFNESSES

In preceding sections, laminate stiffnesses were predicted on the basis of combination of lamina stiffnesses according to classical lamination theory. However, the actual realization of those laminate stiffnesses remains to be demonstrated. The purpose of this section is to display a comparison of predicted laminate stiffnesses with measured laminate stiffnesses. Results obtained by Tsai (Ref. 4-5) and Azzi and Tsai (Ref. 4-6) for two types of laminates, cross-ply and angle-ply laminates, are presented.

4.4.1 Inversion of Stiffness Equations

Before the predicted stiffnesses are compared with experimental stiffnesses, however, a slight reinterpretation of laminate stiffnesses is required. Ordinarily, the resultant forces and moments are written in terms of the middle surface extensional strains and curvatures as

$$\left\{ \frac{N}{M} \right\} = \left[-\frac{A}{B} \begin{array}{c} \vdots \\ \vdots \end{array} \frac{B}{D} - \right] \left\{ \frac{\epsilon^0}{\kappa} \right\} \tag{4.58}$$

However, in most experiments, the deformations are the dependent variables. That is, the loads are applied, and the resulting deformations are measured. Thus, expressions for the middle surface extensional strains and curvatures in terms of the force and moment resultants would be convenient.

The first step in the derivation of the inverse of Eq. (4.58) is to write it in the form

$$N = A\epsilon^0 + B\kappa \tag{4.59}$$

$$M = B\epsilon^0 + D\kappa \tag{4.60}$$

and solve Eq. (4.59) for ϵ^0:

$$\epsilon^0 = A^{-1}N - A^{-1}B\kappa \tag{4.61}$$

whereupon Eq. (4.60) becomes

$$M = BA^{-1}N + (-BA^{-1}B + D)\kappa \tag{4.62}$$

Equations (4.61) and (4.62) can be written as

$$\left\{\frac{\epsilon^0}{M}\right\} = \left[\begin{array}{c|c} A^{-1} & -A^{-1}B \\ \hline BA^{-1} & D - BA^{-1}B \end{array}\right]\left\{\frac{N}{\kappa}\right\} \tag{4.63}$$

or

$$\epsilon^0 = A^*N + B^*\kappa \tag{4.64}$$

$$M = H^*N + D^*\kappa \tag{4.65}$$

where B^* is not equal to H^*. Now solve Eq. (4.65) for κ:

$$\kappa = D^{*-1}M - D^{*-1}H^*N \tag{4.66}$$

and substitute in Eq. (4.64) to get

$$\epsilon^0 = B^*D^{*-1}M + (A^* - B^*D^{*-1}H^*)N \tag{4.67}$$

Thus,

$$\left\{\frac{\epsilon^0}{\kappa}\right\} = \left[\begin{array}{c|c} A^* - B^*D^{*-1}H^* & B^*D^{*-1} \\ \hline -D^{*-1}H^* & D^{*-1} \end{array}\right]\left\{\frac{N}{M}\right\} \tag{4.68}$$

or

$$\left\{\frac{\epsilon^0}{\kappa}\right\} = \left[\begin{array}{c|c} A' & B' \\ \hline H' & D' \end{array}\right]\left\{\frac{N}{M}\right\} \tag{4.69}$$

wherein H' can be shown to be equal to $(B')^T$ by virtue of the symmetry of the A, B, and D matrices and the definitions of the A', B', D', A^*, B^*, and D^* matrices. Of course, H' must equal $(B')^T$ since the matrix of the coefficients in Eq. (4.58) is symmetric so its inverse, the matrix of coefficients in Eq. (4.69), must also be symmetric. Demonstration that the predicted values of A', B', and D' agree with measured values is therefore fully equivalent to verification of the prediction techniques for A, B, and D.

4.4.2 Cross-Ply Laminate Stiffnesses

A cross-ply laminate has N unidirectionally reinforced (orthotropic) layers with principal material directions alternatingly oriented at $0°$ and $90°$ to the laminate coordinate axes. The fiber direction of odd-numbered layers is the x-direction of the laminate. The fiber direction of even-numbered layers is then the y-direction of the laminate. Consider the special, but practical, case of odd-numbered layers with equal thickness and even-numbered layers with equal thickness, but not necessarily the same as that of the odd-numbered layers. Then, two geometrical parameters are important: N, the total number of layers and M, the ratio of the total thickness of odd-numbered layers to the total thickness of even-numbered layers (called the cross-ply ratio). Thus,

$$M = \frac{\displaystyle\sum_{k=\text{odd}} t_k}{\displaystyle\sum_{k=\text{even}} t_k} \tag{4.70}$$

For example, if a five-layered laminate has a lamination or stacking sequence $[t @ 0°/2t @ 90°/t @ 0°/2t @ 90°/t @ 0°]$, then

$$M = \frac{t + t + t}{2t + 2t} = \frac{3}{4} \tag{4.71}$$

Note that the cross-ply ratio, M, has specific meaning only when the layers have alternating $0°$ and $90°$ orientations. If the middle layer of the foregoing example were 2 layers of $0°$ orientation with each layer being $t/2$ thick, then M is easily shown to be one. However, then the layers would not have alternating orientation nor would odd-numbered layers have the same thickness. Thus, more general cross-ply laminates cannot be described by use of the cross-ply ratio, M.

The laminate stiffnesses,

$$A_{ij} = \sum_{k=1}^{N} (\bar{Q}_{ij})_k (z_k - z_{k-1})$$

$$B_{ij} = \frac{1}{2} \sum_{k=1}^{N} (\bar{Q}_{ij})_k (z_k^2 - z_{k-1}^2) \tag{4.72}$$

$$D_{ij} = \frac{1}{3} \sum_{k=1}^{N} (\bar{Q}_{ij})_k (z_k^3 - z_{k-1}^3)$$

can be expressed in terms of M and N for laminates with an odd or even number of layers. In addition, F, the ratio of principal lamina stiffnesses,

$$F = \frac{Q_{22}}{Q_{11}} = \frac{E_2}{E_1} \tag{4.73}$$

is used. Tsai (Ref. 4-5) displayed the following stiffnesses:

Cross-ply laminates with N odd (symmetric)

$$A_{11} = \frac{1}{1 + M} (M + F) t Q_{11}$$

$$A_{12} = t Q_{12}$$

$$A_{22} = \frac{1}{1 + M} (1 + MF) t Q_{11} = \frac{1 + MF}{M + F} A_{11}$$

$$A_{16} = A_{26} = 0$$

$$A_{66} = t Q_{66}$$

(4.74)

$$B_{ij} = 0$$

(4.75)

$$D_{11} = \frac{[(F - 1)P + 1] Q_{11} t^3}{12} = [(F - 1)P + 1] \frac{1 + M}{M + F} \frac{A_{11} t^2}{12}$$

$$D_{12} = \frac{Q_{12} t^3}{12}$$

$$D_{22} = \frac{[(1 - F)P + F] Q_{11} t^3}{12} = [(1 - F)P + F] \frac{1 + M}{M + F} \frac{A_{11} t^2}{12}$$

$$D_{16} = D_{26} = 0$$

$$D_{66} = \frac{Q_{66} t^3}{12}$$

(4.76)

where

$$P = \frac{1}{(1 + M)^3} + \frac{M (N - 3) [M (N - 1) + 2 (N + 1)]}{(N^2 - 1)(1 + M)^3}$$

(4.77)

Cross-ply laminates with N even (antisymmetric)

$$A_{11} = \frac{1}{1 + M} (M + F) Q_{11} t$$

$$A_{12} = Q_{12} t$$

$$A_{22} = \frac{1}{1 + M} (1 + MF) Q_{11} t = \frac{1 + MF}{M + F} A_{11}$$

$$A_{16} = A_{26} = 0$$

$$A_{66} = Q_{66} t$$

(4.78)

$$B_{11} = \frac{M(F - 1)}{N(1 + M)^2} Q_{11} t^2 = \frac{M(F - 1)}{N(1 + M)(M + F)} A_{11} t$$

$$B_{22} = -B_{11}$$

$$B_{12} = B_{16} = B_{26} = B_{66} = 0$$

(4.79)

$$D_{11} = \frac{[(F-1)R+1]Q_{11}t^3}{12} = [(F-1)R+1]\frac{1+M}{M+F}\frac{A_{11}t^2}{12}$$

$$D_{12} = \frac{Q_{12}t^3}{12}$$

$$D_{22} = \frac{[(1-F)R+F]Q_{11}t^3}{12} = [(1-F)R+F]\frac{1+M}{M+F}\frac{A_{11}t^2}{12} \qquad (4.80)$$

$$D_{16} = D_{26} = 0$$

$$D_{66} = \frac{Q_{66}t^3}{12}$$

where

$$R = \frac{1}{1+M} + \frac{8M(M-1)}{N^2(1+M)^3} \qquad (4.81)$$

Observations on cross-ply laminates

The cross-ply laminate stiffnesses are given for symmetric laminates in Eqs. (4.74) through (4.76) and for antisymmetric laminates in Eqs. (4.78) through (4.80). The extensional, coupling, and bending stiffnesses are discussed separately in the following paragraphs.

For both odd- and even-layered cross-ply laminates, the extensional stiffnesses, A_{ij}, are independent of N, the number of layers. However, A_{11} and A_{22} depend on M, the cross-ply ratio, and on F, the stiffness ratio, as shown in Fig. 4-12 and 4-13. For a typical glass fiber-reinforced lamina, $F = .3$ so A_{11} varies from $.65Q_{11}t$ to $.93Q_{11}t$ as M changes from 1 to 10. Similarly, A_{22} varies from A_{11} to $.38A_{11}$ over the same range of M. The stiffnesses A_{12} and A_{66} are independent of M and F. The remaining stiffnesses A_{16} and A_{26} are zero for all cross-ply laminates.

Only cross-ply laminates with an even number of layers have coupling between bending and extension since the B_{ij} are all zero for a cross-ply laminate with an odd number of layers. The coupling stiffnesses B_{11} and B_{22} are plotted as a function of the cross-ply ratio, M, in Fig. 4-14. The number of layers, N, appears in the numerator of the ordinate in Fig. 4-14. Thus, the value of B_{11} obviously decreases as N increases because NB_{11} is constant for a fixed cross-ply ratio. Since N must be even to get any coupling, $N = 2$ corresponds to the largest coupling between bending and extension. One physical interpretation of the coupling stiffness B_{11} is that it is a measure of the location of the neutral (stress-free) surface relative to the laminate middle surface. As a matter of fact, the ordinate in Fig. 4-14 is the fraction of the total laminate thickness, t, that the neutral surface is shifted from the middle surface. The shifting, like B_{11}, is inversely proportional to N so it gets smaller as the number of layers increases.

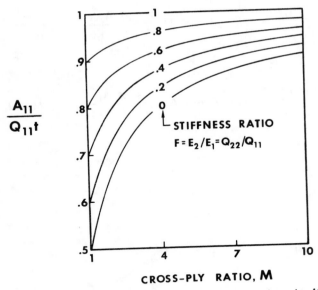

FIG. 4-12. Extensional stiffness, A_{11}, versus cross-ply ratio, M. (*After Tsai, Ref. 4-5.*)

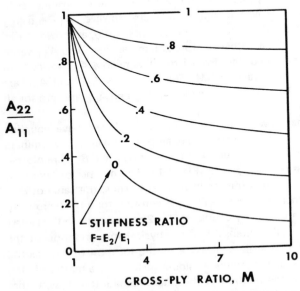

FIG. 4-13. Extensional stiffness, A_{22}, versus cross-ply ratio, M. (*After Tsai, Ref. 4-5.*)

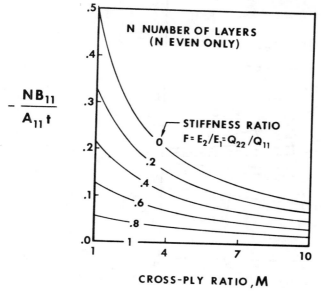

$$-\frac{NB_{11}}{A_{11}t}$$

FIG. 4-14. Coupling stiffness, B_{11}, versus cross-ply ratio, M. (*After Tsai, Ref. 4-5.*)

The bending stiffnesses, D_{ij}, are complicated functions of the number of layers, N, the cross-ply ratio, M, and the stiffness ratio, F. Normalized values of D_{11} and D_{22} are shown in Figs. 4-15 and 4-16 for several values of F and N as a function of M. Extreme values of D_{11} and D_{22} occur when $N = 2$ and $N = 3$ with values for all other N falling in between. The value of D_{11} approaches $A_{11} t^2/12$ and D_{22} approaches $A_{22}t^2/12$ as (1) M gets large, (2) N gets large, or (3) F approaches one. Thus, with certain types of laminate layups, the stiffnesses can approach those for a homogeneous plate- or shell-like element.

4.4.3 Theoretical and Experimental Cross-Ply Laminate Stiffnesses

Two- and three-layered cross-ply laminates were shown to given extrema of behavior in the preceding section. Thus, comparisons between theoretical and experimental stiffnesses for such laminates should be quite revealing. Any agreement for those cases would imply equal or better agreement for cross-ply laminates with more than three layers.

The individual laminae used by Tsai (Ref. 4-5) consist of unidirectional glass fibers in a resin matrix (U.S. Polymeric Co. E-787-NUF) with moduli given in Table 2-3. A series of cross-ply laminates were constructed with $M = 1,2,3,10$ for two-layered laminates and $M = 1,2,5,10$ for three-layered laminates. The laminates were subjected to axial loads and bending moments whereupon surface

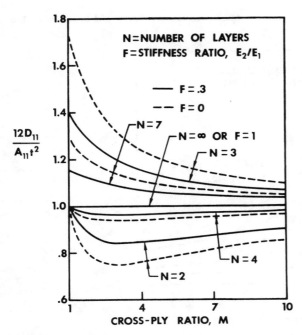

FIG. 4-15. Bending stiffness, D_{11}, versus cross-ply ratio, M.
(*After Tsai, Ref. 4-5.*)

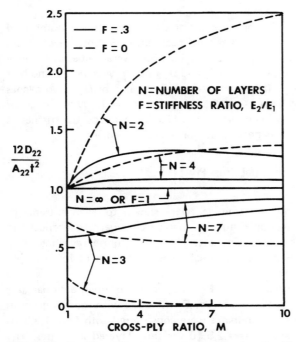

FIG. 4-16. Bending stiffness, D_{22}, versus cross-ply ratio, M.
(*After Tsai, Ref. 4-5.*)

strains were measured. Accordingly, the stiffness relations as strains and curvatures in terms of forces and moments, that is,

$$\left\{ \begin{matrix} \epsilon \\ \kappa \end{matrix} \right\} = \begin{bmatrix} A' & B' \\ B'^T & D' \end{bmatrix} \left\{ \begin{matrix} N \\ M \end{matrix} \right\} \tag{4.82}$$

are natural to use. That is, the theoretical values of A', B', and D' will be compared with the experimentally measured values. Verification of one set of stiffnesses implies verification of the other set since the two sets, A, B, and D and A', B', and D', are inverses of one another as *sets* (that is, $A \neq A'^{-1}$, etc.).

The experiments were performed on two sets of beams with the beam axis at $0°$ and $90°$, respectively, to the fiber direction of the odd-numbered layers. The beams were 1-in. wide, .12-in. thick, and of 6-in. span. Strain rosettes were located on the upper and lower beam surfaces so that the middle surface strains and curvatures can be calculated from simultaneous solution of

$$\left. \begin{matrix} \epsilon_i^0 + \dfrac{t}{2} \kappa_i = \epsilon_i^{\text{upper}} \\[2mm] \epsilon_i^0 - \dfrac{t}{2} \kappa_i = \epsilon_i^{\text{lower}} \end{matrix} \right\} \quad i = 1, 2, 6 \tag{4.83}$$

where t is the beam thickness.

The stiffnesses A'_{11}, $A'_{21} = A'_{12}$, B'_{11}, and B'_{12} were measured after application of pure uniaxial tension, N_1, to a $0°$ beam; the stiffnesses B'_{11}, B'_{21}, D'_{11}, and $D'_{21} = D'_{12}$, after application of pure bending moment, M_1, to a $0°$ beam. The stiffnesses $A'_{12} = A'_{21}$, A'_{22}, B'_{21}, B'_{12}, B'_{22}, $D'_{12} = D'_{21}$, and D'_{22} were measured by similar experiments on a $90°$ beam. A pure twisting test on a $0°$ square plate was used to measure D_{66}. That is, two upward and two downward forces were applied at the four corners of a plate as in Fig. 4-17 whereupon

$$D_{66} = \frac{pL^2}{4w_c} \tag{4.84}$$

where w_c is the corner deflection. The shear stiffness, A_{66}, was not measured.

The experimental stiffnesses for two- and three-layered cross-ply laminates are shown by symbols in Fig. 4-18. The theoretical results are shown by the solid lines in Fig. 4-18. In all cases, the load was kept so low that no strain exceeded 500 microinches per inch. Thus, the behavior was linear and elastic. The agreement between theory and experiment is quite good. Both the qualitative and the quantitative aspects of the theory are verified. Thus, we can claim that the capability to predict cross-ply laminate stiffnesses exists and is quite accurate.

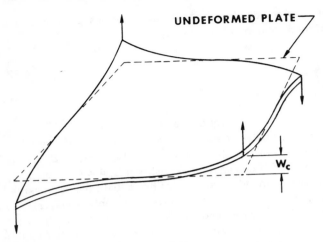

FIG. 4-17. Twisting of a square plate.

4.4.4. Angle-Ply Laminate Stiffnesses

An angle-ply laminate has N unidirectionally reinforced (orthotropic) layers with principal material directions alternatingly oriented at $+\alpha$ and $-\alpha$ to the laminate coordinate axes. The odd-numbered plies are at $-\alpha$ and the even-numbered plies are at $+\alpha$. Consider the special, but practical, case where all layers have the same thickness, that is, regular angle-ply laminates. Then, the laminate behavior can be described by the number of layers, N, the laminae orientation, α, and the laminae stiffnesses, Q_{ij}, in addition to the laminate thickness, t.

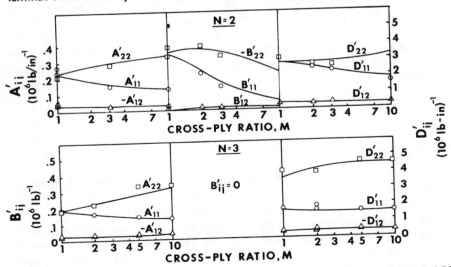

FIG. 4-18. Theoretical and experimental cross-ply laminate stiffnesses. (*After Tsai, Ref. 4-5.*)

The laminate stiffnesses,

$$A_{ij} = \sum_{k=1}^{N} (\bar{Q}_{ij})_k (z_k - z_{k-1})$$

$$B_{ij} = \frac{1}{2} \sum_{k=1}^{N} (\bar{Q}_{ij})_k (z_k^2 - z_{k-1}^2)$$

(4.85)

$$D_{ij} = \frac{1}{3} \sum_{k=1}^{N} (\bar{Q}_{ij})_k (z_k^3 - z_{k-1}^3)$$

can be expressed in terms of N, $\bar{Q}_{ij}$ (in which α is accounted for), and t for laminates with an even number of layers and with an odd number of layers. In both cases, $\bar{Q}_{ij}$ is calculated for $-\alpha$ and

$$\begin{aligned}
\bar{Q}_{11+\alpha} &= \bar{Q}_{11-\alpha} & \bar{Q}_{66+\alpha} &= \bar{Q}_{66-\alpha} \\
\bar{Q}_{12+\alpha} &= \bar{Q}_{12-\alpha} & \bar{Q}_{16+\alpha} &= -\bar{Q}_{16-\alpha} \\
\bar{Q}_{22+\alpha} &= \bar{Q}_{22-\alpha} & \bar{Q}_{26+\alpha} &= -\bar{Q}_{26-\alpha}
\end{aligned}$$

(4.86)

as can be verified by substitution in Eq. (2.80). Tsai (Ref. 4-5) displayed the following stiffnesses:

Angle-ply laminates with N odd (symmetric)

$$A_{11}, A_{12}, A_{22}, A_{66} = t(\bar{Q}_{11}, \bar{Q}_{12}, \bar{Q}_{22}, \bar{Q}_{66})$$

$$A_{16}, A_{26} = \frac{t}{N} (\bar{Q}_{16}, \bar{Q}_{26})$$

(4.87)

$$B_{ij} = 0$$

(4.88)

$$D_{11}, D_{12}, D_{22}, D_{66} = \frac{t^3}{12} (\bar{Q}_{11}, \bar{Q}_{12}, \bar{Q}_{22}, \bar{Q}_{66})$$

$$D_{16}, D_{26} = \frac{t^3}{12} \left(\frac{3N^2 - 2}{N^3} \right) (\bar{Q}_{16}, \bar{Q}_{26})$$

(4.89)

Angle-ply laminates with N even (antisymmetric)

$$A_{11}, A_{12}, A_{22}, A_{66} = t(\bar{Q}_{11}, \bar{Q}_{12}, \bar{Q}_{22}, \bar{Q}_{66})$$

$$A_{16}, A_{26} = 0$$

(4.90)

$$B_{11}, B_{12}, B_{22}, B_{66} = 0$$

$$B_{16}, B_{26} = - \frac{t^2}{2N} (\bar{Q}_{16}, \bar{Q}_{26})$$

(4.91)

$$D_{11}, D_{12}, D_{22}, D_{66} = \frac{t^3}{12} (\bar{Q}_{11}, \bar{Q}_{12}, \bar{Q}_{22}, \bar{Q}_{66})$$

$$D_{16}, D_{26} = 0$$

(4.92)

Observations on angle-ply laminates

The extensional stiffnesses, A_{ij}, are shown in Fig. 4-19 as a function of the lamination angle. The terms A_{11}, A_{12}, A_{22}, and A_{66} are independent of the number of layers, N. However, A_{16} and A_{26} depend on N. When N is odd, they are inversely proportional to N. When N is even, they are zero. Thus, the biggest values of A_{16} and A_{26} occur when $N = 3$. The latter values are shown in Fig. 4-19.

The coupling stiffnesses, B_{ij}, are zero for an odd number of layers, but can be large for an even number of layers. The values of $B_{16} / (tA_{11})$ are shown as a function of lamination angle in Fig. 4-20. Since B_{16} is inversely proportional to N, the largest value of B_{16} occurs when $N = 2$. The quantity plotted can be shown to be

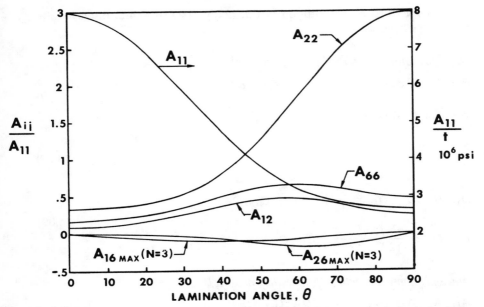

FIG. 4-19. Normalized extensional stiffnesses for a glass/epoxy angle-ply laminate. (After Tsai, Ref. 4-5.)

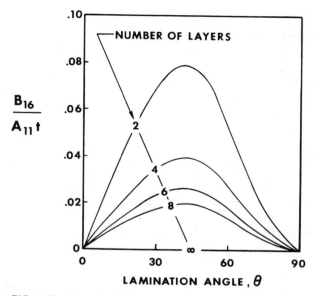

FIG. 4-20. Coupling stiffness B_{16} for a glass/epoxy angle-ply laminate. (*After Tsai, Ref. 4-5.*)

$$\frac{B_{16}}{tA_{11}} = \frac{M_{xy}}{tN_x} \tag{4.93}$$

that is, the ratio of twisting moment to axial extensional force for pure extension, ϵ_x^0. Similarly,

$$\frac{B_{26}}{tA_{22}} = \frac{M_{xy}}{tN_y} \tag{4.94}$$

From Fig. 4-20, apparently the coupling between bending and extension is largest when $\theta = 45°$ (for $N = 2$).

The bending stiffnesses include D_{16} and D_{26} when N is odd, but $D_{16} = D_{26} = 0$ for N even. Since, by virtue of Eq. (4.89), D_{16} and D_{26} are inversely proportional to N, then their maximum value occurs when $N = 3$. Also, D_{16} and D_{26} achieve a maximum for a lamination angle of 45° as shown in Fig. 4-21. Recall that D_{16} and D_{26} are twist coupling terms. In simple bending, the twisting moment induced by the presence of D_{16} and D_{26} is 30 percent of the applied bending moment. This coupling does *not* decrease rapidly as N increases. Thus, approximate solutions in which shear coupling is ignored are not likely to be accurate.

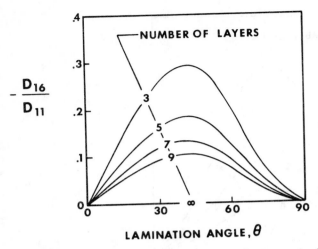

FIG. 4-21. Bending stiffness D_{16} for a glass/epoxy angle-ply laminate. (*After Tsai, Ref. 4-5.*)

4.4.5 Theoretical and Experimental Angle-Ply Laminate Stiffnesses

The laminates and measurement procedures chosen to illustrate comparison between theory and experiment differ from those in Sec. 4.4.3 only in that here angle-ply laminates are treated. A two-layered laminate has the largest B_{16} and B_{26}. A three-layered laminate has the largest A_{16}, A_{26}, D_{16}, and D_{26}. The experiments were conducted on beams with angle plies at $\pm\alpha$ to the beam axis. Note that only half as many specimens are required as for cross-ply laminates because, for example, A_{11} and A_{22} are mirror images of one another about $\theta = 45°$. Since coupling between extensional forces and shearing deformations exists as well as between bending moments and twisting deformations, a complex strain state was anticipated. Thus, three-element strain rosettes were placed on the top and bottom beam surfaces. The shearing strain, γ_{xy}, is then calculated from

$$\gamma_{xy}^{upper} = 2\epsilon_{45°}^{upper} - (\epsilon_x^{upper} + \epsilon_y^{upper})$$

$$\gamma_{xy}^{lower} = 2\epsilon_{45°}^{lower} - (\epsilon_x^{lower} + \epsilon_y^{lower})$$

(4.95)

where $\epsilon_{45°}$ is the third strain (at $45°$ to the x and y axes) in the strain rosette. The middle surface strains and curvatures are calculated from Eq. (4.83).

The actual theoretical and measured stiffnesses are shown in Fig. 4-22. As with cross-ply laminates, very good agreement was obtained. Thus, the predictions of laminate stiffnesses are quite accurate.

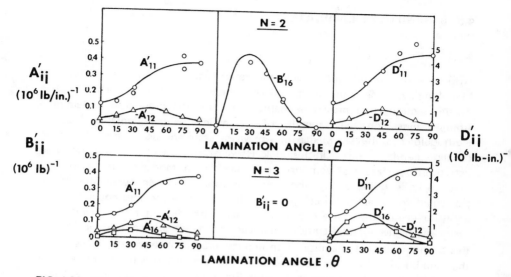

FIG. 4-22. Theoretical and experimental angle-ply laminate stiffnesses. (*After Tsai, Ref.* 4-5.)

4.4.6 Summary Remarks

Experimental measurements of cross-ply and angle-ply laminate stiffnesses were compared with theoretical values of the stiffnesses from lamination theory. Both symmetric and antisymmetric configurations were treated. The number of layers in each configuration corresponded to predictions of the largest coupling stiffnesses A_{16}, A_{26}, B_{ij}, D_{16}, and D_{26} where these stiffnesses existed. Thus, the comparisons between theory and experiment were for worst case conditions. Accordingly, the good agreement obtained lends a high degree of confidence to theoretical predictions of stiffnesses for less severe conditions of coupling.

Problem Set 4.4

Exercise 4.4.1 Prove that $H' = (B')^T$ in Eq. (4.69) by use of matrix algebra.

Exercise 4.4.2 Derive the stiffnesses for regular symmetric cross-ply laminates, that is, derive Eqs. (4.74) - (4.76) for the special case $t_{odd} = t_{even}$. Optional: derive Eqs. (4.74) - (4.76).

Exercise 4.4.3 Derive the stiffnesses for regular antisymmetric cross-ply laminates, that is, derive Eqs. (4.78) - (4.80) for the special case $t_{odd} = t_{even}$ (for which also $M = 1$). Optional: derive Eqs. (4.78) - (4.80).

Exercise 4.4.4 Derive the stiffnesses for symmetric angle-ply laminates given in Eqs. (4.87) - (4.89).

Exercise 4.4.5 Derive the stiffnesses for antisymmetric angle-ply laminates given in Eqs. (4.90) - (4.92).

Exercise 4.4.6 Derive Eqs. (4.93) and (4.94).

4.5 STRENGTH OF LAMINATES

4.5.1 Introduction

With laminate strength, just as with the determination of laminate stiffness, the basic building block is the lamina with its inherent characteristics. The objective of this section is to present a methodology for the prediction of laminate strength based on the strengths of its laminae. Fundamental to such a methodology is the knowledge of the stress state in each lamina, based on concepts developed earlier in this book. However, because of the anisotropic and heterogeneous nature of composite materials, failure modes occur that require new analyses quite unlike those for isotropic homogeneous materials. In particular, for a laminated composite, failure of one layer does not necessarily imply failure of the entire laminate; the laminate may, in fact, be capable of sustaining higher loads despite a significant change in stiffness. An analogy to this phenomenon is the ability of an in-plane loaded plate to carry loads higher than the buckling load, but at an increase in the amount of deformation per unit of load (a decreased stiffness) as shown in Fig. 4-23.

Because of the various characteristics of composite materials, it is difficult to determine a strength theory in which all failure modes and their interactions are properly accounted for. Moreover, the verification of a proposed strength theory is greatly complicated by scatter in measured strengths caused by inconsistent processing techniques (that are mainly unavoidable) and sometimes inappropriate and misleading experimental techniques. Nevertheless, a continuing effort must be made to define strength theories that enable the accurate prediction of composite material strengths. Strength theories are essential to a designer to be able to predict the capability of a structural element under a complex loading state. Such theories must be verified by comparison with measured strengths; subsequently, judgment must be made as to whether the theory adequately represents the physical phenomenon given the inherent experimental difficulties of measuring the phenomenon.

In this section, such an assessment of strength theories for a laminate is made in parallel to the assessment made for an individual lamina in Sec. 2.9. The maximum stress, maximum strain, and maximum distortional energy approaches will be contrasted. Each of the theories is a phenomenological approach; that is, the apparent behavior is studied without reference to what causes it. Thus, such theories can be classed as a macroscopic analysis of strength. Accordingly, the theories are, at best, an imperfect representation of physical reality. However, as will be seen, they are useful and effective in design analysis, some more than others, because they agree with experimental data.

All strength theories for composite materials depend on the strengths in the principal material directions which likely do not coincide with principal stress directions. Therefore, the strength of each lamina in a laminate must be assessed in a coordinate system that is likely different from those of its neighboring

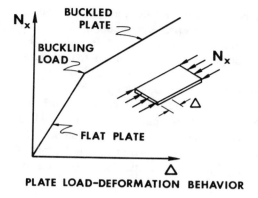

PLATE LOAD-DEFORMATION BEHAVIOR

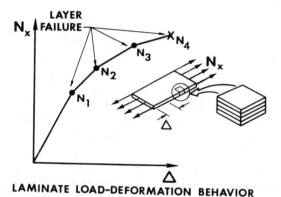

LAMINATE LOAD-DEFORMATION BEHAVIOR

FIG. 4-23. Analogy between buckled plate and laminate load-deformation behavior.

laminae. This coordinate mismatch is but one of the complications that characterizes even a macroscopic strength theory. The main factors that are peculiar to laminate strength analysis are:

- laminae strengths
- laminae stiffnesses
- laminae coefficients of thermal expansion
- laminae orientations
- laminae thicknesses
- stacking sequence
- curing temperature

The thermomechanical properties, thicknesses, and orientations are important in determining the directional characteristics of strength. The stacking sequence affects the bending and coupling stiffnesses and hence the strengths of the laminate. The curing temperature, or operating versus curing temperature, influences the residual stresses that are developed upon cool-down of the laminate from a stress-free elevated temperature curing cycle. In general, if the operating or service temperature is different from the curing temperature, thermal stresses will arise; whether they are called thermal stresses or residual stresses is partly a matter of convenience, but is mainly semantics. In either event, a laminate can be subjected to thermal and mechanical loads with the objective of surviving those loads. A method of strength analysis is required to determine either (1) the maximum loads a given laminate can withstand or (2) the laminate characteristics necessary to withstand a given load. The former is, of course, an analysis situation and the latter a design situation. The design of laminates will be discussed in a subsequent section.

4.5.2 Laminate Strength Analysis Procedure

The analysis of stresses in the laminae of a laminate is a straightforward, but sometimes tedious, task. The reader is presumed to be familiar with the basic lamination principles that were discussed earlier in the chapter. There, the stresses were seen to be a linear function of the applied loads if the laminae exhibit linear elastic behavior. Thus, a single stress analysis suffices to determine the stress field that causes failure of an individual lamina. That is, if all laminae stresses are known, then the stresses in each lamina can be compared with the lamina failure criterion and uniformly scaled upward to determine the load at which failure occurs.

The overall procedure of laminate strength analysis, which at the same time yields the laminate load-deformation behavior, is shown schematically in Fig. 4-24. There, load is taken to mean both forces and moments; similarly, deformations are meant to include both strains and curvatures. The analysis is composed of two different approaches that depend on whether any laminae have failed.

If no laminae have failed, the load must be determined at which the first lamina fails, that is, violates the failure criterion. In the process of this determination, the laminae stresses must be found as a function of the unknown magnitude of loads first in the laminate coordinates and then in the principal material directions. The proportions of load are, of course, specified at the beginning of the analysis. The load parameter is increased until some lamina fails. That lamina is then eliminated, figuratively, from the laminate by assigning zero properties to the failed layer. Actually, because of the matrix manipulations involved in the analysis, the failed lamina properties must not be zero, but rather effectively zero values in order to avoid a singular matrix. The laminate strains are calculated from the known load and the stiffnesses prior to failure of a lamina. The laminate deformations just after failure of a lamina are discussed later.

If one or more laminae have failed, new laminate extensional, coupling, and bending stiffnesses are calculated. Laminae stresses are recalculated to determine

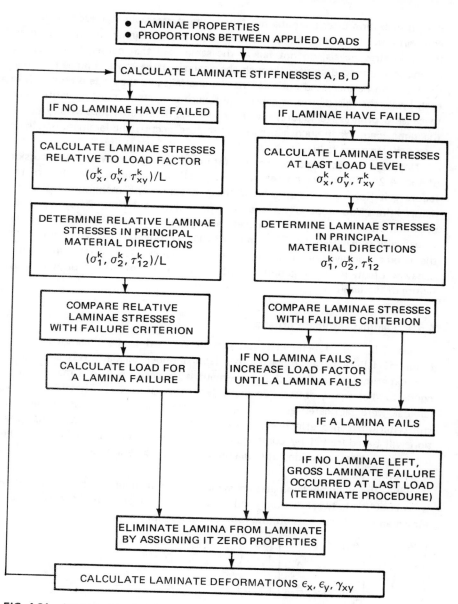

FIG. 4-24. Analysis of laminate strength and load-deformation behavior.

their distribution after a lamina has failed (the stresses will increase to maintain equilibrium). Then it must be verified that the remaining laminae, at their increased stress levels, do not fail at the same load that caused failure of the lamina in the preceding pass through the analysis. If no more laminae fail, then the load can be increased until another lamina fails and the cycle is repeated. In each cycle, the raised stresses caused by failure of a lamina must be verified not to cause a progressive failure, that is, where the laminae all successfully fail at the same load. When such a progressive failure occurs, the laminate is said to have suffered gross failure.

Note that the failure criterion was not mentioned explicitly in the discussion of Fig. 4-24. The entire *procedure* for strength analysis is independent of the failure criterion, but the *results* of the procedure, the maximum loads and deformations, do depend on the specific failure criterion. Also, the load-deformation behavior is piecewise linear owing to the consideration of linear elastic behavior of each lamina. The laminate behavior could be piecewise nonlinear if the laminae behaved in a nonlinear elastic manner. At any rate, the overall behavior of the laminate is nonlinear if one or more laminae fail prior to gross failure of the laminate.

4.5.3 Laminate Strength Criteria

In Sec. 2.9, the Tsai-Hill criterion was determined to be the best representation of failure of an *E*-glass/epoxy lamina under biaxial stress conditions. The experiments used for that assessment were performed in a manner analogous to that for symmetric angle-ply laminates (Ref. 4-7). That is, the same laminae were fabricated as a laminate. The question now is whether the Tsai-Hill criterion is also effective for prediction of laminate strengths. This question will be addressed by comparison of the three lamina failure criteria, the maximum stress criterion, the maximum strain criterion, and the Tsai-Hill criterion.

Since the angle-ply laminates are symmetric about their middle surface, there is no coupling between bending and extension. Therefore, when the laminate is subjected to uniaxial tension, the force-strain relations are

$$\begin{Bmatrix} N_x \\ N_y \\ N_{xy} \end{Bmatrix} = \begin{bmatrix} A_{11} & A_{12} & A_{16} \\ A_{12} & A_{22} & A_{26} \\ A_{16} & A_{26} & A_{66} \end{bmatrix} \begin{Bmatrix} \epsilon_x^0 \\ \epsilon_y^0 \\ \gamma_{xy}^0 \end{Bmatrix} \tag{4.96}$$

whereupon

$$\begin{Bmatrix} \epsilon_x^0 \\ \epsilon_y^0 \\ \gamma_{xy}^0 \end{Bmatrix} = \begin{bmatrix} A_{11} & A_{12} & A_{16} \\ A_{12} & A_{22} & A_{26} \\ A_{16} & A_{26} & A_{66} \end{bmatrix}^{-1} \begin{Bmatrix} N_x \\ N_y \\ N_{xy} \end{Bmatrix} \tag{4.97}$$

From now on, the matrix A^{-1} is called the A' matrix. When $N_x = N_1$ and $N_y = N_{xy} = 0$, the strains are

$$
\left\{
\begin{array}{c}
\epsilon_x^0 \\
\epsilon_y^0 \\
\gamma_{xy}^0
\end{array}
\right\}
=
\begin{bmatrix}
A'_{11} & A'_{12} & A'_{16} \\
A'_{12} & A'_{22} & A'_{26} \\
A'_{16} & A'_{26} & A'_{66}
\end{bmatrix}
\left\{
\begin{array}{c}
N_1 \\
0 \\
0
\end{array}
\right\}
\tag{4.98}
$$

or, more simply,

$$
\begin{aligned}
\epsilon_x^0 &= A'_{11} \, N_1 \\
\epsilon_y^0 &= A'_{12} \, N_1 \\
\gamma_{xy}^0 &= A'_{16} \, N_1
\end{aligned}
\tag{4.99}
$$

The stresses in each layer are obtained by use of the stress-strain relations for a lamina, Eq. (2.79):

$$
\left\{
\begin{array}{c}
\sigma_x \\
\sigma_y \\
\tau_{xy}
\end{array}
\right\}_k
=
\begin{bmatrix}
\bar{Q}_{11} & \bar{Q}_{12} & \bar{Q}_{16} \\
\bar{Q}_{12} & \bar{Q}_{22} & \bar{Q}_{26} \\
\bar{Q}_{16} & \bar{Q}_{26} & \bar{Q}_{66}
\end{bmatrix}_k
\left\{
\begin{array}{c}
A'_{11} \, N_1 \\
A'_{12} \, N_1 \\
A'_{16} \, N_1
\end{array}
\right\}
\tag{4.100}
$$

For a three-layered E-glass/epoxy laminate, the foregoing stresses and strains are substituted in the three strength criteria to determine the maximum value of N_1 that can be sustained without failure of any laminae. (As will be seen later, angle-ply laminates have the behavioral characteristic that *all* the laminae fail simultaneously.) The theoretical and experimental results are plotted for the maximum stress criterion, the maximum strain criterion, and the Tsai-Hill criterion in Figs. 4-25, 4-26, and 4-27, respectively. As in the discussion of the strength of an individual lamina, the results for the maximum stress and maximum strain criteria exhibit cusps that are not borne out experimentally, significant overprediction of the failure stresses, and a rising strength for small angles of lamina orientation away from the direction of the applied force. In contrast, the Tsai-Hill criterion has both the same qualitative and quantitative behavior as the experimental data, has proper failure mode interaction, and reduces for isotropic materials to the von Mises failure criterion. Thus, the Tsai-Hill criterion will be used in subsequent strength predictions in this section. However, one of the other criteria may be more suitable for materials other than E-glass/epoxy.

4.5.4 Thermal and Mechanical Stress Analysis

Mechanical stress analysis, treated earlier in this chapter, does not suffice for analysis of laminates that have been cured at temperatures different from the

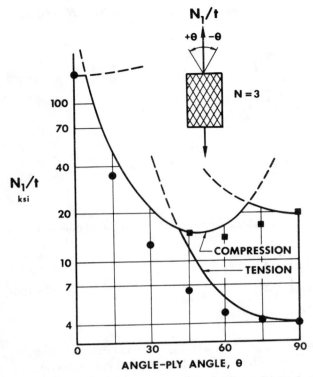

FIG. 4-25. Maximum stress failure criterion. (*After Tsai, Ref. 4-7.*)

design operating temperature. In such cases, thermal stresses arise and must be accounted for. The concepts of mechanical stress analysis will be reiterated in this section along with necessary modifications for thermal stress analysis.

The three-dimensional thermoelastic strain-stress relations are

$$\epsilon_i = S_{ij}\sigma_j + \alpha_i \Delta T \quad i,j = 1, 2, \ldots, 6 \tag{4.101}$$

wherein the total strain, ϵ_i, is the sum of the mechanical strain, $S_{ij}\sigma_j$, and the free thermal strain, $\alpha_i \Delta T$. The three-dimensional stress-strain relations are obtained by inversion:

$$\sigma_i = C_{ij}(\epsilon_j - \alpha_j \Delta T) \quad i,j = 1, 2, \ldots, 6 \tag{4.102}$$

In both Eqs. (4.101) and (4.102), the α_i are the coefficients of thermal expansion and ΔT is the temperature difference. In Eq. (4.102), the terms $C_{ij}\alpha_j \Delta T$ are the thermal stresses if the total strain is zero.

For plane stress on an orthotropic lamina in principal material coordinates,

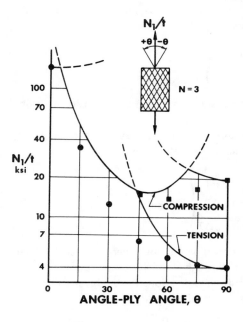

FIG. 4-26. Maximum strain failure criterion. (*After Tsai, Ref. 4-7.*)

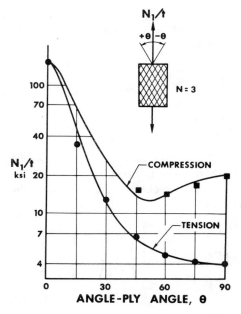

FIG. 4-27. Tsai-Hill failure criterion. (*After Tsai, Ref. 4-7.*)

$$\left\{ \begin{matrix} \sigma_1 \\ \sigma_2 \\ \tau_{12} \end{matrix} \right\} = \begin{bmatrix} Q_{11} & Q_{12} & 0 \\ Q_{12} & Q_{22} & 0 \\ 0 & 0 & Q_{66} \end{bmatrix} \left\{ \begin{matrix} \epsilon_1 & - \alpha_1 \Delta T \\ \epsilon_2 & - \alpha_2 \Delta T \\ \gamma_{12} \end{matrix} \right\} \qquad (4.103)$$

Note that the coefficients of thermal expansion affect only extensional strains, not the shearing strain.

The stresses in laminate coordinates for the kth layer are obtained by transformation of coordinates in the manner of Sec. 2.6 as

$$\left\{ \begin{matrix} \sigma_x \\ \sigma_y \\ \tau_{xy} \end{matrix} \right\}_k = \begin{bmatrix} \bar{Q}_{11} & \bar{Q}_{12} & \bar{Q}_{16} \\ \bar{Q}_{12} & \bar{Q}_{22} & \bar{Q}_{26} \\ \bar{Q}_{16} & \bar{Q}_{26} & \bar{Q}_{66} \end{bmatrix}_k \left\{ \begin{matrix} \epsilon_x & - \alpha_x \Delta T \\ \epsilon_y & - \alpha_y \Delta T \\ \gamma_{xy} & - \alpha_{xy} \Delta T \end{matrix} \right\}_k \qquad (4.104)$$

wherein the appearance of α_{xy} signifies an apparent coefficient of thermal shear. When the linear variation of strain through the thickness, Eq. (4.10), is substituted in Eq. (4.104) and the resulting expressions for the layer stresses are integrated through the thickness, the force resultants are

$$\left\{ \begin{matrix} N_x \\ N_y \\ N_{xy} \end{matrix} \right\} = \begin{bmatrix} A_{11} & A_{12} & A_{16} \\ A_{12} & A_{22} & A_{26} \\ A_{16} & A_{26} & A_{66} \end{bmatrix} \left\{ \begin{matrix} \epsilon_x^0 \\ \epsilon_y^0 \\ \gamma_{xy}^0 \end{matrix} \right\} + \begin{bmatrix} B_{11} & B_{12} & B_{16} \\ B_{12} & B_{22} & B_{26} \\ B_{16} & B_{26} & B_{66} \end{bmatrix} \left\{ \begin{matrix} \kappa_x \\ \kappa_y \\ \kappa_{xy} \end{matrix} \right\} - \left\{ \begin{matrix} N_x^T \\ N_y^T \\ N_{xy}^T \end{matrix} \right\}$$

$$(4.105)$$

in which the A_{ij} and B_{ij} are the usual extensional and coupling stiffnesses defined in Eq. (4.21) and the thermal forces are

$$\left\{ \begin{matrix} N_x^T \\ N_y^T \\ N_{xy}^T \end{matrix} \right\} = \int \begin{bmatrix} \bar{Q}_{11} & \bar{Q}_{12} & \bar{Q}_{16} \\ \bar{Q}_{12} & \bar{Q}_{22} & \bar{Q}_{26} \\ \bar{Q}_{16} & \bar{Q}_{26} & \bar{Q}_{66} \end{bmatrix}_k \left\{ \begin{matrix} \alpha_x \\ \alpha_y \\ \alpha_{xy} \end{matrix} \right\}_k \Delta T \, dz \qquad (4.106)$$

Note that the so-called thermal forces, N^T, are true thermal forces only when the total strains and curvatures are perfectly restrained, that is, zero.

In a similar manner, the moment resultants are obtained by integrating the moment of the stresses through the thickness:

$$\left\{ \begin{matrix} M_x \\ M_y \\ M_{xy} \end{matrix} \right\} = \begin{bmatrix} B_{11} & B_{12} & B_{16} \\ B_{12} & B_{22} & B_{26} \\ B_{16} & B_{26} & B_{66} \end{bmatrix} \left\{ \begin{matrix} \epsilon_x^0 \\ \epsilon_y^0 \\ \gamma_{xy}^0 \end{matrix} \right\} + \begin{bmatrix} D_{11} & D_{12} & D_{16} \\ D_{12} & D_{22} & D_{26} \\ D_{16} & D_{26} & D_{66} \end{bmatrix} \left\{ \begin{matrix} \kappa_x \\ \kappa_y \\ \kappa_{xy} \end{matrix} \right\} - \left\{ \begin{matrix} M_x^T \\ M_y^T \\ M_{xy}^T \end{matrix} \right\}$$

$$(4.107)$$

In which the D_{ij} are the usual bending stiffnesses defined in Eq. (4.21) and the thermal moments are

$$\begin{Bmatrix} M_x^T \\ M_y^T \\ M_{xy}^T \end{Bmatrix} = \int \begin{bmatrix} \bar{Q}_{11} & \bar{Q}_{12} & \bar{Q}_{16} \\ \bar{Q}_{12} & \bar{Q}_{22} & \bar{Q}_{26} \\ \bar{Q}_{16} & \bar{Q}_{16} & \bar{Q}_{66} \end{bmatrix}_k \begin{Bmatrix} \alpha_x \\ \alpha_y \\ \alpha_{xy} \end{Bmatrix}_k \Delta T z dz \qquad (4.108)$$

Actually, only in the restricted case of perfect constraint are the N^T and M^T thermal forces and moments, respectively. However, the force and moment resultants can be rearranged to read

$$\begin{Bmatrix} \bar{N}_x \\ \bar{N}_y \\ \bar{N}_{xy} \end{Bmatrix} = \begin{bmatrix} N_x + N_x^T \\ N_y + N_y^T \\ N_{xy} + N_{xy}^T \end{bmatrix} = \begin{bmatrix} A_{11} & A_{12} & A_{16} \\ A_{12} & A_{22} & A_{26} \\ A_{16} & A_{26} & A_{66} \end{bmatrix} \begin{Bmatrix} \epsilon_x^0 \\ \epsilon_y^0 \\ \gamma_{xy}^0 \end{Bmatrix} + \begin{bmatrix} B_{11} & B_{12} & B_{16} \\ B_{12} & B_{22} & B_{26} \\ B_{16} & B_{26} & B_{66} \end{bmatrix} \begin{Bmatrix} \kappa_x \\ \kappa_y \\ \kappa_{xy} \end{Bmatrix}$$

$$(4.109)$$

$$\begin{Bmatrix} \bar{M}_x \\ \bar{M}_y \\ \bar{M}_{xy} \end{Bmatrix} = \begin{bmatrix} M_x + M_x^T \\ M_y + M_y^T \\ M_{xy} + M_{xy}^T \end{bmatrix} = \begin{bmatrix} B_{11} & B_{12} & B_{16} \\ B_{12} & B_{22} & B_{26} \\ B_{16} & B_{26} & B_{66} \end{bmatrix} \begin{Bmatrix} \epsilon_x^0 \\ \epsilon_y^0 \\ \gamma_{xy}^0 \end{Bmatrix} + \begin{bmatrix} D_{11} & D_{12} & D_{16} \\ D_{12} & D_{22} & D_{26} \\ D_{16} & D_{26} & D_{66} \end{bmatrix} \begin{Bmatrix} \kappa_x \\ \kappa_y \\ \kappa_{xy} \end{Bmatrix}$$

$$(4.110)$$

In the form of Eqs. (4.109) and (4.110), the thermal portion of thermal and mechanical stress problems can be treated as equivalent mechanical loads defined by N^T and M^T in Eqs. (4.106) and (4.108), respectively, in addition to the mechanical loads, N and M.

The fictitious forces and moments, $\bar{N}$ and $\bar{M}$, are subject to the same rules as N and M for problems of mechanical loading only. For example, Eqs. (4.109) and (4.110) can be written as

$$\begin{Bmatrix} \bar{N} \\ \hline \bar{M} \end{Bmatrix} = \begin{bmatrix} A & \vdots & B \\ \hline B & \vdots & D \end{bmatrix} \begin{Bmatrix} \epsilon^0 \\ \hline \kappa \end{Bmatrix} \qquad (4.111)$$

In analogy to Eq. (4.58). Also upon inversion of Eq. (4.111),

$$\begin{Bmatrix} \epsilon^0 \\ \hline \kappa \end{Bmatrix} = \begin{bmatrix} A' & \vdots & B' \\ \hline H' & \vdots & D' \end{bmatrix} \begin{Bmatrix} \bar{N} \\ \hline \bar{M} \end{Bmatrix} \qquad (4.112)$$

in analogy to Eq. (4.69). Thus, a highly advantageous formulation has been achieved.

Upon normal solution of mechanical, or mechanical and thermal, loading problems, the stresses in the laminae can be determined from Eq. (4.103). The laminae stresses are used in the failure criterion to determine the laminate stiffness up to the maximum load the laminate can take. Obviously, lamination theory including thermal effects is essential to the correct description of laminate behavior because of heterogeneity and the natural curing process for fabricating laminates. Interactions between laminae are developed as a result of the manner in which the laminae are placed in the laminate and cured. These interactions will be described and discussed in examples of cross-ply and angle-ply laminates.

4.5.5. Strength of Cross-Ply Laminates

The procedure of strength analysis outlined in Sec. 4.5.2, with the Tsai-Hill failure criterion as justified in Sec. 4.5.3, will be illustrated for cross-ply laminates that have been cured at a temperature above their service or operating temperature (Ref. 4-8). Thus, the thermal effects discussed in Sec. 4.5.4 must be considered as well. For cross-ply laminates, the transformations of lamina properties are trivial, so the strength analysis procedure is readily interpreted.

The particular cross-ply laminate to be examined (Ref. 4-8) has three layers so is symmetric about its middle surface. Thus, no coupling exists between bending and extension. Under the condition $N_x = N$ and all other loads and moments are zero, the stresses in the (symmetric) outer layers are identical. One outer layer is called the 1-layer and has fibers in the x-direction (see Fig. 4-28). The inner layer is called the 2-layer and has fibers in the y-direction. The other outer layer is the 3-layer, but because of symmetry there is no need to refer to it. The cross-ply ratio, M, is .2 so the thickness of the inner layer is ten times that of each of the outer layers.

The properties of the example E-glass/epoxy lamina are

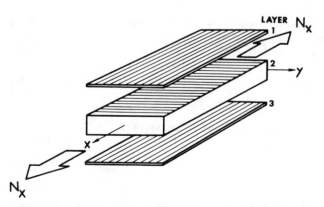

FIG. 4-28. Exploded (unbonded) view of a three-layered $M = .2$ cross-ply laminate under tensile loading.

$$E_1 = 7.8 \times 10^6 \text{ psi} \quad G_{12} = 1.25 \times 10^6 \text{ psi}$$

$$X_t = X_c = 150 \text{ ksi}$$

$$E_2 = 2.6 \times 10^6 \text{ psi} \quad \alpha_1 = 3.5 \times 10^{-6}/°\text{F}$$

$$Y_t = \quad 4 \text{ ksi}$$

$$Y_c = \quad 20 \text{ ksi} \tag{4.113}$$

$$\nu_{12} = \ .25 \quad\quad \alpha_2 = 11.4 \times 10^{-6}/°\text{F}$$

$$S = \quad 6 \text{ ksi}$$

The highest modulus is in the fiber direction, and the highest coefficient of thermal expansion is in the direction perpendicular to the fibers. Moreover, all stiffnesses are regarded as the same in tension as in compression, although the strengths are different.

Pre-failure deformation

The lamina reduced stiffnesses are

$$Q_{11}^{(1)} = Q_{22}^{(2)} = 7.9660 \times 10^6 \text{ psi} \quad Q_{12}^{(1)} = Q_{12}^{(2)} = .6638 \times 10^6 \text{ psi}$$

$$Q_{22}^{(1)} = Q_{11}^{(2)} = 2.6550 \times 10^6 \text{ psi} \quad Q_{66}^{(1)} = Q_{66}^{(2)} = 1.250 \times 10^6 \text{ psi} \tag{4.114}$$

$$Q_{16}^{(1)} = Q_{16}^{(2)} = Q_{26}^{(1)} = Q_{26}^{(2)} = 0$$

and the apparent coefficients of thermal expansion are

$$\alpha_x^{(1)} = \alpha_y^{(2)} = 3.5 \times 10^{-6}/°\text{F}$$

$$\alpha_y^{(1)} = \alpha_x^{(2)} = 11.4 \times 10^{-6}/°\text{F} \tag{4.115}$$

$$\alpha_{xy}^{(1)} = \alpha_{xy}^{(2)} = 0$$

The laminate extensional stiffnesses are

$$A_{11} = 3.5401 \times 10^6 \ t \text{ psi}$$

$$A_{12} = \ .6638 \times 10^6 \ t \text{ psi}$$

$$A_{22} = 7.0809 \times 10^6 \ t \text{ psi} \tag{4.116}$$

$$A_{66} = 1.2500 \times 10^6 \ t \text{ psi}$$

where t is the laminate thickness, that is, if $t = 1''$, then the value of A_{11} would be 3.5401×10^6 lb/in. The inverse extensional stiffnesses are

$$A_{11}' = \ .2875 \times 10^{-6}/(t \text{ psi})$$

$$A_{12}' = -.0270 \times 10^{-6}/(t \text{ psi})$$

$$A_{22}' = \ .1438 \times 10^{-6}/(t \text{ psi}) \tag{4.117}$$

$$A_{66}' = \ .8000 \times 10^{-6}/(t \text{ psi})$$

Thus, all numbers are in hand for calculation of the stresses in the example cross-ply laminate.

Consider a constant temperature of the laminate different from, and relative to, its stress-free curing temperature. Then, the thermal forces are, from Eq. (4.106),

$$N_x^T = 33.1 \, t\Delta T \, \text{psi/}°\text{F}$$
$$N_y^T = 35.0 \, t\Delta T \, \text{psi/}°F$$
$$N_{xy}^T = 0$$

(4.118)

and the thermal moments, from Eq. (4.108), are zero.

By means of rather involved successive substitutions of Eq. (4.109) in (4.112) in (4.10) and finally in (4.104), the stresses in the inner and outer layers can be shown to be

$$\sigma_x^{(1)} = 2.27 \left(\frac{N_x}{t} \right) + 35.5 \, \Delta T \, \text{psi/}°\text{F}$$

$$\sigma_y^{(1)} = .12 \left(\frac{N_x}{t} \right) - 16.0 \, \Delta T \, \text{psi/}°\text{F}$$

(4.119)

$$\tau_{xy}^{(1)} = 0$$

$$\sigma_x^{(2)} = .75 \left(\frac{N_x}{t} \right) - 7.1 \, \Delta T \, \text{psi/}°\text{F}$$

$$\sigma_y^{(2)} = -.024 \left(\frac{N_x}{t} \right) + 3.2 \, \Delta T \, \text{psi/}°\text{F}$$

(4.120)

$$\tau_{xy}^{(2)} = 0$$

The stresses have now been determined as a linear function of the applied loads, N_x and ΔT.

Application of the failure criterion

The failure criterion must be applied to determine the maximum values of N_x and ΔT that can be sustained without failure of any layer. Actually, the failure criterion is applied to each layer separately. For the special orientation of cross-ply laminates, the Tsai-Hill failure criterion for each layer can be expressed as

$$\left(\frac{\sigma_x}{X} \right)^2 - \frac{\sigma_x \sigma_y}{X^2} + \left(\frac{\sigma_y}{Y} \right)^2 + \left(\frac{\tau_{xy}}{S} \right)^2 = 1$$

(4.121)

In the outer layer, since $\tau_{xy} = 0$, the criterion simplifies to

$$\sigma_x^2 - \sigma_x\sigma_y + \left(\frac{X}{Y}\right)^2 \sigma_y^2 = X^2 \tag{4.122}$$

from which, upon substitution of the stresses, results a quadratic equation with solution

$$\frac{N_x}{t} = 110\ \Delta T \text{ psi/}^\circ\text{F} + [57.5Y^2 - 3000\ \Delta T^2 (\text{psi/}^\circ\text{F})^2]^{1/2} \tag{4.123}$$

If the curing temperature is 270°F and the laminate operates at 70°F, $\Delta T = -200^\circ$F, so

$$\frac{N_x}{t} = 6{,}300 \text{ psi} \tag{4.124}$$

Alternatively, if the laminate is cured at room temperature, $\Delta T = 0$, so

$$\frac{N_x}{t} = 30{,}400 \text{ psi} \tag{4.125}$$

In the inner layer, a similar set of steps yields

$$\frac{N_x}{t} \cong 9.6\ \Delta T \text{ psi/}^\circ\text{F} + 5{,}320 \text{ psi} \tag{4.126}$$

from which, if the laminate is cured at 270°F and used at 70°F,

$$\frac{N_x}{t} = 3{,}400 \text{ psi} \tag{4.127}$$

or if cured and used at 70°F,

$$\frac{N_x}{t} = 5{,}320 \text{ psi} \tag{4.128}$$

Obviously, if the laminate is cured at 270°F, the inner layer will fail first. Actually, the inner layer fails for both example curing temperatures, although the outer layer would fail first if the curing temperature were high enough. On the other hand, if the curing temperature were lowered, the laminate would exhibit higher strength. The values of N_x/t for the two curing conditions are the values at failure of the inner layer. Those values correspond to the point labeled N_1 in Fig. 4-23, that is, the so-called "knee" of the load-deformation diagram.

Up to the load corresponding to the knee, the load-deformation diagram is linear and all layers are intact. The axial strain at the knee is

$$\epsilon_x^0 = A_{11}' N_x = .098\%$$ (4.129)

if the residual strains are ignored, that is, ϵ_x^0 is measured from zero load which is not the stress-free state.

Behavior after the first layer fails

After a layer fails, the behavior of the laminate depends on how the mechanical and thermal interactions between layers uncouple. Actually, failure of a layer may not mean that it can no longer carry load. In the present example of a cross-ply laminate, the inner layer at 90° to the x axis has "failed", but, owing to the orientation of the fibers (perpendicular to the main failure-causing stress), the failure should be only a series of cracks *parallel* to the fibers. Thus, stress can still be carried by the inner layer in the fiber direction (y-direction).

The degraded laminate then has stiffnesses based on the original properties of the outer layer and the following properties of the inner layer

$$Q_{11}^{(2)} = 0 \qquad Q_{22}^{(2)} = 7.97 \times 10^6 \text{ psi}$$
$$Q_{12}^{(2)} = 0 \qquad Q_{66}^{(2)} = 0$$ (4.130)

where the zeros are actually a very small number in order to avoid numerical difficulties in a computer analysis. The inverse extensional stiffness matrix of the laminate then has the values

$$A_{11}' = .75 \times 10^{-6}/(t \text{ psi})$$
$$A_{12}' = .01 \times 10^{-6}/(t \text{ psi})$$ (4.131)
$$A_{22}' = .14 \times 10^{-6}/(t \text{ psi})$$

Note that A_{22}' is about the same as in the undegraded state.

The resulting stresses are

$$\sigma_x^{(1)} = 6.00 \frac{N_x}{t}$$

$$\sigma_y^{(1)} = .47 \frac{N_x}{t} - 19.3 \Delta T \text{ psi}/{}^\circ F$$ (4.132)

$$\tau_{xy}^{(1)} = 0$$

$$\sigma_x^{(2)} = 0$$

$$\sigma_y^{(2)} = -.09 \frac{N_x}{t} + 3.9 \Delta T \text{ psi}/{}^\circ F$$ (4.133)

$$\tau_{xy}^{(2)} = 0$$

Obviously, there is no thermal coupling in the x-direction, but the thermal coupling in the y-direction has increased from the undegraded state [compare Eqs. (4.119) and (4.120) with Eqs. (4.132) and (4.133)]. The thermal coupling is so strong under the condition $N_x/t = 3{,}400$ psi and $\Delta T = -200°$F that the outer layers fail by developing cracks parallel to the fibers. This contention can be verified by substituting the resulting stresses in the failure criterion for the outer layer. Thus, as is indicated to be possible in the strength analysis procedure of the right side of Fig. 4-24, more than one lamina fails simultaneously, that is, at the same load.

Behavior after degradation

The laminate is now degraded to the point where the outer layers carry only stress in the x-direction and the inner layers can carry only stress in the y-direction. In both cases, the stress is parallel to the fibers. Thus, the laminate is completely decoupled, both thermally and mechanically. The only nonzero reduced stiffnesses are

$$Q_{11}^{(1)} = Q_{22}^{(2)} = 7.97 \times 10^6 \text{ psi} \tag{4.134}$$

and the associated laminate inverse extensional stiffnesses are

$$A_{11}' = .77 \times 10^{-6}/(t \text{ psi})$$
$$A_{12}' = 0 \tag{4.135}$$
$$A_{22}' = .15 \times 10^{-6}/(t \text{ psi})$$

Accordingly, the only lamina stress that develops is

$$\sigma_x^{(1)} = 6.00 \, \frac{N_x}{t} \tag{4.136}$$

and the resulting laminate extensional stiffness in the x-direction above the knee of the load-deformation curve is

$$\frac{N_x/t}{\epsilon_x^0} = \frac{1}{A_{11}' t} = 1.3 \times 10^6 \text{ psi} \tag{4.137}$$

which is about one-third the undegraded stiffness.

Maximum laminate load

The stage is now set to determine the largest load the laminate can carry. Only the outer layers resist the load N_x after the "knee" of the load-deformation curve. There, the stress in the outer layers is, from Eq. (4.119),

$$\sigma_x^{(1)} = 618 \text{ psi} \tag{4.138}$$

The largest possible value of σ_x under uniaxial conditions is 150 ksi. Thus, the outer layer can be stressed about an additional 149.4 ksi. The corresponding change in the force resultant is obtained from Eq. (4.136) as

$$\frac{\Delta N_x}{t} = \frac{\Delta\sigma_x^{(1)}}{6.00} = 149{,}400 \text{ psi}/6 = 24{,}900 \text{ psi} \tag{4.139}$$

When this change in force resultant is added to the force resultant at the "knee", the largest laminate load is determined to be

$$\frac{N_x}{t} = 3{,}400 \text{ psi} + 24{,}900 \text{ psi} = 28{,}300 \text{ psi} \tag{4.140}$$

which is reasonably close to the experimental maximum load in Fig. 4-29. Note that a "knee" is observed in the experiments. Also, the results of another theory called Netting Analysis are shown in Fig. 4-29; obviously, that theory is incorrect for fiber-reinforced materials. Netting Analysis is based on all load being carried in the fibers. Such a theory is more appropriate for woven fabrics since they have no matrix to carry loads.

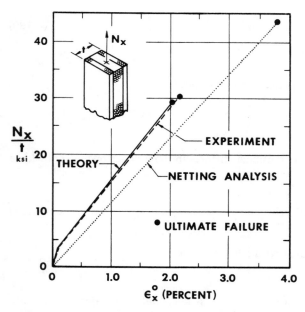

FIG. 4-29. Strength of a cross-ply laminate with $M = .2$.
(After Tsai, Ref. 4-8.)

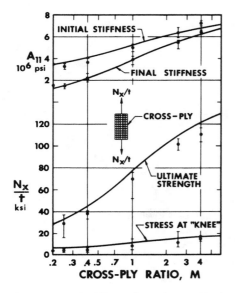

FIG. 4-30. Strength of cross-ply laminates. (*After Tsai, Adams, and Doner, Ref. 4-9.*)

Strength and stiffness for other cross-ply ratios

Theoretical and experimental strengths and stiffnesses of three-layer cross-ply laminates with cross-ply ratios ranging from .2 to 4 are shown in Fig. 4-30. The scatter in the data is partially due to the difficulty of making tensile specimens; the characteristic dog bone shape is formed by routing which often damages the 90° layer.

The predicted strengths are generally somewhat above the measured values. The predicted and observed stiffnesses, both initial (below the knee) and final, are in very good agreement.

4.5.6 Strength of Angle-Ply Laminates

Angle-ply laminates have more complicated stiffness matrices than cross-ply laminates since nontrivial coordinate transformations are involved. However, their behavior will be shown to be a bit simpler than that of cross-ply laminates since no "knee" results in the load-deformation diagram under uniaxial loading. Other than the two preceding differences, the analysis of angle-ply laminates is conceptually the same as that of cross-ply laminates.

The example considered to illustrate the strength analysis procedure is a three-layered laminate with a +15°/−15°/+15° stacking sequence (Ref. 4-8.) The laminae are the same *E*-glass/epoxy as in the cross-ply example. In the laminate coordinates, the transformed reduced stiffnesses are

$$\bar{Q}_{11}^{(1)} = \bar{Q}_{11}^{(2)} = 7.342 \times 10^6 \text{ psi}$$

$$\bar{Q}_{12}^{(1)} = \bar{Q}_{12}^{(2)} = .932 \times 10^6 \text{ psi} \tag{4.141}$$

$$\bar{Q}_{22}^{(1)} = \bar{Q}_{22}^{(2)} = 2.763 \times 10^6 \text{ psi}$$

$$\bar{Q}_{16}^{(1)} = -\bar{Q}_{16}^{(2)} = -1.129 \times 10^6 \text{ psi}$$

$$\bar{Q}_{26}^{(1)} = -\bar{Q}_{26}^{(2)} = -.199 \times 10^6 \text{ psi}$$

$$\bar{Q}_{66}^{(1)} = \bar{Q}_{66}^{(2)} = 1.519 \times 10^6 \text{ psi}$$

(4.141)
(cont'd.)

and the apparent coefficients of thermal expansion are

$$\alpha_x^{(1)} = \alpha_x^{(2)} = 4.029 \times 10^{-6}/^\circ F$$

$$\alpha_y^{(1)} = \alpha_y^{(2)} = 10.870 \times 10^{-6}/^\circ F$$

$$\alpha_{xy}^{(1)} = -\alpha_{xy}^{(2)} = 1.975 \times 10^{-6}/^\circ F$$

(4.142)

The inverse extensional stiffness matrix can be shown to be

$$A' = \begin{bmatrix} .144 & -.0481 & .0336 \\ & .381 & .0047 \\ \text{(symmetric)} & & .6668 \end{bmatrix} \times 10^{-6}/(t \text{ psi})$$

(4.143)

For a constant lamination temperature,

$$N_x^T = 37.5 \, t\Delta T \text{ psi}/^\circ F$$

$$N_y^T = 33.2 \, t\Delta T \text{ psi}/^\circ F$$

$$N_{xy}^T = -1.24 \, t\Delta T \text{ psi}/^\circ F$$

(4.144)

and the thermal moments are zero. When the laminate is subjected to N_x only, the stresses are

$$\sigma_x^{(1)} = .97 \frac{N_x}{t} - .44 \, \Delta T \text{ psi}/^\circ F$$

$$\sigma_y^{(1)} = -.005 \frac{N_x}{t} - .08 \, \Delta T \text{ psi}/^\circ F$$

$$\tau_{xy}^{(1)} = -.10 \frac{N_x}{t} - 1.79 \, \Delta T \text{ psi}/^\circ F$$

(4.145)

$$\sigma_x^{(2)} = 1.05 \frac{N_x}{t} + .89 \, \Delta T \text{ psi}/^\circ F$$

$$\sigma_y^{(2)} = .01 \frac{N_x}{t} + .16 \, \Delta T \text{ psi}/^\circ F$$

$$\tau_{xy}^{(2)} = .20 \frac{N_x}{t} + 3.58 \, \Delta T \text{ psi}/^\circ F$$

(4.146)

Note that the stresses σ_y are very small in comparison to the shearing stresses. Thus, the Tsai-Hill failure criterion can be simplified for this laminate to

$$K_1 \sigma_x^2 + K_2 \sigma_x \tau_{xy} + K_3 \tau_{xy}^2 = X^2 \tag{4.147}$$

in which

$$
\begin{aligned}
K_1 &= \cos^4\theta + 624 \cos^2\theta \sin^2\theta + 1406 \sin^4\theta \\
K_2 &= -(1244 \cos^3\theta \sin\theta + 4386 \cos\theta \sin^3\theta) \\
K_3 &= 625 \cos^4\theta + 4382 \cos^2\theta \sin^2\theta + 625 \sin^4\theta
\end{aligned} \tag{4.148}
$$

The values of K_i for $\theta = -15°$ are

$$K_1 = 46.20 \qquad K_2 = 363.91 \qquad K_3 = 821.00 \tag{4.149}$$

and for $\theta = +15°$ are

$$K_1 = 46.20 \qquad K_2 = -363.91 \qquad K_3 = 821.00 \tag{4.150}$$

Accordingly, in the outer layer, the largest force is

$$\frac{N_x}{t} = 11.14 \,\Delta T \text{ psi/}°\text{F} + 37,400 \text{ psi} \tag{4.151}$$

which if the laminate is cured at $270°$F yields

$$\frac{N_x}{t} = 35,200 \text{ psi} \tag{4.152}$$

Similarly, in the inner layer,

$$\frac{N_x}{t} = 52,600 \text{ psi} \tag{4.153}$$

so the outer layer fails first. Because the remaining inner layer cannot, by itself, withstand the laminate force of 35,200 psi, the inner layer fails immediately after the outer layer. Therefore, the maximum laminate force is

$$\frac{N_x}{t} = 35,200 \text{ psi} \tag{4.154}$$

and there is, as claimed, no knee in the load-deformation behavior.

For other angle-ply lamination angles, similar predicted strengths were obtained and are shown along with experimental results in Fig. 4-31. The agreement is quite good. As further substantiation of the stiffness prediction techniques in Sec. 4.4, theoretical and experimental stiffnesses are plotted in Fig. 4-31 and are also seen to be in very good agreement. However, for

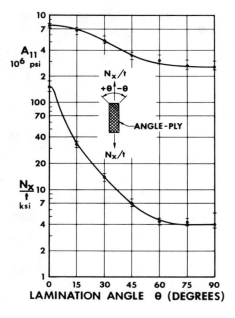

FIG. 4-31. Strength of angle-ply laminates. (*After Tsai, Ref. 4-8.*)

lamination angles around $45°$, the deformations at failure are, in general, several times the predicted deformations because of nonlinear stress-strain behavior. The nonlinear behavior is not unexpected since the large shearing stresses that are developed tend to deform the (nonlinear) matrix of the fiber-reinforced composite more than they deform the fibers. Another interesting observation involves a comparison of the present angle-ply data with the data for a unidirectional lamina at various orientations given in Fig. 2-25 in Sec. 2.9. For angles larger than $0°$ but less than $45°$, the angle-ply has up to about 50 percent higher strength than the unidirectional lamina. However, above $45°$, the unidirectional lamina exhibits higher strength. Such differences result from mechanical and thermal interactions between layers.

4.5.7 Summary Remarks

The strength of special classes of laminated fiber-reinforced composites has been analyzed on the basis of several hypotheses:

- Linear elastic behavior to failure for individual laminae
- Kirchhoff hypothesis of linear strain variation through the laminate thickness (prior to degradation, if any; after degradation, linear only through the lamina thickness).
- Strengths and stiffnesses of the laminae are the same in tension as in compresssion.
- The Tsai-Hill criterion governs failure of a lamina (procedure could involve another criterion).

- Failure of a lamina may mean, for example, only lack of stiffness and strength perpendicular to the fibers with no degradation in the fiber direction.

For cross-ply laminates, a knee in the load-deformation curve occurs after the mechanical and thermal interactions between layers uncouple owing to failure (degradation) of a lamina. The mechanical interactions are due to Poisson effects and, for example, shear strain–normal stress coupling. The thermal interactions are due to different coefficients of thermal expansion in different layers because of different angular orientations of the layers (even though the orthotropic materials can be the same). The interactions are disrupted if the layers in a laminate separate.

For angle-ply laminates, no such knee or change in slope occurs in the load-deformation behavior. Simultaneous failure of all layers occurs.

For two- and three-layered cross-ply and angle-ply laminates of E-glass/ epoxy, Tsai (Ref. 4-8) gives all the stiffnesses, inverse stiffnesses, thermal forces and moments, etc. in a tabular form. The results are obtained for various cross-ply ratios and lamination angles, as appropriate, from a short computer program that could be used for other materials.

In the strength analysis discussed in this section, no account was taken of the possible increase in deflection that occurs when a layer fails. That is, if the laminate is simplistically represented by a set of springs in parallel (one spring represents one lamina) as in Fig. 4-32, then when one spring breaks, the

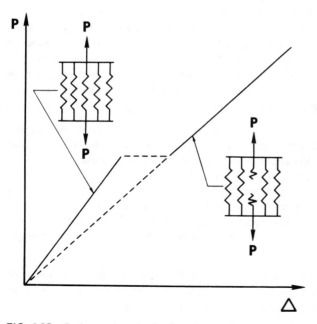

FIG. 4-32. Spring analogy for laminate load deflection behavior.

remaining springs must each have a higher load and hence a higher deflection. Accordingly, a horizontal jump (strain or deflection) occurs in the load-deflection behavior of the laminate as depicted schematically in Fig. 4-32. Such a step-wise load-deflection behavior has not been observed in experiments nor has it been treated, to the author's knowledge, from an analytical point of view. Perhaps, in the experimental procedure, the jump is smoothed out enough that it is difficult to detect. At any rate, the analysis of laminate strength is not altogether satisfactory.

Problem Set 4.5

Exercise 4.5.1 Derive the thermoelastic stress-strain relations for an orthotropic lamina under plane stress, Eq. (4.103).

Exercise 4.5.2 Derive expressions for α_x, α_y, and α_{xy} in Eq. (4.104) as a function of α_1, α_2, and θ and verify that α_{xy} vanishes for isotropic materials.

Exercise 4.5.3 Verify that the thermal force N_x^T for a three-layered cross-ply laminate with $M = .2$ is given by Eq. (4.118).

Exercise 4.5.4 Verify that the stress $\sigma_x^{(1)}$ for a three-layered cross-ply laminate with $M = .2$ is given by Eq. (4.119).

4.6 INTERLAMINAR STRESSES

In classical lamination theory, no account is taken of interlaminar stresses such as σ_z, τ_{zx}, and τ_{zy} which are shown in Fig. 4-33. Rather, only the stresses in the plane of the laminate, σ_x, σ_y, and τ_{xy}, are considered; that is, a plane stress state is assumed. Accordingly, classical lamination theory is incapable of providing predictions of some of the stresses that actually cause failure of a composite material. Interlaminar stresses are one of the failure mechanisms uniquely characteristic of composite materials. Moreover, classical lamination theory implies values of τ_{xy} where it cannot possibly exist, namely at the edge of a laminate. Physical grounds will be used to establish that:

- at the free edges of a laminate (sides of a laminate or holes), the interlaminar shearing stress is very high (perhaps even singular) and would therefore cause the debonding that has been observed in such regions.
- layer stacking sequence changes produce differences in tensile strength of a laminate even though the orientations of each layer do not change (in classical lamination theory, such changes have no effect on the extensional stiffnesses). Interlaminar normal stress, σ_z, changes near the laminate boundaries are believed to provide the answer to such strength differences.

In this section, the interlaminar stresses in the simple case of the free edge of an angle-ply laminate will be analyzed. First, the concept of interlaminar stresses will be described. Next, experimental verification of the theory is offered.

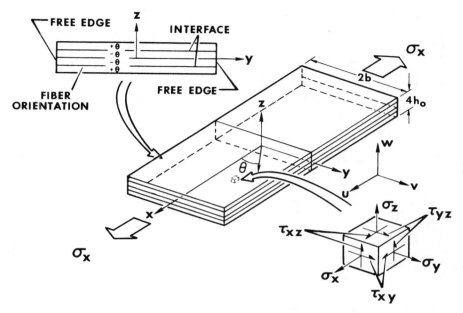

FIG. 4-33. Laminate geometry and stresses. (*After Pipes and Pagano, Ref. 4-10.*)

Finally, the interaction of interlaminar stresses and stacking sequence and their influence on laminate strength will be examined.

4.6.1 Classical Lamination Theory

Consider an angle-ply laminate composed of orthotropic laminae that are symmetrically disposed about the middle surface as shown in Fig. 4-33. Note that there is an even number of layers and that the inner layers have the same angular orientation; this is not, therefore, a regular symmetric angle-ply laminate. Because of the symmetry, there is no coupling between bending and extension. That is, the laminate in Fig. 4-33 can be subjected to N_x and will only extend in the x-direction and contract in the y- and z- directions, but will not bend.

The analysis of such a laminate by use of classical lamination theory revolves about the stress-strain relations of an individual orthotropic lamina in plane stress in principal material directions:

$$\begin{Bmatrix} \sigma_1 \\ \sigma_2 \\ \tau_{12} \end{Bmatrix}_k = \begin{bmatrix} Q_{11} & Q_{12} & 0 \\ Q_{12} & Q_{22} & 0 \\ 0 & 0 & Q_{66} \end{bmatrix}_k \begin{Bmatrix} \epsilon_1^0 \\ \epsilon_2^0 \\ \gamma_{12}^0 \end{Bmatrix} \qquad (4.155)$$

which can be transformed to the laminate axes by use of Eq. (2.80):

$$
\begin{Bmatrix} \sigma_x \\ \sigma_y \\ \tau_{xy} \end{Bmatrix}_k = \begin{bmatrix} \bar{Q}_{11} & \bar{Q}_{12} & \bar{Q}_{16} \\ \bar{Q}_{12} & \bar{Q}_{22} & \bar{Q}_{26} \\ \bar{Q}_{16} & \bar{Q}_{26} & \bar{Q}_{66} \end{bmatrix}_k \begin{Bmatrix} \epsilon_x^0 \\ \epsilon_y^0 \\ \gamma_{xy}^0 \end{Bmatrix}
\tag{4.156}
$$

The extensional stiffnesses of the laminate are then given by

$$
A_{ij} = \sum_{k=1}^{N} (\bar{Q}_{ij})_k (z_k - z_{k-1})
\tag{4.157}
$$

and the force-strain relations are

$$
\begin{Bmatrix} N_x \\ 0 \\ 0 \end{Bmatrix} = \begin{bmatrix} A_{11} & A_{12} & 0 \\ A_{12} & A_{22} & 0 \\ 0 & 0 & A_{66} \end{bmatrix} \begin{Bmatrix} \epsilon_x^0 \\ \epsilon_y^0 \\ \gamma_{xy}^0 \end{Bmatrix}
\tag{4.158}
$$

The membrane strain state is

$$
\epsilon_x^0 = \frac{A_{22} N_x}{A_{11} A_{22} - A_{12}^2}
$$
$$
\epsilon_y^0 = \frac{-A_{12} N_x}{A_{11} A_{22} - A_{12}^2}
\tag{4.159}
$$

There is no γ_{xy}^0 of the laminate. However, there is shearing strain in the principal material coordinates of each lamina in addition to normal strains as is proved by use of Eq. (2.70):

$$
\begin{Bmatrix} \epsilon_1 \\ \epsilon_2 \\ \gamma_{12} \end{Bmatrix}_k = \begin{Bmatrix} \cos^2\theta - \left(\dfrac{A_{12}}{A_{22}}\right)\sin^2\theta \\ \sin^2\theta - \left(\dfrac{A_{12}}{A_{22}}\right)\cos^2\theta \\ -2\cos\theta\sin\theta \left[1 + \left(\dfrac{A_{12}}{A_{22}}\right)\right] \end{Bmatrix}_k \frac{A_{22} N_x}{A_{11} A_{22} - A_{12}^2}
\tag{4.160}
$$

The shearing stresses that correspond to such shearing strains are not physically possible at the edge of a lamina, say $y = b$ of Fig. 4-33.

The free body diagram of each layer of a laminate in Fig. 4-34 is useful in understanding the physical mechanism of shear transfer between layers. There, the fact that τ_{xy} must be zero on a free edge means that the couple caused by

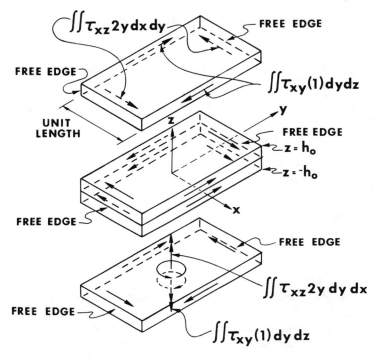

FIG. 4-34. Interlaminar shear stress mechanism. (*After Pipes and Pagano, Ref. 4-10.*)

τ_{xy} acting along the other edges of the free body must be reacted. The only possible reacting couple to satisfy moment equilibrium is caused by τ_{xz} acting on part of the lower face of the layers at the interface with the next layer.

4.6.2 Elasticity Formulation

Rather than a plane stress state, a three-dimensional stress state is considered in the elasticity approach of Pipes and Pagano (Ref. 4-10) to the problem of Sec. 4.6.1. The stress-strain relations for each orthotropic layer in principal material directions are

$$
\begin{Bmatrix} \sigma_1 \\ \sigma_2 \\ \sigma_3 \\ \tau_{23} \\ \tau_{31} \\ \tau_{12} \end{Bmatrix} = \begin{bmatrix} C_{11} & C_{12} & C_{13} & 0 & 0 & 0 \\ C_{12} & C_{22} & C_{23} & 0 & 0 & 0 \\ C_{13} & C_{23} & C_{33} & 0 & 0 & 0 \\ 0 & 0 & 0 & C_{44} & 0 & 0 \\ 0 & 0 & 0 & 0 & C_{55} & 0 \\ 0 & 0 & 0 & 0 & 0 & C_{66} \end{bmatrix} \begin{Bmatrix} \epsilon_1 \\ \epsilon_2 \\ \epsilon_3 \\ \gamma_{23} \\ \gamma_{31} \\ \gamma_{12} \end{Bmatrix}
$$

(4.161)

and can, upon transformation of coordinates in the 1-2 plane, be expressed in laminate coordinates as

$$
\begin{Bmatrix} \sigma_x \\ \sigma_y \\ \sigma_z \\ \tau_{yz} \\ \tau_{zx} \\ \tau_{xy} \end{Bmatrix} = \begin{bmatrix} \bar{C}_{11} & \bar{C}_{12} & \bar{C}_{13} & 0 & 0 & \bar{C}_{16} \\ \bar{C}_{12} & \bar{C}_{22} & \bar{C}_{23} & 0 & 0 & \bar{C}_{26} \\ \bar{C}_{13} & \bar{C}_{23} & \bar{C}_{33} & 0 & 0 & \bar{C}_{36} \\ 0 & 0 & 0 & \bar{C}_{44} & \bar{C}_{45} & 0 \\ 0 & 0 & 0 & \bar{C}_{45} & \bar{C}_{55} & 0 \\ \bar{C}_{16} & \bar{C}_{26} & \bar{C}_{36} & 0 & 0 & \bar{C}_{66} \end{bmatrix} \begin{Bmatrix} \epsilon_x \\ \epsilon_y \\ \epsilon_z \\ \gamma_{yz} \\ \gamma_{zx} \\ \gamma_{xy} \end{Bmatrix}
\tag{4.162}
$$

The strain-displacement relations are

$$
\epsilon_x = u_{,x} \qquad\qquad \epsilon_y = v_{,y} \qquad\qquad \epsilon_z = w_{,z}
$$
$$
\gamma_{yz} = v_{,z} + w_{,y} \qquad \gamma_{zx} = w_{,x} + u_{,z} \qquad \gamma_{xy} = u_{,y} + v_{,x}
\tag{4.163}
$$

where a comma denotes partial differentiation of the principal symbol with respect to the subscript.

If the laminate is subjected to uniform axial extension on ends $x =$ constant, then all the stresses are independent of x. The stress-displacement relations are obtained by substituting the strain-displacement relations, Eq. (4.163), in the stress-strain relations, Eq. (4.162). Next, the stress-displacement relations can be integrated under the condition that all stresses are functions of y and z only to obtain, after imposing symmetry and antisymmetry conditions, the form of the displacement field for the present problem:

$$
u = Kx + U(y,z)
$$
$$
v = V(y,z)
\tag{4.164}
$$
$$
w = W(y,z)
$$

The stress equilibrium equations then reduce to

$$
\tau_{xy,y} + \tau_{zx,z} = 0
$$
$$
\sigma_{y,y} + \tau_{yz,z} = 0
\tag{4.165}
$$
$$
\tau_{yz,y} + \sigma_{z,z} = 0
$$

Upon substitution of the displacement field, Eq. (4.164), in the stress-displacement relations and subsequently in the stress equilibrium differential equations, Eq. (4.165), the displacement equilibrium equations are, for each layer,

$$\bar{C}_{66}U_{,yy} + \bar{C}_{55}U_{,zz} + \bar{C}_{26}V_{,yy} + \bar{C}_{45}V_{,zz} + (\bar{C}_{36} + \bar{C}_{45})W_{,yz} = 0$$

$$\bar{C}_{26}U_{,yy} + \bar{C}_{45}U_{,zz} + \bar{C}_{22}V_{,yy} + \bar{C}_{44}V_{,zz} + (\bar{C}_{23} + \bar{C}_{44})W_{,yz} = 0 \qquad (4.166)$$

$$(\bar{C}_{45} + \bar{C}_{36})U_{,yz} + (\bar{C}_{44} + \bar{C}_{23})V_{,yz} + \bar{C}_{44}W_{,yy} + \bar{C}_{33}W_{,zz} = 0$$

These coupled second-order partial differential equations do not admit a closed-form solution. Accordingly, the approximate numerical technique of finite differences is employed. First, however, the boundary conditions must be stipulated in order to complete the formulation of the problem. Symmetry of the laminate about several planes permits reduction of the region of consideration to a quarter of the laminate cross section in the yz plane at any value of x as shown in Fig. 4-35. There, along the upper surface,

$$\tau_{yz} = 0 \qquad \sigma_z = 0 \qquad \tau_{xz} = 0 \tag{4.167}$$

along the outer edge,

$$\tau_{xy} = 0 \qquad \sigma_y = 0 \qquad \tau_{yz} = 0 \tag{4.168}$$

along the middle surface, $z = 0$, since U and V must be symmetric and W antisymmetric,

$$U_{,z}(y, 0) = 0$$

$$V_{,z}(y, 0) = 0 \tag{4.169}$$

$$W(y, 0) = 0$$

and along the line $y = 0$, since U and V must be antisymmetric and W symmetric,

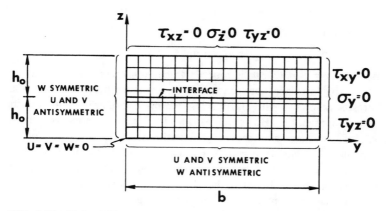

FIG. 4-35. Finite difference representation and boundary conditions. (*After Pipes and Pagano, Ref. 4-10.*)

$$U(0, z) = 0$$
$$V(0, z) = 0 \tag{4.170}$$
$$W_{,y}(0, z) = 0$$

At the corner $(b, 2h_0)$ of the region, five stress conditions apparently govern the behavior. However, the problem would be overspecified if all five conditions were imposed at the same time. Rather, three are specified and, subsequently, the remaining two are seen to be satisfied thereby acting as a built-in verification of the numerical results. Numerical experimentation revealed that the choice of the three conditions is immaterial; the remaining two are always satisfied.

The numerical solution, as mentioned earlier, was obtained by the finite difference method. The two regions (layers) indicated in Fig. 4-35 are represented by a series of regularly spaced material points as shown. At each point, the differential equations are approximated by finite difference operators (central difference operators inside the region with forward and backward difference operators being used at the boundaries). At the interface between layers, the continuity conditions for $U, V, W, \sigma_z, \tau_{xz}$, and τ_{yz} are approximately satisfied by locating material points symmetrically about the interface.

The resulting finite difference equations constitute a set of linear nonhomogeneous algebraic equations. Since there are three dependent variables, the number of equations in the set is three times the number of material points. Obviously, if a large number of points is required to accurately represent the continuous elastic body, a digital computer is essential. Even with the use of a computer, advantage must be taken of the banded nature of the coefficient matrix of the equations; otherwise, even computer solution times would be prohibitive.

4.6.3 Elasticity Solution Results

For a high modulus graphite/epoxy composite material with

$$E_1 = 20.0 \times 10^6 \text{ psi} \qquad G_{12} = G_{23} = G_{31} = .85 \times 10^6 \text{ psi}$$
$$\tag{4.171}$$
$$E_2 = E_3 = 2.1 \times 10^6 \text{ psi} \qquad \nu_{12} = \nu_{23} = \nu_{31} = .21$$

in a laminate with $b = 8h_0$ (width is four times the thickness), the stresses σ_x, τ_{xy}, and τ_{xz} at the interface between layers $(z = h_0)$ are shown in Fig. 4-36. There, the stresses predicted with classical lamination theory are obtained in the center of the cross-section. However, as the free edge is approached, σ_x decreases, τ_{xy} goes to zero, and, most significantly, τ_{xz} increases from zero to infinity (a singularity exists at $y = \pm b$). By use of other laminate geometries, the width of the region in which the stresses differ from those of classical lamination theory has been shown to be about the thickness of the laminate, $4h_0$. Thus, the deviation from classical lamination theory can be regarded as a boundary layer or edge effect. One laminate thickness away from the edge, classical lamination theory is expected to be valid.

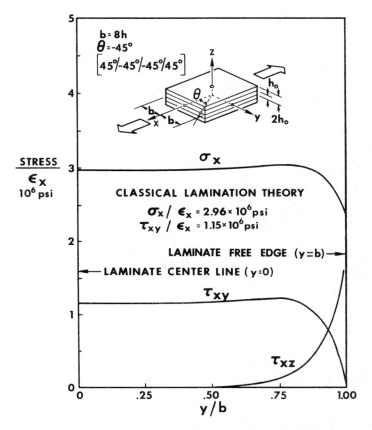

FIG. 4-36.   Stresses at the interface. (*After Pipes and Pagano, Ref. 4-10.*)

The interlaminar shear stress, τ_{xz}, has a distribution over the cross-section thickness as shown by several profiles at various distances from the middle of the laminate in Fig. 4-37. Stress values that have been extrapolated from the numerical data are shown by dashed lines. The value of τ_{xz} is zero at the upper surface of the laminate and at the middle surface. The maximum value for any profile always occurs at the interface between layers. The largest value of τ_{xz} occurs, of course, at the intersection of the free edge with the interface between layers and appears to be a singularity although such a contention cannot be proved by use of a numerical technique.

For other values of θ for a four-layered laminate, an indication of the variation of τ_{xz} with θ is shown in Fig. 4-38. The values plotted correspond to τ_{xz} at the material point nearest the interface between layers on the free edge and not, of course, to the probable singular value at the interface itself. Note that τ_{xz} is zero at $\theta = 60°$ as well as at the expected $0°$ and $90°$. Plots like Fig. 4-38 would have different values for materials other than graphite/epoxy.

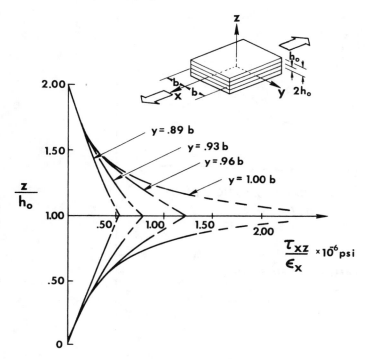

FIG. 4-37. Interlaminar shear stress distribution through the laminate thickness. (*After Pipes and Pagano, Ref. 4-10.*)

4.6.4 Experimental Confirmation of Interlaminar Stresses

Experiments (Ref. 4-11) were performed to confirm Pipes and Pagano's solution for interlaminar stresses. The Moiré technique was utilized to examine the surface displacements of the symmetric angle-ply laminate under axial extension in Sec. 4.6.1. The Moiré technique depends on an optical phenomenon of fringes caused by relative displacement of two sets of arrays of lines. One array is placed on the specimen and the other nearby. A fringe is the locus of points with the same component of displacement normal to the direction of the array lines.

The stress-strain response of long, flat graphite/epoxy specimens was linear to fracture. At various load levels, the Moiré fringes were photographed as typified by the left half of Fig. 4-39. On the right half of Fig. 4-39 is shown a schematic representation of the S-shaped Moiré fringes. The axial displacements determined by the Moiré fringe analysis are shown along with the elasticity solution of Pipes and Pagano in Fig. 4-40. Obviously, the agreement is quite good; thus, the physical existence of interlaminar stresses has been demonstrated.

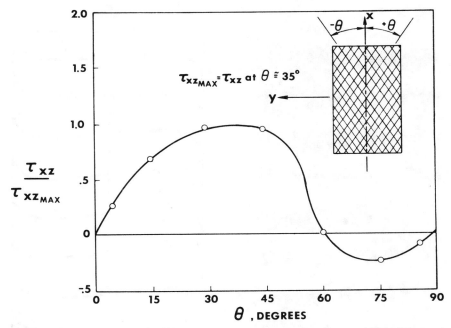

FIG. 4-38. Interlaminar shear stress as a function of fiber orientation. (*After Pipes and Pagano, Ref. 4-10.*)

4.6.5 Implications of Interlaminar Stresses

The existence of interlaminar stresses means that laminated composites can delaminate near free edges whether they be at the edge of a plate, around a hole. or at the ends of a tubular configuration used to obtain material properties. In the latter case, delamination at the ends of the tube could cause premature failure so must be considered in specimen design.

That the interlaminar stresses can be affected by the laminate stacking sequence (arrangement of laminae, that is, $[+45/-45/+15/-15]_s$ versus $[+15/-15/+45/-45]_s$) is significant to design analysts. Pagano and Pipes (Ref. 4-12) hypothesized that the interlaminar normal stress, σ_z, can be changed from tension to compression by changing the stacking sequence. Their work was motivated by observations of Foye and Baker (Ref. 4-13) of fatigue strengths differing by about 25,000 psi for $\pm 15°$, $\pm 45°$ symmetric angle-ply laminates when the positions of the $\pm 15°$ laminae and the $\pm 45°$ laminae were reversed. Other data on static strength reveals qualitatively similar differences. However, classical lamination theory extensional stresses are unaffected by stacking sequence (bending stresses are excluded from the discussion since there is no coupling between bending and extension owing to middle surface symmetry. Foye and Baker observed delamination and stated that progressive delamination

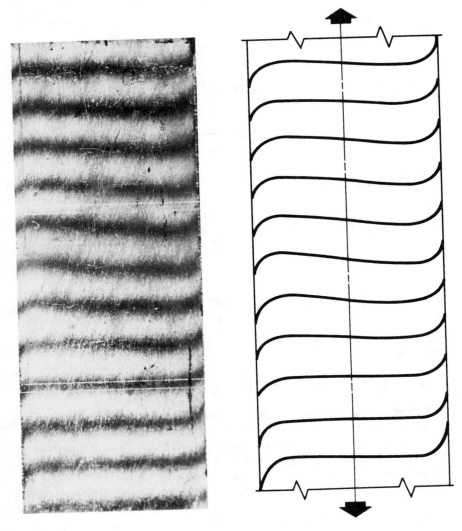

FIG. 4-39. Moiré fringe pattern. (*After Pipes and Daniels, Ref. 4-11.*)

was the failure mode in fatigue. The contention of Pagano and Pipes that the interlaminar normal stress, σ_z, is responsible for delamination seems quite reasonable in view of the following analysis.

Consider the free body diagram of a symmetric eight-layered laminate cross section in Fig. 4-41. The laminate is subjected to load in the x-direction as in Fig. 4-33. Recall that the mechanism of interlaminar stress transfer involves a balance of moments of stresses. First, since τ_{xy} is zero on a free edge, but finite elsewhere, its moment must be balanced by the moment of interlaminar shear

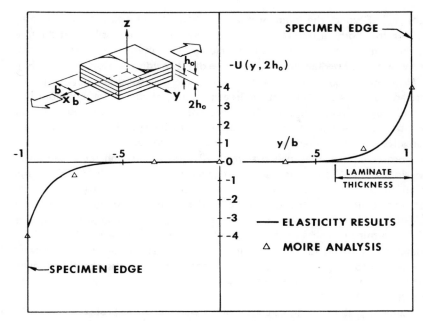

FIG. 4-40. Axial displacement distribution at the laminate surface, $Z = 2h_0$. (*After Pipes and Daniel, Ref. 4-11.*)

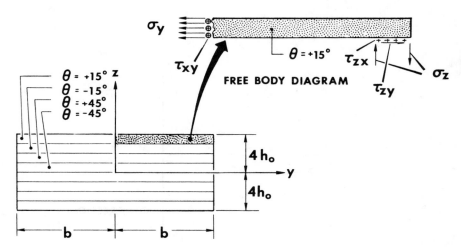

FIG. 4-41. Interlaminar stresses in top layer. (*After Pagano and Pipes, Ref. 4-12.*)

stresses, τ_{xz}. Second, the moment due to σ_z must balance the moment due to σ_y. In the free body, a tensile σ_y in the 15° layer implies a tensile σ_z at the free edge; the converse holds for a compressive σ_y. The interlaminar normal stress, σ_z, is hypothesized by Pagano and Pipes to exhibit the distribution shown in Fig. 4-42. Note that σ_z goes to zero in the region where classical lamination theory applies and perhaps to infinity at the free edge; it is, of course, self-equilibrating. If the 45° layers were placed on the outside of the laminate, a compressive σ_y would be predicted with classical lamination theory; thus, σ_z would be compressive and the laminate would not tend to delaminate.

Accordingly, they reasoned that σ_z is distributed through the thickness as shown in Fig. 4-43 for two stacking sequences, $[15°/-15°/45°/-45°]_{\subseteq}$ and $[15°/45°/-45°/-15°]_{\subseteq}$. Obviously, the latter sequence should provide a greater strength than the former because of less tendency to delaminate. The sequence $[45°/-45°/15°/-15°]_{\subseteq}$ should, by similar reasoning, lead to compressive stresses that are the mirror images of the tensile stresses of the $[15°/-15°/45°/-45°]_{\subseteq}$ laminate and be much stronger. The interlaminar shear stresses in the two cases can be shown to be essentially the same if not identical. Thus, the only conclusion to be drawn is that the interlaminar normal stress, σ_z, may be the key to the success of a laminate.

In summary, there are three classes of interlaminar stress problems:

- $\pm\theta$ laminates exhibit only shear coupling (no Poisson mismatch between layers), so τ_{xz} is the only nonzero interlaminar stress.
- 0°/90° laminates exhibit only a Poisson mismatch between layers (no shear coupling), so τ_{yz} and σ_z are the only nonzero interlaminar stresses.

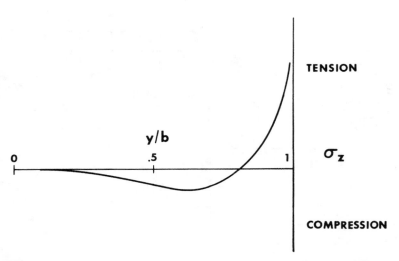

FIG. 4-42. Distribution of interlaminar stress vs. y. (*After Pagano and Pipes, Ref. 4-12.*)

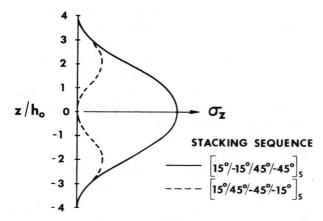

FIG. 4-43. Distribution of interlaminar normal stress in boundary layer region vs. z. (*After Pagano and Pipes, Ref. 4-12.*)

- combinations of the above, for example, $\pm\theta_1$, $\pm\theta_2$ laminates, exhibit both shear coupling and Poisson mismatch between layers, so have τ_{xz}, τ_{yz}, and σ_z interlaminar stresses.

Problem Set 4.6

Exercise 4.6.1   Demonstrate that use of classical lamination theory leads to

$$\frac{\sigma_x}{\epsilon_x} = 2.96 \times 10^6 \text{ psi} \qquad \frac{\tau_{xy}}{\epsilon_x} = 1.15 \times 10^6 \text{ psi}$$

as stresses in each layer of the four-layered graphite/epoxy laminate discussed in Sec. 4.6.3. Disregard the sign of τ_{xy}. What is σ_y?

Exercise 4.6.2   Obtain the displacements

$$u = -(C_1 z + C_2)y + (C_4 y + C_5 z + C_6)x + U(y, z)$$

$$v = (C_1 z + C_2)x - \frac{C_4 x^2}{2} + V(y, z)$$

$$w = -C_1 xy + C_7 x - \frac{C_5 x^2}{2} + C_8 + W(y, z)$$

by integration of the stress-displacement relations when the stresses are functions of y and z only. These displacements result before the various symmetry conditions are applied to obtain Eq. (4.164). (Hint: see Timoshenko and Goodier, *Theory of Elasticity*, 3d ed., McGraw-Hill, New York, pp. 240 and 280.)

4.7 DESIGN OF LAMINATES

In a design situation, the loads to be carried are generally known. The difference between analysis, which has been the principal concern of this book, and design is as follows. *Design* is the altering of dimensions and materials in a configuration to carry specified loads whereas *analysis* is the determination of the load a specific configuration will take. Actually, in design, analysis techniques can be used in an iterative fashion to come to a suitable design; hence, the term design analysis has evolved. More recently, optimization concepts have received wide attention: given design variables such as geometry, stacking sequence, laminae orientation, etc., what is the lightest laminate that will sustain the design loads?

The key to the design of efficient laminates is to resist both the *magnitude* and the *directional nature* of the loads without over-design in either respect. That is, the laminate is tailored to just meet specific requirements. Isotropic materials are usually inefficient because excess strength and stiffness is inevitably available in some direction. By appropriate consideration of the loads and their directions, a laminate can be constructed of individual laminae in such a manner as to just resist those loads and no more (with, of course, an appropriate factor of safety). For example, a cross-ply laminate can be used to resist loads in the principal directions 1 and 2 where N_1 is resisted by A_{11}, A_{12}, D_{11}, and D_{12} and N_2 is resisted by A_{12}, A_{22}, D_{12}, and D_{22} if the laminate is symmetric. In more complex situations where shearing forces and twisting moments are important, angle-ply laminates may be required in order to obtain the necessary shearing and twisting stiffnesses. Other design factors become evident when the strength characteristics are considered.

The analytical tools to accomplish laminate design are twofold. First, invariant stiffness concepts as developed by Tsai and Pagano (Ref. 4-14) are used as an aid to understanding how stiff laminates of arbitrary orientation are and how that stiffness can be varied. Second, structural optimization techniques as described by Schmit (Ref. 4-15) are used to provide a decision-making process for variation of laminate design parameters. This duo of techniques is particularly well suited to composite materials design because the simultaneous possibility and necessity to tailor the material to meet structural requirements exists to a degree not seen in isotropic materials.

The topic of invariant stiffness concepts and their usage will be discussed in Sec. 4.7.1 through 4.7.3, and the important subject of laminate joints will be briefly described in Sec. 4.7.4. Optimization techniques will not be covered in order to avoid raising the mathematical level of the book.

4.7.1 Invariant Laminate Stiffness Concepts

The topic of invariant transformed reduced stiffnesses of orthotropic and anisotropic laminae was introduced in Sec. 2.7. There, the rearrangement of stiffness transformation equations by Tsai and Pagano (Ref. 4-14) was shown to to be quite advantageous. In particular, certain invariant components of the lamina stiffnesses become apparent and are helpful in determining how the lamina stiffnesses change with transformation to nonprincipal material directions.

The invariant stiffness concepts for a lamina will now be extended to a laminate. All results in this and succeeding subsections on invariant laminate stiffnesses were obtained by Tsai and Pagano (Ref. 4-14). The laminate is composed of orthotropic laminae with arbitrary orientations and thicknesses. The stiffnesses of the laminate in the xy plane can be written in the usual manner as

$$(A_{ij}, B_{ij}, D_{ij}) = \int \bar{Q}_{ij}(1, z, z^2)\,dz \tag{4.172}$$

where the $\bar{Q}_{ij}$ are constant in each layer, but vary from layer to layer. The values of the $\bar{Q}_{ij}$ are given in Table 2-2 when U_6 and U_7 are zero. For examples,

$$(A_{11}, B_{11}, D_{11})$$
$$= \int [U_1(1, z, z^2) + U_2 \cos 2\theta\,(1, z, z^2) + U_3 \cos 4\theta\,(1, z, z^2)]\,dz \tag{4.173}$$

When all orthotropic laminae are of the same material, the constants U_1, U_2, and U_3 can be brought outside the integral:

$$(A_{11}, B_{11}, D_{11})$$
$$= U_1\left(t, 0, \frac{t^3}{12}\right) + U_2\int \cos 2\theta\,(1, z, z^2)\,dz + U_3\int \cos 4\theta\,(1, z, z^2)\,dz \tag{4.174}$$

The final result is given in Table 4-6 along with the values for all the stiffnesses. There, the $V_{i(A,B,D)}$ are

$$V_{0(A,B,D)} = \left(t, 0, \frac{t^3}{12}\right)$$
$$V_{1(A,B,D)} = \int \cos 2\theta\,(1, z, z^2)\,dz$$
$$V_{2(A,B,D)} = \int \sin 2\theta\,(1, z, z^2)\,dz \tag{4.175}$$
$$V_{3(A,B,D)} = \int \cos 4\theta\,(1, z, z^2)\,dz$$
$$V_{4(A,B,D)} = \int \sin 4\theta\,(1, z, z^2)\,dz$$

TABLE 4-6. Laminate stiffnesses as a function of lamina properties

Stiffness	$V_{0(A,B,D)}$	$V_{1(A,B,D)}$	$V_{2(A,B,D)}$	$V_{3(A,B,D)}$	$V_{4(A,B,D)}$
(A_{11},B_{11},D_{11})	U_1	U_2	0	U_3	0
(A_{22},B_{22},D_{22})	U_1	$-U_2$	0	U_3	0
(A_{12},B_{12},D_{12})	U_4	0	0	$-U_3$	0
(A_{66},B_{66},D_{66})	U_5	0	0	$-U_3$	0
$2(A_{16},B_{16},D_{16})$	0	0	$-U_2$	0	$-2U_3$
$2(A_{26},B_{26},D_{26})$	0	0	$-U_2$	0	$2U_3$

Because each layer is macroscopically homogeneous in its own region of space, the integrals in Eq. (4.175) further simplify to summations:

$$V_{iA} = \sum_{k=1}^{N} W_k (z_{k+1} - z_k)$$

$$V_{iB} = \frac{1}{2} \sum_{k=1}^{N} W_k (z_{k+1}^2 - z_k^2) \tag{4.176}$$

$$V_{iD} = \frac{1}{3} \sum_{k=1}^{N} W_k (z_{k+1}^3 - z_k^3)$$

in which the z_i are defined in Eq. (4.21), N is the number of layers, and

$$W_k = \begin{cases} \cos 2\theta_k, i = 1 \\ \sin 2\theta_k, i = 2 \\ \cos 4\theta_k, i = 3 \\ \sin 4\theta_k, i = 4 \end{cases} \tag{4.177}$$

wherein θ_k is the orientation of the 1-direction in the kth lamina from the laminate x axis. The stiffnesses in Table 4-6 are for a laminate with N layers of a single orthotropic material with various laminae principal material property orientations.

The stiffnesses in Table 4-6 are *not* transformed stiffnesses in analogy to Table 2-2. That is, the x axis of the laminate is fixed relative to the θ_k of each lamina. However, the transformed stiffnesses can be obtained by rotating the entire laminate through angle ϕ, that is, by substituting $(\theta - \phi)$ for θ in Eq. (4.174). For example,

$$\bar{A}_{11} = U_1 t + U_2 \int \cos 2(\theta - \phi) dz + U_3 \int \cos 4(\theta - \phi) dz \tag{4.178}$$

Then, by use of the trigonometric identities for subtraction of two angles

$$\cos (\alpha - \beta) = \cos \alpha \cos \beta + \sin \alpha \sin \beta \tag{4.179}$$

and the fact that ϕ, and hence its trigonometric functions, are independent of z, we see that

$$\bar{A}_{11} = U_1 t + U_2 V_{1A} \cos 2\phi + U_2 V_{2A} \sin 2\phi + U_3 V_{3A} \cos 4\phi$$
$$+ U_3 V_{4A} \sin 4\phi \tag{4.180}$$

TABLE 4-7. Transformation equations for A_{ij}

Transformed extensional stiffness	Constant	$\cos 2\phi$	$\sin 2\phi$	$\cos 4\phi$	$\sin 4\phi$
$\bar{A}_{11}$	$U_1 V_{0A}$	$U_2 V_{1A}$	$U_2 V_{2A}$	$U_3 V_{3A}$	$U_3 V_{4A}$
$\bar{A}_{22}$	$U_1 V_{0A}$	$-U_2 V_{1A}$	$-U_2 V_{2A}$	$U_3 V_{3A}$	$U_3 V_{4A}$
$\bar{A}_{12}$	$U_4 V_{0A}$	0	0	$-U_3 V_{3A}$	$-U_3 V_{4A}$
$\bar{A}_{66}$	$U_5 V_{0A}$	0	0	$-U_3 V_{3A}$	$-U_3 V_{4A}$
$2\bar{A}_{16}$	0	$U_2 V_{2A}$	$-U_2 V_{1A}$	$2U_3 V_{4A}$	$-2U_3 V_{3A}$
$2\bar{A}_{26}$	0	$U_2 V_{2A}$	$-U_2 V_{1A}$	$-2U_3 V_{4A}$	$2U_3 V_{3A}$

where V_{iA} are defined in Eq. (4.176). The transformed extensional stiffnesses, $\bar{A}_{ij}$, are given in Table 4-7. The transformed coupling stiffnesses, $\bar{B}_{ij}$, and the transformed bending stiffnesses, $\bar{D}_{ij}$, have the same form as in Table 4-7 except the V_{iA} are replaced by V_{iB} and V_{iD}, respectively.

The form of the transformation relations for the A_{ij}, B_{ij}, and D_{ij} in Table 4-7 is identical to that for the anisotropic Q_{ij} in Table 2-2. Exercises in which additional points are made about the invariants are given at the end of the section.

4.7.2 Special Results for Invariant Laminate Stiffnesses

Several special laminates will be examined to help understand the summations $V_{i(A,B,D)}$ in Eq. (4.176). Recall first from mathematics that the integral of an odd (antisymmetric) function over an interval symmetric about the origin is zero. Similarly, the integral of an even (symmetric) function over the same interval is finite. That is,

$$\int_{-z}^{+z} (\text{odd function})\, dz = 0, \qquad \int_{-z}^{+z} (\text{even function})\, dz = \text{finite} \qquad (4.181)$$

where z is the coordinate perpendicular to the plane of the laminate. We will consider laminates with laminae orientations that are: (1) odd functions of z; (2) even functions of z; (3) random functions of z; and (4) increments of π/N where N is the number of equal thickness layers. Examples of these laminates are shown in Fig. 4-44.

First, for laminae orientations with θ_k that are an odd function of z, as illustrated by the 2-layer angle-ply laminate with $\pm\alpha$ orientation in Fig. 4-44a, the following integrands in $V_{i(A,B,D)}$, Eq. (4.175), are odd:

$$\cos p\theta\,(z) \qquad \sin p\theta\,(1, z^2)$$

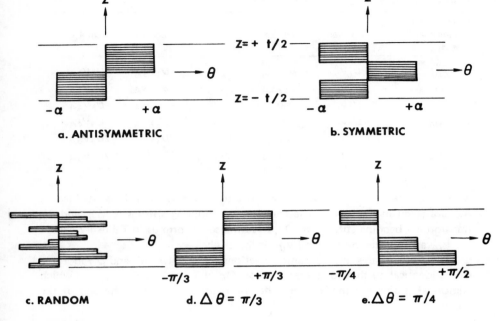

FIG. 4-44. Laminate examples.

and the following integrands are even:

$$\cos p\theta \, (1, z^2) \qquad \sin p\theta \, (z)$$

where p is 2 or 4. Thus, the following summations vanish:

$$V_{2A} = V_{4A} = V_{1B} = V_{3B} = V_{2D} = V_{4D} = 0 \tag{4.182}$$

Accordingly, the following stiffnesses in Table 4-6 are zero:

$$A_{16} = A_{26} = B_{11} = B_{22} = B_{12} = B_{66} = D_{16} = D_{26} = 0 \tag{4.183}$$

The extensional and bending stiffnesses are those of an orthotropic material, but the coupling stiffnesses are not all zero (B_{16} and B_{26} remain).

Next, for laminae orientations with θ_k that are an even function of z (symmetric), as illustrated by the three-layered laminate in Fig. 4-44b, the following integrands in $V_{i(A,B,D)}$, Eq. (4.175), are odd:

$$\cos p\theta \, (z) \qquad \sin p\theta \, (z)$$

and the following integrands are even

$$\cos p\theta \, (1, z^2) \qquad \sin p\theta \, (1, z^2)$$

Thus, the following summations vanish:

$$V_{1B} = V_{2B} = V_{3B} = V_{4B} = 0 \qquad (4.184)$$

Accordingly, all the coupling stiffnesses, B_{ij}, in Table 4-6 vanish. The A_{ij} and D_{ij} are those of an anisotropic material.

   If the laminae orientation is a random function of z as in Fig. 4-44c, define $\overline{V}_i$ as the spatial average of the individual $V_{i(A,B,D)}$ (they all will be treated alike):

$$\overline{V}_i = \frac{1}{\pi} \int_{-\pi/2}^{\pi/2} V_i\, d\theta = \frac{1}{\pi} \int_{-\pi/2}^{\pi/2} \int_{-t/2}^{t/2} \left\{ \begin{array}{c} \cos p\theta \\ \sin p\theta \end{array} \right\} (1,z,z^2)\, dz\, d\theta \qquad (4.185)$$

where p is even. Interchange the order of integration to get

$$\overline{V}_i = \frac{1}{\pi} \int_{-t/2}^{t/2} \int_{-\pi/2}^{\pi/2} \left\{ \begin{array}{c} \cos p\theta \\ \sin p\theta \end{array} \right\} d\theta\, (1,z,z^2)\, dz \qquad (4.186)$$

which is zero. When all the $V_{i(A,B,D)}$ are zero, only the constant terms remain in the stiffnesses. Moreover, the laminate is macroscopically isotropic because now

$$A_{11} = A_{22} = U_1 t \qquad A_{12} = U_4 t \qquad A_{66} = U_5 t \qquad A_{16} = A_{26} = 0$$

$$\qquad (4.187)$$

$$B_{ij} = 0 \qquad\qquad D_{ij} = \frac{A_{ij}t^2}{12}$$

and $A_{11} - A_{12} = 2A_{66}$. Although the laminate is macroscopically isotropic, it is still inhomogeneous so the stress distribution is discontinuous and different from that of an isotropic material.

   For a laminate of N equal thickness layers ($N > 2$) with orientation angles differing by π/N as in Fig. 4-44d and e, the summation for V_{1A} is

$$V_{1A} = \left(\cos \frac{2\pi}{N} + \cos \frac{4\pi}{N} + \cdots + \cos 2\pi \right) \frac{t}{N} \qquad (4.188)$$

but

$$\cos x + \cos 2x + \cdots + \cos nx = \frac{\sin\left(n + \dfrac{1}{2}\right)x}{2 \sin \dfrac{x}{2}} - \frac{1}{2} \qquad (4.189)$$

which for $x = 2\pi/N$ is zero. Also,

$$V_{3A} = \left(\sin \frac{2\pi}{N} + \sin \frac{4\pi}{N} + \cdots + \sin 2\pi \right) \frac{t}{N} \qquad (4.190)$$

and

$$\sin x + \sin 2x + \cdots + \sin nx = \frac{\sin \dfrac{1+n}{2} x \sin \dfrac{n}{2} x}{\sin \dfrac{x}{2}} \tag{4.191}$$

which for $x = 2\pi/N$ is zero. Similarly, $V_{2A} = 0$ because the expression in Eq. (4.189) vanishes for $x = 4\pi/N$ and, as well, $V_{4A} = 0$. Thus, since the variable terms are zero, the A_{ij} are isotropic and are given in Eq. (4.187). However, the B_{ij} are not zero so the laminates in this class are not extensionally isotropic but are called quasi-isotropic. This class of laminates occurs for laminae stacking sequences of $0/\pm\pi/3$; $\pi/2/\pi/4/0/-\pi/4$; etc. Other more complicated lamination sequences have isotropic B_{ij} or isotropic D_{ij}.

A final result of interest is the integral of the area under the transformed stiffness versus angle of rotation curve from $\phi = 0$ to $\phi = 2\pi$, that is, one complete revolution of the laminate:

$$\int_0^{2\pi} \overline{A}_{ij} \, d\phi \tag{4.192}$$

The integral

$$\int_0^{2\pi} \left\{ \begin{array}{c} \cos p\phi \\ \sin p\phi \end{array} \right\} d\phi \tag{4.193}$$

is zero when p is an integer so only the constant terms contribute to Eq. (4.178) which is then independent of ϕ. The average values of the integral are the isotropic A_{ij} in Eq. (4.187) obtained for randomly oriented laminates and extensionally quasi-isotropic laminates. Those A_{ij} contain U_1, U_4, and U_5, but we showed in Exercise 2.7.3 that U_4 is dependent on U_1 and U_5. Thus, U_1 and U_5 appear to be a measure of orthotropic laminates as well as of orthotropic materials. That is, because the integral of $\overline{A}_{ij}$ is constant irrespective of the lamination sequence of laminae orientations, there are constant measures of the laminate, namely U_1 and U_5, which are related to the area under the $\overline{A}_{ij}$ versus ϕ curve. Similarly, the area under the $\overline{B}_{ij}$ versus ϕ curve can be shown to be zero and that under the $\overline{D}_{ij}$ versus ϕ curve is constant. These results will be put to use in the next subsection.

4.7.3 Use of Invariant Laminate Stiffnesses in Design

Two basic invariants, U_1 and U_5, were shown in the previous subsection to be the basic indicators of average laminate stiffnesses. For isotropic materials, these invariants reduce to $U_1 = Q_{11}$ and $U_5 = Q_{66}$, the extensional stiffness and shear stiffness. Accordingly, Tsai and Pagano (Ref. 4-14) suggested the orthotopic invariants U_1 and U_5 be called the isotropic stiffness and isotropic shear rigidity, respectively. They observed that these "isotropic properties" are a realistic measure of the minimum stiffness capability of composite laminates. These isotropic properties can be compared directly to properties of isotropic materials as well as to properties of other orthotropic laminates. Obviously, the comparison criterion is more complex than for isotropic materials because now we have two measures, U_1 and U_5, instead of the usual isotropic stiffness U_1 or E_1. Comparison of values of U_1 alone is not fair because of the degrading influence of the usually low values of U_5 for a composite material.

The optimization or design of a laminate can be performed with the aid of the isotropic stiffnesses. Start with a laminate of unidirectional layers for which $A_{ij} = Q_{ij}t$. If some of the laminae orientations are changed from $0°$, then the new values of A_{ij} will be given by the relations of Table 4-6. The actual $\overline{A}_{ij}$ will vary with rotation ϕ in accordance with Table 4-7. However, that variation is always about the isotropic values. For example, variation of $\overline{A}_{11}$ and $\overline{A}_{66}$ with ϕ for boron/epoxy (properties are given in Table 2-3) is shown in Fig. 4-45. The unidirectional and isotropic laminate values are both shown in addition to results for two cross-ply and two angle-ply laminates. The areas under all curves in Fig. 4-45 are obviously all the same. Thus, if the cross-ply laminate with $M = 1$ and the angle-ply laminate of the same thickness with $\alpha = 45°$ are combined, the resulting laminate is extensionally isotropic (this is the case of the four-layered quasi-isotropic laminate with difference in orientation of $45°$). Note, however, even though $\overline{A}_{11}$ and $\overline{A}_{66}$ are constant irrespective of rotation, they are not related in the same manner as true isotropic material properties E and G.

Tsai and Pagano (Ref. 4-14) further define the isotropic stiffness and shear rigidity to be

$$U_1 = \overline{E} \qquad U_5 = \overline{G} \tag{4.194}$$

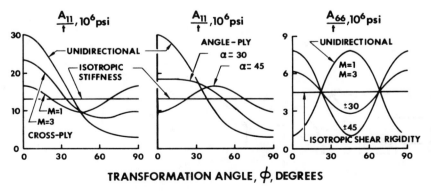

FIG. 4-45.   Transformed laminate stiffnesses.

and show for highly orthotropic composite materials such as boron/epoxy and graphite/epoxy that

$$\bar{E} \cong \frac{3}{8}E_1 + \frac{5}{8}E_2$$

$$\bar{G} \cong \frac{1}{8}E_1 + \frac{1}{4}E_2$$

(4.195)

Thus, the usual emphasis on the value of E_1 is misplaced. Obviously, the value of E_2 enters the representative average properties quite strongly. These approximations are quite accurate as can be verified by simple calculations.

4.7.4 Laminate Joints

High stiffnesses and strengths can be attained for composite laminates. However, these characteristics are quite different from those of ordinary materials to which we will generally want to fasten composite laminates. Often, the full strength and stiffness characteristics of the laminate cannot be transferred through the joint without a significant weight penalty. Thus, the topic of joints or other fastening devices is critical to the successful use of composite materials.

The purpose of this subsection is to familiarize the reader with some of the basic characteristics and problems of composite laminate joints. The specific design of a joint is much too complex for an introductory textbook such as this. The published state-of-the-art of laminate joint design is summarized in the *Structural Design Guide For Advanced Composite Applications* (Ref. 4-16) and

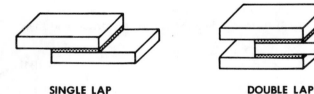

SINGLE LAP   DOUBLE LAP

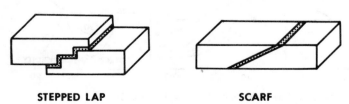

STEPPED LAP   SCARF

FIG. 4-46. Bonded joints.

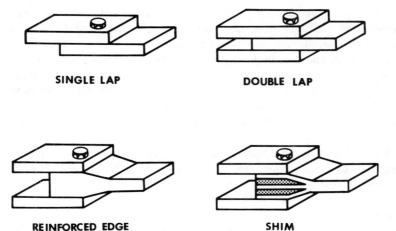

SINGLE LAP   DOUBLE LAP

REINFORCED EDGE   SHIM

FIG. 4-47. Bolted joints.

Military Handbook 17A, Plastics for Aerospace Vehicles, Part 1., Reinforced Plastics (Ref. 4-17). Further developments will be found in the technical literature and revisions of the two preceding references.

The two major classes of laminate joints are bonded joints as in Fig. 4-46 and bolted joints as in Fig. 4-47. Often, the two classes are combined, for example, the bonded-bolted joint in Fig. 4-48. Joints involving composite materials are usually bonded because of the natural presence of resin in the composite and as well bolted for reasons discussed later. Several characteristics of fiber-reinforced composite materials render them more susceptible to joint problems than conventional metals. These characteristics are weakness in in-plane shear, transverse tension, interlaminar shear, and bearing strength relative to the primary asset of a lamina, the strength and stiffness in the fiber direction.

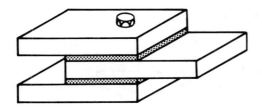

FIG. 4-48. Bonded-bolted double lap joint.

Bonded joints

Goland and Reissner (Ref. 4-18; this classic paper is referred to by nearly every researcher in bonded joints) studied the stresses in bonded single lap joints for two important limiting cases: (1) a bond layer so thin that it has no contribution to the joint flexibility, and (2) a bond layer so thick that it is the primary contributor to the joint flexibility. They considered the shearing and normal stresses in the bond layer as well as those in the joined plates. For fiber-reinforced composites, the thick bond layer approach of Goland and Reissner is more appropriate than the thin bond layer approach because of the presence of epoxy resin in the composite and the thickness of the bond relative to the joined pieces. They found, for equal thickness isotropic plates, that the bond layer shear stress has nearly uniform distribution except for a large concentration near the end of the joint. The bond stress perpendicular to the bond layer also has high values near the joint edge although not nearly as high as the inflexible bond case.

Berg (Ref. 4-19) analyzed a bonded double lap joint. He suggested interleaving the materials of a lap joint to reduce the high stresses that otherwise occur where the layers meet.

Bolted joints

The principal failure modes of bolted joints are (1) bearing failure of the material as in the elongated bolt hole of Fig. 4-49a, (2) tension failure of the material in the reduced cross section through the bolt hole in Fig. 4-49b; (3) shear-out or cleavage failure of the material (actually transverse tension failure of the material) as in Figs. 4-49c and d, and (4) bolt failures (mainly shear failures). Of course, combinations of these failures do occur.

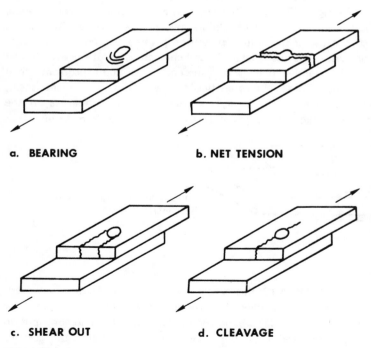

a. **BEARING**   b. **NET TENSION**

c. **SHEAR OUT**   d. **CLEAVAGE**

FIG. 4-49. Bolted joint failures.

One of the ways to increase the bearing strength of a joint is to use metal inserts as in the shim joint of Fig. 4-47. Another way is to thicken a section of the composite laminate as in the reinforced edge joint in Fig. 4-47.

Net tension failures can be avoided or delayed by increased joint flexibility to spread the load transfer over several lines of bolts. Composite materials are generally more brittle than conventional metals so loads are not easily redistributed around a stress concentration such as a bolt hole. Simultaneously, shear lag effects due to discontinuous fibers lead to difficult design problems around bolt holes. A possible solution is to put a ductile composite such as S-glass/epoxy in a strip of several times the bolt diameter in line with the bolt rows.

Bonded-bolted joints

Bonded-bolted joints have better performance than either bonded or bolted joints. The bonding results in reduction of the normal tendency of a bolted joint to shear out. The bolting decreases the likelihood of a bonded joint debonding in an interfacial shear mode. The usual mode of failure for a bonded-bolted joint is either a tension failure through a section including a fastener or an interlaminar shear failure in the composite or a combination of both.

Bonded-bolted joints have good load distribution and are generally designed so that the bolts take all the load. Then, the bolts would take all the load after the bond breaks. The bond provides a change in failure mode and a sizable margin against fatigue failure.

An example of a complex bonded-bolted joint used in the box beam of a folding aircraft wing is shown in Fig. 4-50. There, a basic structure of

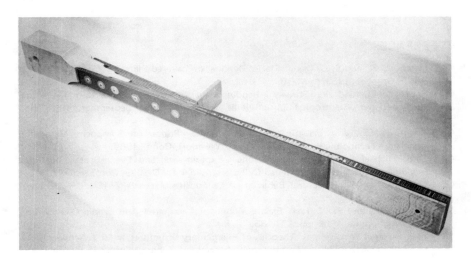

FIG. 4-50. Complex bonded-bolted joint. (*Courtesy of Vought Systems Division, LTV Aerospace Corporation. The fabrication of this joint was sponsored by the Nonmetallic Branch of the Air Force Materials Laboratory.*)

graphite/epoxy and boron/epoxy layers over honeycomb is attached to an aluminum forging. The honeycomb is gradually replaced by graphite/epoxy as the joint is approached. Then, titanium sheaves are successively introduced to increase the bearing strength of the joint. Finally, the combined graphite/epoxy and titanium composite is stair-stepped over the Christmas tree-like aluminum forging. The titanium sheaves are bonded to the graphite/epoxy with a film adhesive. The graphite/epoxy is bonded to the aluminum forging with a paste adhesive. The entire joint is then bolted together.

Problem Set 4.7

Exercise 4.7.1 Show that $\bar{A}_{11} + \bar{A}_{22} + 2\bar{A}_{12}$ is invariant under rotation about the z axis, that is, that

$$\bar{A}_{11} + \bar{A}_{22} + 2\bar{A}_{12} = A_{11} + A_{22} + 2A_{12}$$

irrespective of angle of rotation, ϕ. Also, relate this invariant to the reduced stiffness invariant $Q_{11} + Q_{22} + 2Q_{12}$.

Exercise 4.7.2 Show that $A_{66} - \bar{A}_{12}$ is invariant under rotation about the z axis, that is, that

$$\bar{A}_{66} - \bar{A}_{12} = A_{66} - A_{12}$$

Also, relate this invariant to the reduced stiffness invariant $Q_{66} - Q_{12}$.

Exercise 4.7.3 What is the value of the coupling stiffness invariants $\bar{B}_{11} + \bar{B}_{22} + 2\bar{B}_{12}$ and $\bar{B}_{66} - \bar{B}_{12}$?

Exercise 4.7.4 Relate the bending stiffness invariants $\bar{D}_{11} + \bar{D}_{22} + 2\bar{D}_{12}$ and $\bar{D}_{66} - \bar{D}_{12}$ to the reduced stiffness invariants and the extensional stiffness invariants.

REFERENCES

4-1   Pister, K. S., and S. B. Dong: Elastic Bending of Layered Plates, *J. Eng. Mech. Div.*, *ASCE*, October, 1959, pp. 1-10.

4-2   Reissner, E., and Y. Stavsky: Bending and Stretching of Certain Types of Heterogeneous Aelotropic Elastic Plates, *J. Appl. Mech.*, September, 1961, pp. 402-408.

4-3   Ashton, J. E., J. C. Halpin, and P. H. Petit: "Primer on Composite Materials: Analysis," Technomic Publishing Company, Westport, Conn., 1969.

4-4   Card, Michael F., and Robert M. Jones: Experimental and Theoretical Results for Buckling of Eccentrically Stiffened Cylinders, *NASA* TN D-3639, October, 1966.

4-5   Tsai, Stephen W.: Structural Behavior of Composite Materials, *NASA* CR-71, July, 1964.

4-6   Azzi, V. D., and S. W. Tsai: Elastic Moduli of Laminated Anisotropic Composites, *Exp. Mech.*, June, 1965, pp. 177-185.

4-7   Tsai, Stephen W.: Strength Theories of Filamentary Structures, in R. T. Schwartz and H. S. Schwartz (eds.), "Fundamental Aspects of Fiber Reinforced Plastic Composites," Wiley Interscience, New York, 1968.

4-8   Tsai, Stephen W.: Strength Characteristics of Composite Materials, *NASA* CR-224, April, 1965.

4-9 Tsai, Stephen W., Donald F. Adams, and Douglas R. Doner: Analysis of Composite Structures, *NASA* CR-620, November, 1966.

4-10 Pipes, R. Byron, and N. J. Pagano: Interlaminar Stresses in Composite Laminates Under Uniform Axial Extension, *J. Composite Materials*, October, 1970, pp. 538-548.

4-11 Pipes, R. Byron, and I. M. Daniel: Moiré Analysis of the Interlaminar Shear Edge Effect in Laminated Composites, *J. Composite Materials*, April, 1971, pp. 255-259.

4-12 Pagano, N. J., and R. Byron Pipes: The Influence of Stacking Sequence on Laminate Strength, *J. Composite Materials*, January, 1971, pp. 50-57.

4-13 Foye, R. L., and D. J. Baker: Design of Orthotropic Laminates, *AIAA/ASME 11th Conf. Structures, Structural Dynamics, and Materials*, Denver, Colorado, April, 1970.

4-14 Tsai, Stephen W., and Nicholas J. Pagano: Invariant Properties of Composite Materials, in S. W. Tsai, J. C. Halpin, and Nicholas J. Pagano (eds.), "Composite Materials Workshop," Technomic Publishing Co., Stamford, Connecticut, 1968, pp. 233-253. Also AFML-TR-67-349, March, 1968.

4-15 Schmit, Lucien A., Jr.: The Structural Synthesis Concept and Its Potential Role in Design with Composites, in F. W. Wendt, H. Liebowitz, and N. Perrone (eds.), "Mechanics of Composite Materials," *Proc. 5th Symp. Naval Structural Mechanics, 1967*, Pergamon, New York, 1970, pp. 553-582.

4-16 Structural Design Guide for Advanced Composite Applications, Air Force Materials Laboratory, Advanced Composites Division, January, 1971.

4-17 Plastics for Aerospace Vehicles, Part 1, Reinforced Plastics, *Military Handbook* 17A, January, 1971.

4-18 Goland, M., and E. Reissner: The Stresses in Cemented Joints, *J. Appl. Mech.,* March, 1944, pp. A-17–A-27.

4-19 Berg, K. R.: Problems in the Design of Joints and Attachments, in F. W. Wendt, H. Liebowitz, and N. Perrone (eds.), "Mechanics of Composite Materials," *Proc. 5th Symp. Naval Structural Mechanics, 1967*, Pergamon, New York, 1970, pp. 467-479.

Chapter 5
BENDING, BUCKLING, AND VIBRATION OF LAMINATED PLATES

5.1 INTRODUCTION

Laminated plates are one of the simplest and most widespread practical applications of composite laminates. Beams are, of course, simpler. However, such essentially one-dimensional structural elements do not display well the two-dimensional capabilities and characteristics of composite laminates.

The objective in this chapter is to demonstrate the effect of the various coupling stiffnesses (B_{ij}, A_{16}, A_{26}, D_{16}, and D_{26}) on the bending, buckling, and vibration behavior of laminated plates. The study of these effects is the logical culmination of a course on the mechanics of fiber-reinforced composite materials. The objective does not include a complete study of laminated plate theory. Instead, some of the important laminated plate theory results are examined so that the physical significance of the stiffnesses is appreciated. The theory of laminated plates with associated solution techniques is an eminently suitable topic for a follow-on course. A more complete cataloguing and classification of laminated plate problems is found in Ashton and Whitney's "Theory of Laminated Plates" (Ref. 5-1).

The theoretical considerations underlying the basic theory of laminated plates are discussed in Sec. 5.2. Then, the differential equations and associated boundary conditions governing the bending, buckling, and vibration behavior of laminated plates are displayed along with a brief discussion of possible solution techniques. Next, solutions for the various laminate configurations described in Sec. 4.3 are displayed in Secs. 5.3 through 5.5 for bending, buckling, and vibration problems.

A simply supported rectangular plate is used consistently in all sections to illustrate the kinds of results that can be obtained, i.e., the influence of the various stiffnesses on laminated plate behavior. In addition, only the simplest types of loading will be studied in order to avoid the solution difficulties inherent to complex loadings. Accordingly, in the interest of simplicity, just the bare thread of laminated plate results will be displayed.

Specially orthotropic plates and plates with multiple specially orthotropic layers that are symmmetric about the plate middle surface have, as has already been noted in Sec. 4.3, force and moment resultants in which there is no bending-extension coupling nor any shear or twist coupling, that is,

$$
\left\{ \begin{array}{c} N_x \\ N_y \\ N_{xy} \end{array} \right\} = \begin{bmatrix} A_{11} & A_{12} & 0 \\ A_{12} & A_{22} & 0 \\ 0 & 0 & A_{66} \end{bmatrix} \left\{ \begin{array}{c} \epsilon_x^{\circ} \\ \epsilon_y^{\circ} \\ \gamma_{xy}^{\circ} \end{array} \right\}
\tag{5.1}
$$

$$
\left\{ \begin{array}{c} M_x \\ M_y \\ M_{xy} \end{array} \right\} = \begin{bmatrix} D_{11} & D_{12} & 0 \\ D_{12} & D_{22} & 0 \\ 0 & 0 & D_{66} \end{bmatrix} \left\{ \begin{array}{c} \kappa_x \\ \kappa_y \\ \kappa_{xy} \end{array} \right\}
\tag{5.2}
$$

For plate problems, whether the specially orthotropic laminate has a single layer or multiple layers is essentially immaterial; it need be characterized only by D_{11}, D_{12}, D_{22}, and D_{66} in Eq. (5.2). That is, because there is no bending-extension coupling, the force-strain relations, Eq. (5.1), are not used in plate analysis. However, note that force-strain relations are needed in shell analysis because of the differences in deformation of plates and shells.

Often, because specially orthotropic laminates are virtually as easy to analyze as isotropic plates, other laminates are regarded as, or approximated by, specially orthotropic laminates. This approximation will be studied by comparison of results for each type of laminate with and without the various stiffnesses that distinguish it from a specially orthotropic laminate. Specifically, the importance of the twist coupling terms D_{16} and D_{26} will be examined for symmetric angle-ply laminates. Then, bending-extension coupling will be analyzed for antisymmetric cross-ply and angle-ply laminates and compared with the specially orthotropic approximation in which the B_{ij} are ignored. These comparisons will be made successively for bending, buckling, and vibration of simply supported plates in Secs. 5.3, 5.4, and 5.5, respectively.

Finally, the significance of the various coupling stiffnesses is summarized in Sec. 5.6.

5.2 GOVERNING EQUATIONS FOR BENDING, BUCKLING, AND VIBRATION OF LAMINATED PLATES

5.2.1 Basic Restrictions and Assumptions

Most of the restrictions and assumptions on which laminated plate theory is based have been utilized in Chap. 4. However, for completeness, they will be reiterated here. The seemingly dual terminology of restrictions and assumptions is used because the terms have basically different meanings. *Restrictions* are limitations on the use of the theory that are *obviously* either satisfied or they are not. For example, a theory for square plates does not apply to round plates.

Assumptions are limitations on the theory that have a nature of uncertainty to them. For example, stresses perpendicular to the surface of a plate are commonly assumed to be small enough to be regarded as zero, or assumed to be zero; however, we do not know for sure just how small the stresses are unless we appeal to a more accurate theory. In summary, the difference between restrictions and assumptions is that restrictions involve the *known* and assumptions involve the *unknown* (about which we wish to speculate). The following restrictions and assumptions provide further opportunity to clarify the difference between the two classifications, but mainly to build a firm foundation for the study of laminated plate theory. The geometry, forces, and moments for a plate are shown in Figs. 5-1, 5-2, and 5-3, respectively. Recall that the laminate geometry is shown in Fig. 4-5.

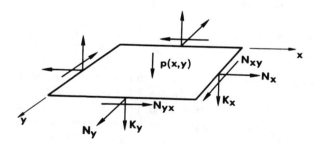

FIG. 5-1. Plate geometry.

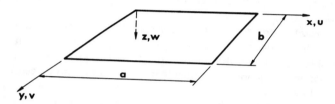

FIG. 5-2. Plate forces.

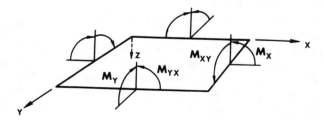

FIG. 5-3. Plate moments.

Restrictions

- Each layer is orthotropic (but the principal material directions of each layer need not be aligned with the plate axes), linear elastic, and of constant thickness (whereupon the entire plate is of constant thickness).
- The plate thickness is very small compared to its length and width (such a configuration is commonly called a thin plate, although the name plate itself connotes such a geometry).
- No body forces exist.

Assumptions

- Stresses acting in the xy plane (the plane of the plate) dominate the plate behavior. Then, σ_z, τ_{xz}, and τ_{yz} are assumed to be zero such that an approximate state of *plane stress* is said to exist (wherein only σ_x, σ_y, and τ_{xy} are considered).
- The *Kirchhoff assumption* of negligible transverse shear strains, γ_{xz} and γ_{yz}, and negligible transverse normal strain, ϵ_z, constitutes a statement of nondeformable normals to the middle surface although there is an inherent, but commonly ignored, conflict with the assumption of zero transverse normal stress, σ_z.
- Displacements u, v, and w are small compared to the plate thickness (generally, although not necessarily, indicative of small deflection theory).
- Strains, ϵ_x, ϵ_y, and γ_{xy}, are small compared to unity (small strain theory).
- Rotatory inertia terms are negligible.

Consequences

- If transverse shear strains are ignored or are assumed to be zero, then transverse shear stresses are also zero throughout the plate by virtue of the stress-strain relations. On the other hand, even if nothing is said about the transverse shear strains, we still know that the transverse shear stresses are zero on both the upper and lower plate surfaces if there is no shear loading. Commonly, in classical plate theory, the transverse shear strains are regarded as zero, yet transverse shear stresses are calculated from equilibrium considerations. Such procedures will be ignored in this book for the sake of simplicity in presenting only a demonstration of the effect of coupling terms.
- By virtue of the Kirchhoff assumption, the remaining strains, ϵ_x, ϵ_y, and γ_{xy}, as well as the displacements, u and v, are a linear function of the transverse coordinate z. Moreover, the stresses are accordingly a linear, but discontinuous, function of the transverse coordinate z. Both of these results are shown schematically in Fig. 4-2.
- As the restriction to thin plates is relaxed, the assumption of plane stress, $\sigma_z = \tau_{xz} = \tau_{yz} = 0$, becomes worse.

Note that none of the assumptions involve fiber-reinforced composite materials explicitly. Instead, only the restriction to orthotropic materials is significant. Therefore, what follows is essentially a classical plate theory. Actually, interlaminar stresses cannot be entirely disregarded in laminated plates, but this refinement will not be treated in this book other than what was studied in Sec. 4.6. Transverse shear effects will be addressed briefly in Sec. 6.5.

5.2.2 Equilibrium Equations for Laminated Plates

The equilibrium differential equations in terms of the force and moment resultants derived in Chap. 4 are

$$N_{x,x} + N_{xy,y} = 0 \tag{5.3}$$

$$N_{xy,x} + N_{y,y} = 0 \tag{5.4}$$

$$M_{x,xx} + 2M_{xy,xy} + M_{y,yy} = -p \tag{5.5}$$

where a comma denotes differentiation of the principal symbol with respect to the subscript. In this form, the equilibrium equations are merely those of classical plate theory. When the stipulation of a laminated plate is introduced by the explicit use of the force and moment resultants in Eqs. (4.19) and (4.20) and the strain and change of curvature definitions in Eqs. (4.11) and (4.12), then the equilibrium equations, Eqs. (5.3) to (5.5), become (upon dropping the zero subscript used to denote middle surface displacements)

$$A_{11}u_{,xx} + 2A_{16}u_{,xy} + A_{66}u_{,yy} + A_{16}v_{,xx} + (A_{12} + A_{66})v_{,xy} + A_{26}v_{,yy}$$
$$-B_{11}w_{,xxx} - 3B_{16}w_{,xxy} - (B_{12} + 2B_{66})w_{,xyy} - B_{26}w_{,yyy} = 0 \tag{5.6}$$

$$A_{16}u_{,xx} + (A_{12} + A_{66})u_{,xy} + A_{26}u_{,yy} + A_{66}v_{,xx} + 2A_{26}v_{,xy} + A_{22}v_{,yy}$$
$$-B_{16}w_{,xxx} - (B_{12} + 2B_{66})w_{,xxy} - 3B_{26}w_{,xyy} - B_{22}w_{,yyy} = 0 \tag{5.7}$$

$$D_{11}w_{,xxxx} + 4D_{16}w_{,xxxy} + 2(D_{12} + 2D_{66})w_{,xxyy} + 4D_{26}w_{,xyyy}$$
$$+ D_{22}w_{,yyyy} - B_{11}u_{,xxx} - 3B_{16}u_{,xxy} - (B_{12} + 2B_{66})u_{,xyy} - B_{26}u_{,yyy}$$
$$-B_{16}v_{,xxx} - (B_{12} + 2B_{66})v_{,xxy} - 3B_{26}v_{,xyy} - B_{22}v_{,yyy} = p \tag{5.8}$$

Obvious and sometimes drastic simplifications result when the laminate is symmetric about the middle surface ($B_{ij} = 0$), specially orthotropic (all the terms with 16 and 26 subscripts vanish in addition to the B_{ij}), homogeneous ($B_{ij} = 0$ and $D_{ij} = A_{ij}t^2/12$), or isotropic. In all those cases, Eqs. (5.6) and (5.7) are uncoupled from Eq. (5.8). That is, Eq. (5.8) contains derivatives of the normal displacement w only and Eqs. (5.6) and (5.7) contain both u and v but not w. Accordingly, only Eq. (5.8) must be solved to determine the deflections of a plate with the aforementioned simplifications. The more general case of unsymmetrical laminates requires the solution of the coupled equations, Eqs. (5.6-5.8).

Boundary conditions used to be thought of as a choice between simply supported, clamped, or free edges if all classes of elastically restrained edges are neglected. The real situation for laminated plates is more complex because now there are actually *four* types of boundary conditions that can be called simply supported edges. Similarly, there are four kinds of clamped edges. These boundary conditions can be concisely described as a displacement or derivative of a displacement or function thereof is equal to some prescribed value (often zero) denoted by an overbar at the edge:

$$
\begin{array}{lll}
u_n = \bar{u}_n & \text{or} & N_n = \bar{N}_n \\
u_t = \bar{u}_t & \text{or} & N_{nt} = \bar{N}_{nt} \\
w,_n = \bar{w},_n & \text{or} & M_n = \bar{M}_n \\
w = \bar{w} & \text{or} & M_{nt,t} + Q_n = \bar{K}_n
\end{array}
\tag{5.9}
$$

in n and t coordinates where n is the direction normal to the edge and t is the direction tangent to the edge. Also, Q_n is the shear force and K_n is the well-known Kirchhoff force of classical plate theory (see Timoshenko and Woinowsky-Krieger, Ref. 5-2). For example, on the edge $x = 0$ in Figs. 5-1 and 5-2

$$
u = 0 \quad \text{or} \quad N_x = 0
\tag{5.10}
$$

The eight possible types of simply supported (prefix S) and clamped (prefix C) edge boundary conditions (combinations of the conditions in Eq. (5.9)) are commonly classified as (see Almroth, Ref. 5-3)

$$
\begin{array}{llllll}
S1: & w = 0 & M_n = 0 & u_n = \bar{u}_n & u_t = \bar{u}_t \\
S2: & w = 0 & M_n = 0 & N_n = \bar{N}_n & u_t = \bar{u}_t \\
S3: & w = 0 & M_n = 0 & u_n = \bar{u}_n & N_{nt} = \bar{N}_{nt} \\
S4: & w = 0 & M_n = 0 & N_n = \bar{N}_n & N_{nt} = \bar{N}_{nt}
\end{array}
\tag{5.11}
$$

$$
\begin{array}{llllll}
C1: & w = 0 & w,_n = 0 & u_n = \bar{u}_n & u_t = \bar{u}_t \\
C2: & w = 0 & w,_n = 0 & N_n = \bar{N}_n & u_t = \bar{u}_t \\
C3: & w = 0 & w,_n = 0 & u_n = \bar{u}_n & N_{nt} = \bar{N}_{nt} \\
C4: & w = 0 & w,_n = 0 & N_n = \bar{N}_n & N_{nt} = \bar{N}_{nt}
\end{array}
\tag{5.12}
$$

whereupon a rectangular plate can be characterized as having any one of the above eight conditions on each of its four edges. The range of possibilities is, therefore, enormous (12 possible conditions on each of the four edges if the free edge conditions are included). The simplest cases to analyze naturally involve like types of boundary conditions on opposite, if not all, edges. The emphasis in this book will be on plates with four simply supported edges so cases will be chosen from those in Eq. (5.11). Note that simply supported edges have no rotational restraint, but when this simplified terminology is used, the specific in-plane conditions are not determined. Obviously, the in-plane conditions must be specified by stating, for example, that a solution is being obtained for $S1$ boundary conditions.

5.2.3 Buckling Equations for Laminated Plates

A plate buckles when the in-plane load gets so large that the originally flat equilibrium state is no longer stable and the plate deflects into a nonflat configuration. The load at which the departure from the flat state takes place is called the buckling load.

Analysis of plates buckling under in-plane loading involves solution of an eigenvalue problem as opposed to the boundary value problem of equilibrium analysis. The distinctions between boundary value problems and eigenvalue problems are too involved to treat here. Instead, the buckling differential equations governing the buckling behavior from a membrane prebuckled state (prebuckling deformations are ignored) are

$$\delta N_{x,\,x} + \delta N_{xy,\,y} = 0 \qquad (5.13)$$

$$\delta N_{xy,\,x} + \delta N_{y,\,y} = 0 \qquad (5.14)$$

$$\delta M_{x,\,xx} + 2\delta M_{xy,\,xy} + \delta M_{y,\,yy}$$
$$+ \bar{N}_x \delta w,_{xx} + 2\bar{N}_{xy}\delta w,_{xy} + \bar{N}_y \delta w,_{yy} = 0 \qquad (5.15)$$

where the δ denotes a variation of the principal symbol from its value in the prebuckled equilibrium state. Thus, the terms $\delta N_x, \ldots \delta M_x, \ldots$ are variations of forces and moments, respectively, from a membrane prebuckling equilibrium state. The terms δw and, by implication, δu and δv are variations in displacement from the same flat prebuckled state. In appearance, the buckling differential equations resemble the equilibrium differential equations except for the variational notation. Note that if the prebuckling state is a membrane, then $\delta w = w$. Also note that the applied in-plane loads $\bar{N}_x$, $\bar{N}_{xy}$, and $\bar{N}_y$ enter the mathematical formulation of the eigenvalue problem as coefficients of the curvatures rather than as "loads" on the right-hand side of the equation. The essence of the eigenvalue problem is to determine the smallest applied loads, $\bar{N}_x$, etc. that cause buckling. An important consequence of this type of problem is that the magnitude of the deformations after buckling cannot be determined without resort to large deflection considerations; i.e., the deformations are indeterminate when only Eqs. (5.13) to (5.15) are available.

The variations in force and moment resultants are

$$
\begin{Bmatrix} \delta N_x \\ \delta N_y \\ \delta N_{xy} \end{Bmatrix} = \begin{bmatrix} A_{11} & A_{12} & A_{16} \\ A_{12} & A_{22} & A_{26} \\ A_{16} & A_{26} & A_{66} \end{bmatrix} \begin{Bmatrix} \delta \epsilon_x^\circ \\ \delta \epsilon_y^\circ \\ \delta \gamma_{xy}^\circ \end{Bmatrix} + \begin{bmatrix} B_{11} & B_{12} & B_{16} \\ B_{12} & B_{22} & B_{26} \\ B_{16} & B_{26} & B_{66} \end{bmatrix} \begin{Bmatrix} \delta \kappa_x \\ \delta \kappa_y \\ \delta \kappa_{xy} \end{Bmatrix} \tag{5.16}
$$

$$
\begin{Bmatrix} \delta M_x \\ \delta M_y \\ \delta M_{xy} \end{Bmatrix} = \begin{bmatrix} B_{11} & B_{12} & B_{16} \\ B_{12} & B_{22} & B_{26} \\ B_{16} & B_{26} & B_{66} \end{bmatrix} \begin{Bmatrix} \delta \epsilon_x^\circ \\ \delta \epsilon_y^\circ \\ \delta \gamma_{xy}^\circ \end{Bmatrix} + \begin{bmatrix} D_{11} & D_{12} & D_{16} \\ D_{12} & D_{22} & D_{26} \\ D_{16} & D_{26} & D_{66} \end{bmatrix} \begin{Bmatrix} \delta \kappa_x \\ \delta \kappa_y \\ \delta \kappa_{xy} \end{Bmatrix} \tag{5.17}
$$

where the variations in extensional strains and changes in curvature are related to the variations in displacements by

$$ \delta \epsilon_x^\circ = \delta u,_x \qquad \delta \epsilon_y^\circ = \delta v,_y \qquad \delta \gamma_{xy}^\circ = \delta u,_y + \delta v,_x \tag{5.18} $$

$$ \delta \kappa_x = -\delta w,_{xx} \qquad \delta \kappa_y = -\delta w,_{yy} \qquad \delta \kappa_{xy} = -2\delta w,_{xy} \tag{5.19} $$

The buckling differential equations can be expressed in terms of the variations in displacements by substituting the variations in extensional strains and curvatures, Eqs. (5.18) and (5.19), in the variations in force and moment resultants, Eqs. (5.16) and (5.17), and subsequently in Eqs. (5.13) to (5.15). The resulting equations take on a form similar to the corresponding equilibrium equations, Eqs. (5.6) to (5.8). Just as for equilibrium problems, buckling of generally laminated plates has coupling between bending and extension. However, some special laminates exhibit no coupling; hence, their buckling loads are obtained by solution of only Eq. (5.15) or its variation of deflections equivalent.

The boundary conditions for buckling problems are applied only to the buckling deformations since the prebuckling deformations are assumed to be a membrane state. One of the distinguishing features of an eigenvalue problem is that all the boundary conditions are homogeneous, i.e., zero. Thus, during buckling, the simply supported edge and clamped edge boundary conditions are

$$
\begin{aligned}
S1: &\quad \delta w = 0 \quad \delta M_n = 0 \quad \delta u_n = 0 \quad \delta u_t = 0 \\
S2: &\quad \delta w = 0 \quad \delta M_n = 0 \quad \delta N_n = 0 \quad \delta u_t = 0 \\
S3: &\quad \delta w = 0 \quad \delta M_n = 0 \quad \delta u_n = 0 \quad \delta N_{nt} = 0 \\
S4: &\quad \delta w = 0 \quad \delta M_n = 0 \quad \delta N_n = 0 \quad \delta N_{nt} = 0
\end{aligned} \tag{5.20}
$$

$$
\begin{aligned}
C1: &\quad \delta w = 0 \quad \delta w,_n = 0 \quad \delta u_n = 0 \quad \delta u_t = 0 \\
C2: &\quad \delta w = 0 \quad \delta w,_n = 0 \quad \delta N_n = 0 \quad \delta u_t = 0
\end{aligned} \tag{5.21}
$$

$C3$: $\delta w = 0$ $\delta w,_n = 0$ $\delta u_n = 0$ $\delta N_{nt} = 0$

$C4$: $\delta w = 0$ $\delta w,_n = 0$ $\delta N_n = 0$ $\delta N_{nt} = 0$

$$(5.21)$$
(cont'd.)

The boundary conditions could be different for each edge of a plate, so the number of combinations of possible boundary conditions is enormous, as it was with equilibrium problems.

5.2.4 Vibration Equations for Laminated Plates

As with plate buckling, plate vibration, or oscillation about a state of static equilibrium, is an eigenvalue problem. The objective of the analysis is to determine the frequencies and mode shapes in which laminated plates vibrate. The magnitude of the deformations in a particular mode, however, is indeterminate since this is an eigenvalue problem. The governing vibration differential equations are obtained from the buckling differential equations by adding an acceleration term to the right-hand side of Eq. (5.15) and reinterpreting all variations to be during vibration about an equilibrium state (no problem is presented since the variations during buckling were from the equilibrium state, too.):

$$\delta N_{x,x} + \delta N_{xy,y} = 0 \qquad (5.22)$$

$$\delta N_{xy,x} + \delta N_{y,y} = 0 \qquad (5.23)$$

$$\delta M_{x,xx} + 2\delta M_{xy,xy} + \delta M_{y,yy}$$
$$+ \bar{N}_x \delta w,_{xx} + 2\bar{N}_{xy}\delta w,_{xy} + \bar{N}_y \delta w,_{yy} = \rho \delta w,_{tt} \qquad (5.24)$$

where ρ is the mass per unit area of the plate.

The variations in forces and moments during vibration are given by Eqs. (5.16) and (5.17). The prestress state (equilibrium stress state) is specified by $\bar{N}_x$, $\bar{N}_y$, and $\bar{N}_{xy}$.

As with both the plate bending and buckling problems, plate vibrations include coupling between bending and extension when the plate is unsymmetrically laminated. For symmetrically laminated plates, the coupling vanishes, and the vibration problem reduces to solution of Eq. (5.24) alone since rotatory inertia terms are ignored. Irrespective of the lamination characteristics, the boundary conditions are the same as for the buckling problem.

Alternatively, both the buckling and vibration problems can be formulated as a vibration problem with buckling loads being determined when the vibration frequency is equated to zero.

5.2.5 Solution Techniques

Many techniques exist for solution of the equilibrium, buckling, and vibration problems formulated in the preceding subsections. The techniques range from fortuitous exact solutions that are obtained essentially by "observation" or examination through numerical approximations such as finite element and finite difference approaches to the various approximate energy methods such as Rayleigh-Ritz and Galerkin. Since the objective here is to demonstrate the importance of the coupling stiffnesses, only those solution techniques necessary for fruitful illustrations will be used.

A prominent adjunct of many of the techniques is separation of variables. In that method, the deflection variables or variation in deflection variables are arbitrarily separated into functions of plate coordinate x alone times functions of y alone. Wang (Ref. 5-4) determined that separation of variables leads to exact solutions for some classes of plate problems, but does not for others, i.e., the deflections are not always separable. A specific example of an approximate use of separation of variables due to Ashton (Ref. 5-5) will be discussed in Sec. 5.3.2. Other exact uses of the method abound throughout Sec. 5.3 through 5.5.

5.3 DEFLECTION OF SIMPLY SUPPORTED LAMINATED PLATES UNDER DISTRIBUTED LATERAL LOAD

Consider the general class of laminated rectangular plates that are simply supported along edges $x = 0$, $x = a$, $y = 0$, and $y = b$ and subjected to a distributed lateral load, $p(x,y)$, in Fig. 5-4. The lateral load can be expanded in a double Fourier series:

$$p(x, y) = \sum_{m=1}^{\infty} \sum_{n=1}^{\infty} p_{mn} \sin \frac{m\pi x}{a} \sin \frac{n\pi y}{b} \qquad (5.25)$$

Many different types of loading can easily be represented by Eq. (5.25). For example, a uniform load, p_0, is given by

$$p(x, y) = \sum_{m=1,3,..}^{\infty} \sum_{n=1,3,..}^{\infty} \frac{16p_0}{\pi^2} \frac{1}{mn} \sin \frac{m\pi x}{a} \sin \frac{n\pi y}{b} \qquad (5.26)$$

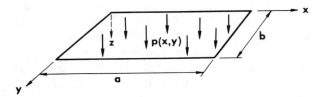

FIG. 5-4. Simply supported laminated rectangular plate under distributed lateral load.

Numerous other examples are discussed by Timoshenko and Woinowsky-Kreiger (Ref. 5-2).

Some of the various types of lamination possible, specially orthotropic, symmetric angle-ply, antisymmetric cross-ply, and antisymmetric angle-ply, will be analyzed for the loading given in Eq. (5.25). The results will be compared to determine the influence of the twist-coupling stiffnesses (D_{16} and D_{26}) and the bending-extension coupling stiffnesses (B_{ij}). All the plate edges are simply supported; however, as has been observed in Sec. 5.2, such a specification is still ambiguous. Thus, pay special attention to the precise formulation for the boundary conditions in each of the cases discussed.

5.3.1 Specially Orthotropic Laminates

A specially orthotropic laminate has either a single layer of a specially orthotropic material or multiple specially orthotropic layers that are symmetrically arranged about the laminate middle surface. In both cases, the laminate stiffnesses consist solely of $A_{11}, A_{12}, A_{22}, A_{66}, D_{11}, D_{12}, D_{22}$, and D_{66}. That is, neither shear or twist coupling nor bending-extension coupling exists. Thus, for plate problems, the lateral deflections are described by only one differential equation of equilibrium:

$$D_{11} w_{,xxxx} + 2(D_{12} + 2D_{66}) w_{,xxyy} + D_{22} w_{,yyyy} = p(x, y) \tag{5.27}$$

subject to the boundary conditions of a simply supported edge which for this simple laminate are

$$\begin{aligned} x = 0, a: \quad & w = 0 \quad M_x = -D_{11} w_{,xx} - D_{12} w_{,yy} = 0 \\ y = 0, b: \quad & w = 0 \quad M_y = -D_{12} w_{,xx} - D_{22} w_{,yy} = 0 \end{aligned} \tag{5.28}$$

Note that because the in-plane deformations, u and v, are not present in the differential equation, the simply supported edge boundary condition takes on an especially simple form as compared to Eq. (5.11).

The solution to this fourth-order partial differential equation and associated boundary conditions is remarkably simple. As with isotropic plates, the solution can easily be verified as

$$w = \sum_{m=1}^{\infty} \sum_{n=1}^{\infty} a_{mn} \sin \frac{m\pi x}{a} \sin \frac{n\pi y}{b} \tag{5.29}$$

That is, Eq. (5.29) satisfies the differential equation, Eq. (5.27), and the boundary conditions, Eq. (5.28), so is the exact solution if

$$a_{mn} = \frac{\dfrac{p_{mn}}{\pi^4}}{D_{11}\left(\dfrac{m}{a}\right)^4 + 2(D_{12} + 2D_{66})\left(\dfrac{m}{a}\right)^2\left(\dfrac{n}{b}\right)^2 + D_{22}\left(\dfrac{n}{b}\right)^4} \tag{5.30}$$

In particular, for a uniform lateral load, the solution is easily shown to be

$$w = \frac{16 p_0}{\pi^6} \sum_{m=1,3,5,\ldots}^{\infty} \sum_{n=1,3,5,\ldots}^{\infty}$$

$$\times \frac{(1/mn) \sin (m\pi x/a) \sin (n\pi y/b)}{D_{11} (m/a)^4 + 2(D_{12} + 2D_{66})(m/a)^2 (n/b)^2 + D_{22} (n/b)^4} \tag{5.31}$$

Once the deflections are known, the stresses are straightforwardly obtained by substitution in the stress-strain relations, Eq. (4.13), after the strains are found from Eq. (4.9). Note that the solution in Eq. (5.31) is expressed in terms of only the laminate stiffnesses D_{11}, D_{12}, D_{22}, and D_{66}. This solution will not be plotted here, but will be used as a baseline solution in the following subsections and plotted there in comparison with more complicated results.

5.3.2 Symmetric Angle-Ply Laminates

Symmetric angle-ply laminates were described in Sec. 4.3.2 and found to be characterized by a full matrix of extensional stiffnesses as well as bending stiffnesses (but of course no bending-extension coupling stiffnesses because of middle surface symmetry). The new facet of this type of laminate as opposed to specially orthotropic laminates is the appearance of the twist coupling stiffnesses D_{16} and D_{26} (the shear coupling stiffnesses A_{16} and A_{26} do not affect a plate problem when the laminate is symmetric). The governing differential equation of equilibrium is

$$D_{11} w_{,xxxx} + 4D_{16} w_{,xxxy} + 2(D_{12} + 2D_{66}) w_{,xxyy}$$

$$+ 4D_{26} w_{,xyyy} + D_{22} w_{,yyyy} = p(x, y) \tag{5.32}$$

subject to the simply supported edge boundary conditions

$$x = 0, a: \quad w = 0 \quad M_x = -D_{11} w_{,xx} - D_{12} w_{,yy} - 2D_{16} w_{,xy} = 0$$

$$y = 0, b: \quad w = 0 \quad M_y = -D_{12} w_{,xx} - D_{22} w_{,yy} - 2D_{26} w_{,xy} = 0 \tag{5.33}$$

As with the specially orthotropic laminate, the simply supported edge boundary condition cannot be further distinguished by the character of the in-plane boundary conditions on u and v since the latter do not appear in any plate problem for a symmetric laminate.

The solution to the governing differential equation, Eq. (5.32), is not as simple as for specially orthotropic laminates because of the presence of D_{16} and D_{26}. The Fourier expansion of the deflection w, Eq. (5.29), is an example of separation of variables. However, because of the terms involving D_{16} and D_{26}, the expansion does not satisfy the governing differential equation. Thus, the variables are not actually separable. Moreover, the deflection expansion also does not satisfy the boundary conditions, Eq. (5.33), again because of the terms involving D_{16} and D_{26}.

Ashton (Ref. 5-5) solved this problem approximately by recognizing that the differential equation, Eq. (5.32), is but one result of the equilibrium requirement of making the total potential energy of the mechanical system stationary relative to the independent variable w. An alternative method is to express the total potential energy in terms of the deflections and their derivatives. Specifically, Ashton approximated the deflection by the Fourier expansion in Eq. (5.29) and substituted it in the expression for the total potential energy, V:

$$V = \frac{1}{2} \iint [D_{11}(w,_{xx})^2 + 2D_{12}w,_{xx}w,_{yy} + D_{22}(w,_{yy})^2 + 4D_{66}(w,_{xy})^2$$

$$+ 4D_{16}w,_{xx}w,_{xy} + 4D_{26}w,_{yy}w,_{xy} - 2pw] \, dxdy \quad (5.34)$$

where the term involving pw is the work of the external forces, namely the lateral load p and the remainder of V is the strain energy of the plate. The energy for an unsymmetrically laminated plate is displayed by Ashton and Whitney (Ref. 5-1). Basic energy principles and their application to applied mechanics, particularly structural mechanics, are discussed in the classical book by Langhaar (Ref. 5-6).

If enough terms are taken in the deflection expansion, the approximate energy converges to the exact energy as long as the geometric boundary conditions ($w = 0$ and $w,_x = 0$) are satisfied even if the natural boundary conditions ($M_n = \bar{M}_n$ and $N_n = \bar{N}_n$) are not satisfied. This method is the well-known Rayleigh-Ritz method when the energy is made stationary relative to the coefficients of the deflection expansion according to the principle of stationary potential energy. The resulting equations are a set of simultaneous linear algebraic equations that can be solved numerically with the aid of a digital computer. Note that, for the simply supported edge boundary condition, only one geometric boundary condition, $w = 0$, exists. Also, only one natural boundary condition, $M_n = 0$, exists. The double sine series deflection function, Eq. (5.29), satisfies the geometric boundary condition, but not the natural boundary condition. Thus, it is an acceptable deflection approximation for the Rayleigh-Ritz method. However, the convergence of the method may be slow because the natural boundary condition is not satisfied exactly.

Ashton (Ref. 5-5) used 49 terms (up through $m = 7$ and $n = 7$) in the deflection approximation, Eq. (5.29), to obtain for a uniformly loaded square plate with $D_{22}/D_{11} = 1$, $(D_{12} + 2D_{66})/D_{11} = 1.5$, and $D_{16}/D_{11} = D_{26}/D_{11} = -.5$ a maximum deflection (at the center) of

$$w_{max} = \frac{.00425a^4p}{D_{11}} \quad (5.35)$$

whereas if D_{16} and D_{26} are ignored, that is, the symmetric angle-ply is approximated by a specially orthotropic laminate with $D_{22}/D_{11} = 1$, $(D_{12} + 2D_{66})/D_{11} = 1.5$, and $D_{16} = D_{26} = 0$, then the maximum deflection is

$$w_{max} = \frac{.00324 a^4 p}{D_{11}} \tag{5.36}$$

Thus, the error in ignoring the twist coupling terms is about 24 percent, certainly not a negligible error. Hence, the specially orthotropic laminate is an unacceptable approximation to a symmetric angle-ply laminate. Recognize, however, that Ashton's Rayleigh-Ritz results are also approximate since only a finite number of terms were used in the deflection approximation. Thus, a comparison of his results with an exact solution would lend more confidence to the rejection of the specially orthotropic approximation.

Ashton (Ref. 5-7) observed that skew (parallelogram-shaped) isotropic plates under uniform distributed load $\bar{p}_0$ as shown in Fig. 5-5 are governed by the

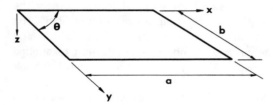

FIG. 5-5. Skew plate geometry.

equilibrium differential equation

$$w_{,xxxx} - 4 \cos\theta \ w_{,xxxy} + 2(1 + 2 \cos^2\theta) w_{,xxyy}$$

$$- 4 \cos\theta \ w_{,xyyy} + w_{,yyyy} = \frac{\bar{p}_0 \sin^4\theta}{D} \tag{5.37}$$

with simply supported edge boundary conditions

$$\begin{aligned} x = 0, a: &\quad w = 0 \quad w_{,xx} - 2 \cos\theta \ w_{,xy} = 0 \\ y = 0, b: &\quad w = 0 \quad w_{,yy} - 2 \cos\theta \ w_{,xy} = 0 \end{aligned} \tag{5.38}$$

The essence of Ashton's contribution is that he identified the skew plate stiffnesses as being a transformation of the symmetric angle-ply stiffnesses, or, more generally, the anisotropic bending stiffnesses, that is,

$$D_{22} = D_{11} = D \qquad \frac{D_{12} + 2D_{66}}{D_{11}} = (1 + 2 \cos^2\theta)$$

$$\frac{D_{16}}{D_{11}} = \frac{D_{26}}{D_{11}} = -\cos\theta \qquad p = \bar{p} \sin^4\theta \tag{5.39}$$

The stiffnesses in Eq. (5.39) are equivalent to the stiffnesses of an equivalent orthotropic material with principal material axes of orthotropy at 45° to the plate sides. The orthotropic bending stiffnesses of the equivalent material can be shown to be

$$D'_{11} = D(1 + 2 \cos \theta + \cos^2 \theta)$$

$$D'_{22} = D(1 - 2 \cos \theta + \cos^2 \theta) \qquad (5.40)$$

$$D'_{12} + 2D'_{66} = D \sin^2 \theta$$

where, of course, D'_{16} and D'_{26} are zero since these are principal material directions for an orthotropic material. Values of D'_{ij} and D_{ij} are given in Table 5-1 for several values of the equivalent skew angle θ. From Eq. (5.40), as θ gets smaller, D'_{11} gets larger, D'_{22} gets smaller, and, most importantly, D_{16} gets larger, that is, the plate becomes more anisotropic.

Since exact solutions for skew plates are readily available, Ashton was able to get some exact solutions for anisotropic rectangular plates by the special identification process outlined in the preceding paragraph. Specifically, values for the center deflection of a square plate are shown in Table 5-2. There, the exact solution is shown along with the Rayleigh-Ritz solution and the specially orthotropic solution. For the case already discussed where $D_{22}/D_{11} = 1$, $(D_{12} + 2D_{66})/D_{11} = 1.5$, and $D_{16}/D_{11} = D_{26}/D_{11} = -.5$, the exact solution is

TABLE 5-1. Equivalent bending stiffness ratios *

Equivalent skew angle θ	$\dfrac{D'_{22}}{D'_{11}}$	$\dfrac{D'_{12} + 2D'_{66}}{D'_{11}}$	$\dfrac{D'_{16}}{D'_{11}}$	$\dfrac{D_{22}}{D_{11}}$	$\dfrac{D_{12} + 2D_{66}}{D_{11}}$	$\dfrac{D_{16}}{D_{11}}$
90°	1.000	1.000	0.	1.	1.000	0.
80°	.495	.702	0.	1.	1.061	−.174
63°	.141	.376	0.	1.	1.412	−.454
60°	.111	.333	0.	1.	1.500	−.500
54°	.0675	.260	0.	1.	1.690	−.587

*After Ashton, Ref. 5-7.

TABLE 5-2. Maximum deflection coefficients, K, for exact, specially orthotropic, and Rayleigh-Ritz solutions *

Equivalent skew angle θ	Exact solution K	Specially orthotropic solution K	Rayleigh-Ritz solution K
90°	.00406	.00406	.00406
80°	.00411	.00394	.00408
63°	.00444	.00336	.00422
60°	.00452	.00324	.00425
54°	.00476	.00301	.00430

*After Ashton, Ref. 5-7.

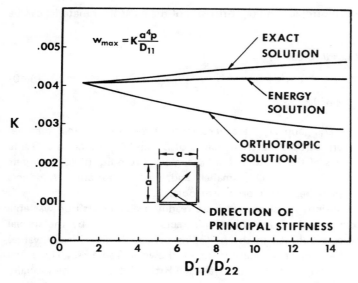

FIG. 5-6. Deflection coefficient versus principal stiffness ratio. (After Ashton, Ref. 5-7.)

$$w_{max} = \frac{.00452a^4p}{D_{11}} \tag{5.41}$$

Thus, the Rayleigh-Ritz solution is 6 percent in error whereas the specially orthotropic solution is 28 percent in error.

The deflection results for the three approaches are plotted in Fig. 5-6 as a function of the ratio of the principal stiffnesses, D'_{11}/D'_{22}, which gets large as θ decreases. Thus, the larger the D_{16} and D_{26}, the smaller θ becomes and hence the more inaccurate both the Rayleigh-Ritz approach and the specially orthotropic approximation become.

5.3.3 Antisymmetric Cross-Ply Laminates

Antisymmetric cross-ply laminates were described in Sec. 4.3.3 and found to have extensional stiffnesses A_{11}, A_{12}, $A_{22} = A_{11}$, and A_{66}, bending-extension coupling stiffnesses, B_{11} and $B_{22} = -B_{11}$, and bending stiffnesses D_{11}, D_{12}, $D_{22} = D_{11}$, and D_{66}. The new terms here in comparison to a specially orthotropic laminate are B_{11} and B_{22}. Because of this coupling, the equilibrium differential equations are coupled:

$$A_{11}u_{,xx} + A_{66}u_{,yy} + (A_{12} + A_{66})v_{,xy} - B_{11}w_{,xxx} = 0 \tag{5.42}$$

$$(A_{12} + A_{66})u_{,xy} + A_{66}v_{,xx} + A_{11}v_{,yy} + B_{11}w_{,yyy} = 0 \tag{5.43}$$

$$D_{11}(w,_{xxxx} + w,_{yyyy}) + 2(D_{12} + 2D_{66})w,_{xxyy}$$

$$-B_{11}(u,_{xxx} - v,_{yyy}) = p \quad (5.44)$$

Whitney and Leissa (Ref. 5-8) chose to solve the problem for simply supported edge boundary condition S2:

$$x = 0, a: \quad w = 0 \quad M_x = B_{11}u,_x - D_{11}w,_{xx} - D_{12}w,_{yy} = 0 \quad (5.45)$$

$$v = 0 \quad N_x = A_{11}u,_x + A_{12}v,_y - B_{11}w,_{xx} = 0 \quad (5.46)$$

$$y = 0, b: \quad w = 0 \quad M_y = -B_{11}v,_y - D_{12}w,_{xx} - D_{11}w,_{yy} = 0 \quad (5.47)$$

$$u = 0 \quad N_y = A_{12}u,_x + A_{11}v,_y + B_{11}w,_{yy} = 0 \quad (5.48)$$

and observed that the deflections

$$u = \sum_{m=1}^{\infty} \sum_{n=1}^{\infty} A_{mn} \cos \frac{m\pi x}{a} \sin \frac{n\pi y}{b}$$

$$v = \sum_{m=1}^{\infty} \sum_{n=1}^{\infty} B_{mn} \sin \frac{m\pi x}{a} \cos \frac{n\pi y}{b} \quad (5.49)$$

$$w = \sum_{m=1}^{\infty} \sum_{n=1}^{\infty} C_{mn} \sin \frac{m\pi x}{a} \sin \frac{n\pi y}{b}$$

satisfy the three governing differential equations and the boundary conditions so are the exact solution (the form of which need not be repeated here).

If the lateral load is but one term of the Fourier series, that is,

$$p = p_0 \sin \frac{\pi x}{a} \sin \frac{\pi y}{b} \quad (5.50)$$

then the normalized maximum deflection of a rectangular antisymmetric cross-ply laminated graphite/epoxy plate is plotted in Fig. 5-7 for 2,4,6, and an infinite number of layers. The infinite number of layers case corresponds to the specially orthotropic solution in which the coupling between bending and extension is ignored. For a two-layered plate, neglect of coupling results in an underprediction of the deflection by 64 percent; that is, the actual deflection is nearly three times the specially orthotropic approximation! The effect of coupling between bending and extension on the deflections obviously dies out quite rapidly as the number of layers increases irrespective of the plate aspect ratio, a/b. That is, the specially orthotropic solution is rapidly approached. However, only when there are more than six layers can coupling be ignored without significant error.

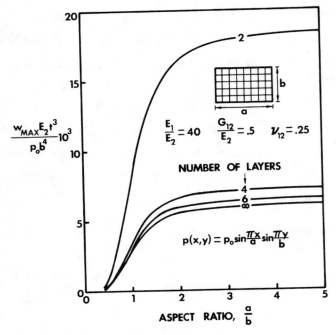

FIG. 5-7. Maximum deflection of a rectangular antisymmetric cross-ply laminated plate under sinusoidal transverse load.

For laminated composite materials, the effect of coupling between bending and extension on plate deflections depends essentially on the orthotropic modulus ratio, E_1/E_2. Values of G_{12}/E_2 and ν_{12} are fixed because the influence of their variation on the deflections is small compared to that of E_1/E_2. At $E_1/E_2 = 1$ in Fig. 5-8, the effect of coupling is nonexistent as expected. As E_1/E_2 increases, the effect of coupling between bending and extension increases. Thus, the deflections in a two-layered boron/epoxy plate ($E_1/E_2 = 10$) are not as much larger than the specially orthotropic approximation as in a graphite/epoxy plate ($E_1/E_2 = 40$).

5.3.4 Antisymmetric Angle-Ply Laminates

Antisymmetric angle-ply laminates were described in Sec. 4.3.3 and found to have extensional stiffnesses A_{11}, A_{12}, A_{22}, and A_{66}, bending-extension coupling stiffnesses B_{16} and B_{26}, and bending stiffnesses D_{11}, D_{12}, D_{22}, and D_{66}. Thus, this laminate exhibits a different type of bending-extension coupling than does the antisymmetric cross-ply laminate. The governing differential equations of equilibrium are

$$A_{11}u_{,xx} + A_{66}u_{,yy} + (A_{12} + A_{66})v_{,xy} - 3B_{16}w_{,xxy} - B_{26}w_{,yyy} = 0$$

$$(5.51)$$

$$(A_{12} + A_{66})u_{,xy} + A_{66}v_{,xx} + A_{22}v_{,yy} - B_{16}w_{,xxx} - 3B_{26}w_{,xyy} = 0$$

$$(5.52)$$

$$D_{11}w_{,xxxx} + 2(D_{12} + 2D_{66})w_{,xxyy} + D_{22}w_{,yyyy}$$
$$- B_{16}(3u_{,xxy} + v_{,xxx}) - B_{26}(u_{,yyy} + 3v_{,xyy}) = p \quad (5.53)$$

Whitney (Refs. 5-9 and 5-10) chose to solve the problem for simply supported edge boundary condition $S3$ (recall that $S2$ was used for antisymmetric cross-ply laminates):

$$x = 0, a: \quad w = 0 \quad M_x = B_{16}(u_{,y} + v_{,x}) - D_{11}w_{,xx} - D_{12}w_{,yy} = 0 \quad (5.54)$$

$$u = 0 \quad N_{xy} = A_{66}(u_{,y} + v_{,x}) - B_{16}w_{,xx} - B_{26}w_{,yy} = 0 \quad (5.55)$$

$$y = 0, b: \quad w = 0 \quad M_y = B_{26}(u_{,y} + v_{,x}) - D_{12}w_{,xx} - D_{22}w_{,yy} = 0 \quad (5.56)$$

$$v = 0 \quad N_{xy} = A_{66}(u_{,y} + v_{,x}) - B_{16}w_{,xx} - B_{26}w_{,yy} = 0 \quad (5.57)$$

He then observed that the deflections

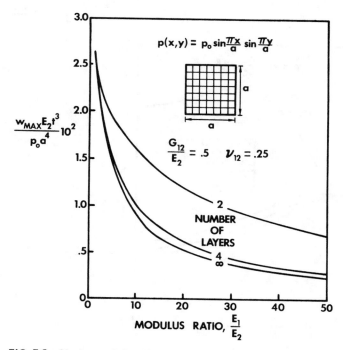

FIG. 5-8. Maximum deflection of a square antisymmetric cross-ply laminated plate under sinusoidal transverse load. (*After Whitney and Leissa, Ref. 5-8.*)

$$u = \sum_{m=1}^{\infty} \sum_{n=1}^{\infty} A_{mn} \sin \frac{m\pi x}{a} \cos \frac{n\pi y}{b}$$

$$v = \sum_{m=1}^{\infty} \sum_{n=1}^{\infty} B_{mn} \cos \frac{m\pi x}{a} \sin \frac{n\pi y}{b} \qquad (5.58)$$

$$w = \sum_{m=1}^{\infty} \sum_{n=1}^{\infty} C_{mn} \sin \frac{m\pi x}{a} \sin \frac{n\pi y}{b}$$

identically satisfy the governing differential equations and boundary conditions so are the exact solution.

Results for a graphite/epoxy laminate for which

$$\frac{E_1}{E_2} = 40 \qquad \frac{G_{12}}{E_2} = .5 \qquad \nu_{12} = .25 \qquad (5.59)$$

are shown in Fig. 5-9 as a function of angle-ply angle for the loading

$$p = p_0 \sin \frac{\pi x}{a} \sin \frac{\pi y}{a} \qquad (5.60)$$

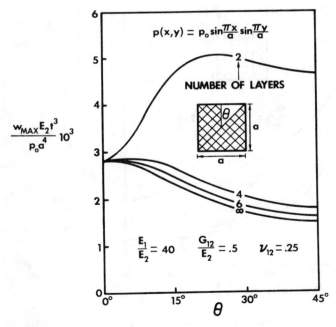

FIG. 5-9. Maximum deflection of a square antisymmetric angle-ply laminated plate under sinusoidal transverse load.

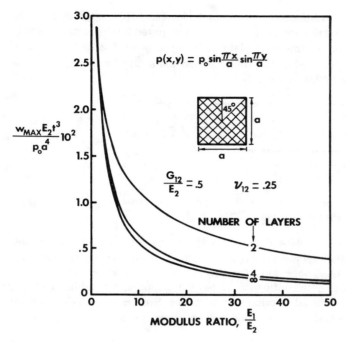

FIG. 5-10. Maximum deflection of a square antisymmetric angle-ply laminated plate under sinusoidal transverse load.

Clearly, coupling is quite significant for two-layered laminates, but rapidly decreases as the number of layers increases. For a fixed laminate thickness, the bending-extension coupling stiffnesses

$$(B_{16}, B_{26}) = (Q_{16}, Q_{26}) \, \frac{t^2}{2N} \qquad\qquad (5.61)$$

obviously decrease as N increases so the source of the change in the influence of coupling is evident.

Results for a square plate under sinusoidal transverse load with a variable modulus ratio and a ±45° lamination angle are shown in Fig. 5-10. There, the effect of coupling between bending and extension on deflections is significant for all modulus ratios except those quite close to $E_1/E_2 = 1$.

5.4 BUCKLING OF SIMPLY SUPPORTED LAMINATED PLATES UNDER IN-PLANE LOAD

Consider the general class of laminated rectangular plates that are simply supported along edges $x = 0$, $x = a$, $y = 0$, and $y = b$ and subjected to uniform in-plane force in the x-direction as in Fig. 5-11. Other more complicated loads and boundary conditions could be treated. However, the importance of the various stiffnesses in buckling problems is well-illustrated by this simple loading. More comprehensive treatment of plate buckling in general is given by Timoshenko and Gere (Ref. 5-11) and of laminated plate buckling in particular is given by Ashton and Whitney (Ref. 5-1).

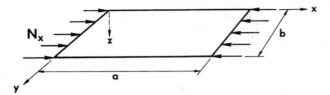

FIG. 5-11. Simply supported laminated rectangular plate under uniform uniaxial in-plane compression.

The buckling load will be determined for plates with various laminations; specially orthotropic, symmetric angle-ply, antisymmetric cross-ply, and antisymmetric angle-ply. The results for the different lamination types will be compared to ascertain the influence of twist coupling and bending-extension coupling. As with the deflection problems in Sec. 5.3, different simply supported edge boundary conditions will be used in the several problems presented.

5.4.1 Specially Orthotropic Laminates

A specially orthotropic laminate has either a single layer of a specially orthotropic material or multiple specially orthotropic layers that are symmetrically arranged about the laminate middle surface. In both cases, the laminate stiffnesses consist solely of $A_{11}, A_{12}, A_{22}, A_{66}, D_{11}, D_{12}, D_{22}$, and D_{66}. That is, neither shear or twist coupling nor bending-extension coupling exists. Then, for plate problems, the buckling loads are described by only one buckling differential equation:

$$D_{11}\delta w_{,xxxx} + 2(D_{12} + 2D_{66})\delta w_{,xxyy} + D_{22}\delta w_{,yyyy} + \bar{N}_x \delta w_{,xx} = 0$$

$$(5.62)$$

subject to the simply supported edge boundary conditions

$$x = 0, a: \quad \delta w = 0 \quad \delta M_x = -D_{11}\delta w_{,xx} - D_{12}\delta w_{,yy} = 0$$
$$y = 0, b: \quad \delta w = 0 \quad \delta M_y = -D_{12}\delta w_{,xx} - D_{22}\delta w_{,yy} = 0$$

$$(5.63)$$

Note that because of the absence of the variations in in-plane displacements, δu and δv, the boundary conditions are much simpler than the general case in Eq. (5.20).

The solution to this fourth-order partial differential equation and associated homogeneous boundary conditions is every bit as simple as the analogous deflection problem in Sec. 5.3.1. The boundary conditions are satisfied by the variation in lateral displacement (for plates, δw actually is the buckle displacement since $w = 0$ in the membrane prebuckling state; however, δu and δv are variations from a nontrivial equilibrium state. Hence, we retain the more rigorous variational notation consistently):

$$\delta w = A_{mn} \sin \frac{m\pi x}{a} \sin \frac{n\pi y}{b} \tag{5.64}$$

where m and n are the number of buckle half wavelengths in the x- and y-directions, respectively. In addition, the governing differential equation is satisfied by Eq. (5.64) if

$$\bar{N}_x = \pi^2 \left[D_{11} \left(\frac{m}{a}\right)^2 + 2(D_{12} + 2D_{66})\left(\frac{n}{b}\right)^2 + D_{22} \left(\frac{n}{b}\right)^4 \left(\frac{a}{m}\right)^2 \right] \tag{5.65}$$

The smallest value of $\bar{N}_x$ obviously occurs when $n = 1$, so the buckling load expression further reduces to

$$\bar{N}_x = \pi^2 \left[D_{11} \left(\frac{m}{a}\right)^2 + 2(D_{12} + 2D_{66}) \frac{1}{b^2} + D_{22} \frac{1}{b^4} \left(\frac{a}{m}\right)^2 \right] \tag{5.66}$$

The smallest value of $\bar{N}_x$ for various m is not obvious, but varies for different values of the stiffnesses and the plate aspect ratio, a/b.

For example, if $D_{11}/D_{22} = 10$ and $(D_{12} + 2D_{66})/D_{22} = 1$, then Eq. (5.66) becomes

$$\bar{N}_x = \pi^2 D_{22} \left[10\left(\frac{m}{a}\right)^2 + \frac{2}{b^2} + \left(\frac{a}{m}\right)^2 \frac{1}{b^4} \right] \tag{5.67}$$

which is plotted in Fig. 5-12 versus the plate aspect ratio. There, for small plate aspect ratios (that is, $a/b < 2.5$), the plate buckles into a single half-wave in the x-direction. For example, the buckling load of a square plate is

$$\bar{N}_x = \frac{13\pi^2 D_{22}}{b^2} \tag{5.68}$$

As the plate aspect ratio increases, the plate buckles into more and more buckle half-waves in the x-direction and has an $\bar{N}_x$ versus a/b curve that gets ever flatter and, in fact, approaches

$$\bar{N}_x = \frac{8.32456\pi^2 D_{22}}{b^2} \tag{5.69}$$

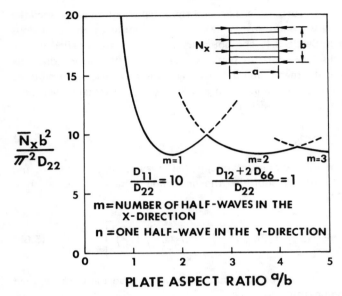

FIG. 5-12. Buckling loads for rectangular specially orthotropic laminated plates under uniform compression, $\bar{N}_x$.

For other materials, other families of curves such as in Fig. 5-12 are obtained with correspondingly different buckling loads and different points of change from one buckling mode to another.

5.4.2 Symmetric Angle-Ply Laminates

Symmetric angle-ply laminates were found in Sec. 4.3.2 to be characterized by a full matrix of extensional stiffnesses as well as bending stiffnesses, but have no coupling between bending and extension. The principal difference between these laminates and specially orthotropic laminates is the introduction here of the twist coupling stiffnesses D_{16} and D_{26} (the shear coupling stiffnesses A_{16} and A_{26} are immaterial for buckling of a symmetrically laminated plate since the governing differential equations are uncoupled). Accordingly, the governing buckling differential equation is

$$D_{11}\delta w,_{xxxx} + 4D_{16}\delta w,_{xxxy} + 2(D_{12} + 2D_{66})\delta w,_{xxyy}$$
$$+ 4D_{26}\delta w,_{xyyy} + D_{22}\delta w,_{yyyy} + \bar{N}_x \delta w,_{xx} = 0 \quad (5.70)$$

subject to the simply supported edge boundary conditions

$$x = 0, a: \quad \delta w = 0 \quad \delta M_x = -D_{11}\delta w,_{xx} - D_{12}\delta w,_{yy} - 2D_{16}\delta w,_{xy} = 0$$
$$y = 0, b: \quad \delta w = 0 \quad \delta M_y = -D_{12}\delta w,_{xx} - D_{22}\delta w,_{yy} - 2D_{26}\delta w,_{xy} = 0$$
$$(5.71)$$

The presence of D_{16} and D_{26} in the governing differential equation and boundary conditions makes a closed form solution impossible. That is, in analogy to bending of symmetric angle-ply laminated plates, the variation in lateral displacement, δw, cannot be separated into a function of x alone times a function of y alone such as in Eq. (5.64). However, again in analogy to bending of symmetrically laminated angle-ply plates, an approximate Rayleigh-Ritz solution can be obtained (Ref. 5-12) (or equivalently a Galerkin solution as presented by Chamis (Ref. 5-13)) by substituting the variation in lateral displacement expression

$$\delta w = \sum_{m=1}^{\infty} \sum_{n=1}^{\infty} A_{mn} \sin \frac{m\pi x}{a} \sin \frac{n\pi y}{b} \tag{5.72}$$

into the expression for the second variation of the total potential energy and subsequently making it stationary relative to the A_{mn}. Note that Eq. (5.72) satisfies the geometric boundary conditions of the problem ($\delta w = 0$ on all edges), but not the natural boundary conditions ($\delta M_n =$ on all edges) so the results may converge slowly toward the actual solution.

The Rayleigh-Ritz procedure (or alternatively the Galerkin procedure) is only incidental to the present objectives so we shall be satisfied to report the results for several laminates of boron/epoxy with $E_1/E_2 = 10$, $G_{12}/E_2 = .3$, and $\nu_{12} = .3$. Normalized buckling loads for three laminates, 20 layers at $+\theta$, 20 layers alternating at $\pm\theta$, and the specially orthotropic approximation, are plotted in Fig. 5-13. The Rayleigh-Ritz curves are for 49 terms ($m = 7$ and $n = 7$). Experimental results by Mandell (Ref. 5-14) are also shown in Fig. 5-13;

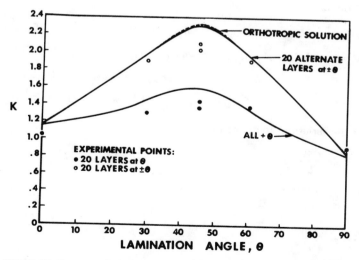

FIG. 5-13. Buckling loads for rectangular symmetric angle-ply plates under uniform compression, $\overline{N}_x$. (After Ashton and Whitney, Ref. 5-1.)

obviously, the agreement between theory and experiment is very satisfactory. Apparently, twist coupling is just as important for buckling problems as it is for bending problems. The principal influence of twist coupling is to lower the buckling load from what would be obtained with the specially orthotropic approximation. Hence, the specially orthotropic approximation is unconservative in design applications.

To the author's knowledge, there are, in contrast to the bending problem, no skew plate analogy results for plate buckling problems.

5.4.3 Antisymmetric Cross-Ply Laminates

Antisymmetric cross-ply laminates were found in Sec. 4.3.3 to have extensional stiffnesses A_{11}, A_{12}, $A_{22} = A_{11}$, and A_{66}, bending-extension coupling stiffnesses B_{11} and $B_{22} = -B_{11}$, and bending stiffnesses D_{11}, D_{12}, $D_{22} = D_{11}$, and D_{66}. The new terms here in comparison to a specially orthotropic laminate are B_{11} and B_{22}. Because of this bending-extension coupling, the buckling differential equations are coupled:

$$A_{11}\delta u,_{xx} + A_{66}\delta u,_{yy} + (A_{12} + A_{66})\delta v,_{xy} - B_{11}\delta w,_{xxx} = 0 \qquad (5.73)$$

$$(A_{12} + A_{66})\delta u,_{xy} + A_{66}\delta v,_{xx} + A_{11}\delta v,_{yy} + B_{11}\delta w,_{yyy} = 0 \qquad (5.74)$$

$$D_{11}(\delta w,_{xxxx} + \delta w,_{yyyy}) + 2(D_{12} + 2D_{66})\delta w,_{xxyy}$$
$$- B_{11}(\delta u,_{xxx} - \delta v,_{yyy}) + \bar{N}_x\delta w,_{xx} = 0 \quad (5.75)$$

Jones (Ref. 5-15) chose to solve the problem for simply supported edge boundary condition $S2$:

$$x = 0, a: \quad \delta w = 0 \quad \delta M_x = B_{11}\delta u,_x - D_{11}\delta w,_{xx} - D_{12}\delta w,_{yy} = 0 \quad (5.76)$$
$$\delta v = 0 \quad \delta N_x = A_{11}\delta u,_x + A_{12}\delta v,_y - B_{11}\delta w,_{xx} = 0 \quad (5.77)$$
$$y = 0, b: \quad \delta w = 0 \quad \delta M_y = -B_{11}\delta v,_y - D_{12}\delta w,_{xx} - D_{11}\delta w,_{yy} = 0 \quad (5.78)$$
$$\delta u = 0 \quad \delta N_y = A_{12}\delta u,_x + A_{11}\delta v,_y + B_{11}\delta w,_{yy} = 0 \quad (5.79)$$

and verified that the variations in deflections

$$\delta u = \bar{u} \cos\frac{m\pi x}{a} \sin\frac{n\pi y}{b}$$

$$\delta v = \bar{v} \sin\frac{m\pi x}{a} \cos\frac{n\pi y}{b} \qquad (5.80)$$

$$\delta w = \bar{w} \sin\frac{m\pi x}{a} \sin\frac{n\pi y}{b}$$

satisfy both the boundary conditions and the governing differential equations exactly if

$$\bar{N}_x = \left(\frac{a}{m\pi}\right)^2 \left(T_{33} + \frac{2T_{12}T_{23}T_{13} - T_{22}T_{13}^2 - T_{11}T_{23}^2}{T_{11}T_{22} - T_{12}^2}\right) \tag{5.81}$$

where

$$T_{11} = A_{11}\left(\frac{m\pi}{a}\right)^2 + A_{66}\left(\frac{n\pi}{b}\right)^2$$

$$T_{12} = (A_{12} + A_{66})\left(\frac{m\pi}{a}\right)\left(\frac{n\pi}{b}\right)$$

$$T_{13} = -B_{11}\left(\frac{m\pi}{a}\right)^3$$

$$T_{22} = A_{11}\left(\frac{n\pi}{b}\right)^2 + A_{66}\left(\frac{m\pi}{a}\right)^2 \tag{5.82}$$

$$T_{23} = B_{11}\left(\frac{n\pi}{b}\right)^3$$

$$T_{33} = D_{11}\left[\left(\frac{m\pi}{a}\right)^4 + \left(\frac{n\pi}{b}\right)^4\right] + 2(D_{12} + 2D_{66})\left(\frac{m\pi}{a}\right)^2\left(\frac{n\pi}{b}\right)^2$$

Note that if B_{11} is zero, then T_{13} and T_{23} are also zero so Eq. (5.81) reduces to the specially orthotropic solution, Eq. (5.65), if $D_{11} = D_{22}$. Because T_{11}, T_{12}, and T_{22} are functions of both m and n, no simple conclusion can be drawn about the value of n at buckling as could be done for specially orthotropic laminates where n was determined to be one. Instead, Eq. (5.81) is a complicated function of both m and n. At this point, recall the discussion in Sec. 3.5.3 about the difference between finding a minimum of a function of discrete variables versus a function of continuous variables. We have already seen that plates buckle with a small number of buckles. Consequently, the lowest buckling load must be found in Eq. (5.81) by a searching procedure involving *integer* values of m and $\bar{n}$ (Ref. 5-16) and not by equating to zero the first partial derivatives of $\bar{N}_x$ with respect to m and n.

Results for graphite/epoxy antisymmetric laminates for which $E_1/E_2 = 40$, $G_{12} = .5$, and $\nu_{12} = .25$ are shown in Fig. 5-14. There, the buckling load is normalized with respect to the plate width b and plate stiffness D_{22} for various rectangular plate aspect ratios and for various numbers of layers. The solution for an infinite number of layers (an orthotropic laminate) is shown as a limiting case of zero coupling between bending and extension. The coupling is extremely significant for a small number of layers. For two layers at a plate aspect ratio of one, the overestimate of buckling resistance is 183 percent if the orthotropic approximation is made. From another point of view, the resistance is 65 percent less than calculated by use of the orthotropic approximation. For four layers, the analogous numbers are 19 percent overestimate and 16 percent reduction,

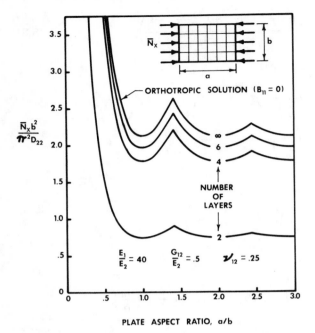

FIG. 5-14. Buckling loads for rectangular antisymmetric cross-ply laminated plates under uniform uniaxial compression $\bar{N}_x$. (*After Jones, Ref. 5-15.*)

respectively. For six layers, the numbers drop to 8 percent overestimate and 7 percent reduction, respectively. Obviously, the effect of coupling between bending and extension dies out rapidly as the number of layers increases for an antisymmetric laminate. However, for fewer than six layers, the effect cannot be ignored.

When other composite materials are considered, the effect of coupling on the buckling load depends essentially on the modulus ratio, E_1/E_2, as shown in Fig. 5-15. There, the buckling load is normalized by the buckling load of an orthotropic plate ($B_{11} = 0$) for square plates. Values of G_{12}/E_2 and ν_{12} are fixed because their influence on the buckling load is small compared to that of E_1/E_2. As the modulus ratio decreases from the graphite/epoxy value of 40, the influence of coupling between bending and extension decreases slowly. As noted previously, the reduction in buckling load of a square two-layered graphite/ epoxy plate from the orthotropic value is about 65 percent. For a square boron/epoxy plate, the reduction is about 43 percent. From the design analysis point of view, the orthotropic solution is too high by 183 percent for a graphite/epoxy plate and by 74 percent for an analogous boron/epoxy plate. Obviously, coupling between bending and extension is extremely important when the plate has only two layers. However, the influence of coupling dies out quite rapidly as the number of layers increases. For example, the reduction in buckling load for a six-layered graphite/epoxy plate is only about 7 percent and about 5 percent for a boron/epoxy plate.

5.4.4 Antisymmetric Angle-Ply Laminates

Antisymmetric angle-ply laminates were found in Sec. 4.3.3 to have extensional stiffnesses A_{11}, A_{12}, A_{22}, and A_{66}, bending-extension coupling stiffnesses B_{16} and B_{26}, and bending stiffnesses D_{11}, D_{12}, D_{22}, and D_{66}. Thus, this type of laminate exhibits a different kind of bending-extension coupling than does the antisymmetric cross-ply laminate examined in Sec. 5.4.3. The coupled buckling differential equations are

$$A_{11}\delta u_{,xx} + A_{66}\delta u_{,yy} + (A_{12} + A_{66})\delta v_{,xy}$$
$$- 3B_{16}\delta w_{,xxy} - B_{26}\delta w_{,yyy} = 0 \quad (5.83)$$

$$(A_{12} + A_{66})\delta u_{,xy} + A_{66}\delta v_{,xx} + A_{22}\delta v_{,yy}$$
$$- B_{16}\delta w_{,xxx} - 3B_{26}\delta w_{,xyy} = 0 \quad (5.84)$$

$$D_{11}\delta w_{,xxxx} + 2(D_{12} + 2D_{66})\delta w_{,xxyy} + D_{22}\delta w_{,yyyy}$$
$$- B_{16}(3\delta u_{,xxy} + \delta v_{,xxx}) - B_{26}(\delta u_{,yyy} + 3\delta v_{,xyy}) + \bar{N}_x\delta w_{,xx} = 0$$
$$(5.85)$$

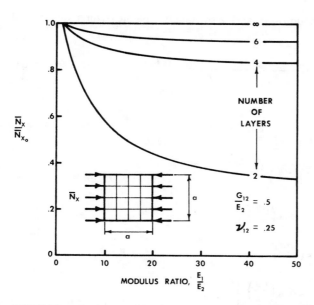

FIG. 5-15.   Relative uniaxial buckling loads of square antisymmetric cross-ply laminated plates. (*After Jones, Ref. 5-15.*)

Whitney (Refs. 5-9 and 5-10) chose to solve the problem for simply supported edge boundary condition $S3$ (note that this boundary condition differs significantly from the $S2$ condition used for antisymmetric cross-ply laminates):

$x = 0, a$:

$$\delta w = 0 \quad \delta M_x = B_{16}(\delta v,_x + \delta u,_y) - D_{11}\delta w,_{xx} - D_{12}\delta w,_{yy} = 0 \quad (5.86)$$

$$\delta u = 0 \quad \delta N_{xy} = A_{66}(\delta v,_x + \delta u,_y) - B_{16}\delta w,_{xx} - B_{26}\delta w,_{yy} = 0 \quad (5.87)$$

$y = 0, b$:

$$\delta w = 0 \quad \delta M_y = B_{26}(\delta v,_x + \delta u,_y) - D_{12}\delta w,_{xx} - D_{22}\delta w,_{yy} = 0 \quad (5.88)$$

$$\delta v = 0 \quad \delta N_{xy} = A_{66}(\delta v,_x + \delta u,_y) - B_{16}\delta w,_{xx} - B_{26}\delta w,_{yy} = 0 \quad (5.89)$$

He then observed that the variations in displacement

$$\delta u = \bar{u} \sin \frac{m\pi x}{a} \cos \frac{n\pi y}{b}$$

$$\delta v = \bar{v} \cos \frac{m\pi x}{a} \sin \frac{n\pi y}{b} \quad (5.90)$$

$$\delta w = \bar{w} \sin \frac{m\pi x}{a} \sin \frac{n\pi y}{b}$$

satisfy the boundary conditions and also satisfy the governing differential equations exactly if

$$\bar{N}_x = \left(\frac{a}{m\pi}\right)^2 \left(T_{33} + \frac{2T_{12}T_{23}T_{13} - T_{22}T_{13}^2 - T_{11}T_{23}^2}{T_{11}T_{22} - T_{12}^2}\right) \quad (5.91)$$

where

$$T_{11} = A_{11}\left(\frac{m\pi}{a}\right)^2 + A_{66}\left(\frac{n\pi}{b}\right)^2$$

$$T_{12} = (A_{12} + A_{66})\left(\frac{m\pi}{a}\right)\left(\frac{n\pi}{b}\right)$$

$$T_{13} = -\left[3B_{16}\left(\frac{m\pi}{a}\right)^2 + B_{26}\left(\frac{n\pi}{b}\right)^2\right]\left(\frac{n\pi}{b}\right)$$

$$T_{22} = A_{22}\left(\frac{n\pi}{b}\right)^2 + A_{66}\left(\frac{m\pi}{a}\right)^2 \quad (5.92)$$

$$T_{23} = -\left[B_{16}\left(\frac{m\pi}{a}\right)^2 + 3B_{26}\left(\frac{n\pi}{b}\right)^2\right]\left(\frac{m\pi}{a}\right)$$

$$T_{33} = D_{11}\left(\frac{m\pi}{a}\right)^4 + 2(D_{12} + 2D_{66})\left(\frac{m\pi}{a}\right)^2\left(\frac{n\pi}{b}\right)^2 + D_{22}\left(\frac{n\pi}{b}\right)^4$$

Note that if B_{16} and B_{26} are zero, then T_{13} and T_{23} are also zero so Eq. (5.91) reduces to the specially orthotropic solution, Eq. (5.65). The character of Eq. (5.91) is the same as that of Eq. (5.81) for antisymmetric cross-ply laminates so the remarks in Sec. 5.4.3 are equally applicable here.

Numerical results for graphite/epoxy composites with $E_1/E_2 = 40$, $G_{12}/E_2 = .5$, and $\nu_{12} = .25$ in square plates are shown in Fig. 5-16. The influence of bending-extension coupling is to reduce the buckling load for two-layered plates from the many layered result (which is the specially orthotropic solution of Sec. 5.4.1). At $45°$, the reduction is by about a factor of 2/3; perhaps more significant is the fact that use of the specially orthotropic approximation leads to a predicted buckling load that is three times the actual buckling load! Obviously, the specially orthotropic approximation is highly unconservative for antisymmetric angle-ply laminates with less than six layers. For six layers, the error in the buckling load is about 7 percent. Thus, the bending-extension coupling effect dies out rapidly as the number of layers increases. This conclusion is valid for other materials as represented in Fig. 5-17 in the manner of Sec. 5.4.3.

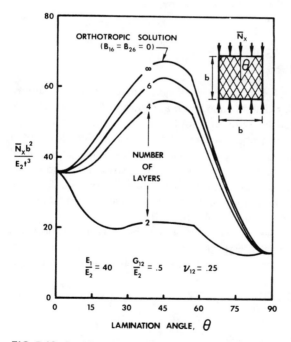

FIG. 5-16. Buckling loads for square antisymmetric angle-ply laminated plates under uniform uniaxial compression, $\bar{N}_x$. (After Jones, Morgan, and Whitney, Ref. 5-17.)

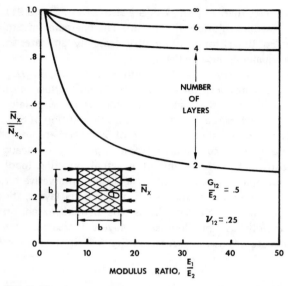

FIG. 5-17.   Relative uniaxial buckling loads of square antisymmetric angle-ply laminated plates.

5.5 VIBRATION OF SIMPLY SUPPORTED LAMINATED PLATES

Consider the general class of laminated rectangular plates that are simply supported along edges $x = 0, x = a, y = 0$, and $y = b$ as shown in Fig. 5-18. The nature of the free vibrations of such a structural configuration about an equilibrium state will be addressed in this section according to the governing differential equations and boundary conditions discussed in Sec. 5.2. Other more complicated boundary conditions and the effect of an equilibrium stress state could be considered. However, in consonance with the objectives of this book, those topics are left for further study. A more comprehensive treatment of laminated plate vibrations is provided by Ashton and Whitney (Ref. 5-1).

The free vibration frequencies and mode shapes will be determined for plates with various laminations: specially orthotropic, symmetric angle-ply, antisymmetric cross-ply, and antisymmetric angle-ply. The results for the different types

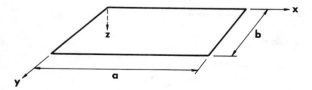

FIG. 5-18. Simply supported laminated rectangular plate.

of lamination will be compared to determine the influence of twist coupling and bending-extension coupling on the vibration behavior. As with the deflection problems in Sec. 5.3 and the buckling problems in Sec. 5.4, different simply supported edge boundary conditions will be used in the several problems presented.

5.5.1 Specially Orthotropic Laminates

A specially orthotropic laminate has either a single layer of a specially orthotropic material or multiple specially orthotropic layers that are symmetrically arranged about the laminate middle surface. In both cases, the laminate stiffnesses consist solely of $A_{11}, A_{12}, A_{22}, A_{66}, D_{11}, D_{12}, D_{22},$ and D_{66}. That is, neither shear or twist coupling nor bending-extension coupling exists. Then, for plate problems, the vibration frequencies and modes are described by a single vibration differential equation:

$$D_{11} \delta w_{,xxxx} + 2(D_{12} + 2D_{66}) \delta w_{,xxyy} + D_{22} \delta w_{,yyyy} + \rho \delta w_{,tt} = 0$$

$$(5.93)$$

subject to the simply supported edge boundary conditions

$$x = 0, a: \quad \delta w = 0 \quad \delta M_x = -D_{11} \delta w_{,xx} - D_{12} \delta w_{,yy} = 0$$

$$y = 0, b: \quad \delta w = 0 \quad \delta M_y = -D_{12} \delta w_{,xx} - D_{22} \delta w_{,yy} = 0 \qquad (5.94)$$

The free vibration of an elastic continuum is harmonic in time so Ashton and Whitney (Ref. 5-1) chose a harmonic solution

$$\delta w(x, y, t) = (A \cos \omega t + B \sin \omega t) \delta w(x, y) \tag{5.95}$$

and observed that the problem has now been separated into time and spatial variations according to Eq. (5.95). The resulting differential equation and boundary conditions are satisfied by

$$\delta w(x, y) = \sin \frac{m\pi x}{a} \sin \frac{n\pi y}{b} \tag{5.96}$$

if

$$\omega^2 = \frac{\pi^4}{\rho} \left[D_{11} \left(\frac{m}{a}\right)^4 + 2(D_{12} + 2D_{66}) \left(\frac{m}{a}\right)^2 \left(\frac{n}{b}\right)^2 + D_{22} \left(\frac{n}{b}\right)^4 \right] \tag{5.97}$$

where the various frequencies, ω, correspond to different mode shapes (different values of m and n in Eq. (5.96) so accordingly different shapes of w). The fundamental frequency (lowest frequency) is obviously obtained when m and n are both one.

For a specially orthotropic square plate with $D_{11}/D_{22} = 10$ and $(D_{12} + 2D_{66}) = 1$, the four lowest frequencies are displayed in Table 5-3 along with the

TABLE 5-3. Normalized vibration frequencies for specially orthotropic and isotropic simply supported square plates

Mode	Specially orthotropic			Isotropic		
	m	n	k	m	n	k
1st	1	1	3.60555	1	1	2
2nd	1	2	5.83095	1	2	5
3rd	1	3	10.44031	2	1	5
4th	2	1	13	2	2	8

four lowest frequencies of an isotropic plate. In Table 5-3, the factor k is defined by

$$\omega = \frac{k\pi^2}{b^2} \sqrt{D_{22}/\rho} \tag{5.98}$$

where $D_{22} = D$ for an isotropic plate. The corresponding mode shapes are shown in Fig. 5-19 where the node lines (lines of zero deflection with time) are indicated by dashed lines. The significant observation is that the specially orthotropic plate has a different set of four lowest frequencies than does the

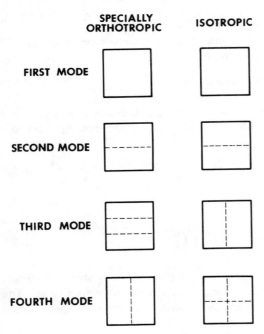

FIG. 5-19. Vibration mode shapes for simply supported square specially orthotropic and isotropic plates.

isotropic plate. That is, a directional preference is exhibited by the specially orthotropic plate as evidenced by the $m = 1$, $n = 3$ mode having a lower frequency than the $m = 2$, $n = 1$ mode. On the other hand, the isotropic plate has the same frequency for both the $m = 2$, $n = 1$ mode and the $m = 1$, $n = 2$ mode.

5.5.2 Symmetric Angle-Ply Laminates

Symmetric angle-ply laminates were found in Sec. 4.3.2 to be characterized by a full matrix of extensional stiffnesses as well as bending stiffnesses, but have no coupling between bending and extension. The principal difference between these laminates and specially orthotropic laminates is the introduction here of the twist coupling stiffnesses D_{16} and D_{26} (the shear coupling stiffnesses A_{16} and A_{26} are immaterial for vibration of a symmetrically laminated plate since the governing differential equations are uncoupled). Accordingly, the governing vibration differential equation is

$$D_{11}\delta w,_{xxxx} + 4D_{16}\delta w,_{xxxy} + 2(D_{12} + 2D_{66})\delta w,_{xxyy}$$
$$+ 4D_{26}\delta w,_{xyyy} + D_{22}\delta w,_{yyyy} + \rho\delta w,_{tt} = 0 \quad (5.99)$$

subject to the simply supported edge boundary conditions at all time

$$x = 0, a: \quad \delta w = 0 \quad \delta M_x = -D_{11}\delta w,_{xx} - D_{12}\delta w,_{yy} - 2D_{16}\delta w,_{xy} = 0$$
$$y = 0, b: \quad \delta w = 0 \quad \delta M_y = -D_{12}\delta w,_{xx} - D_{22}\delta w,_{yy} - 2D_{26}\delta w,_{xy} = 0$$

$$(5.100)$$

The presence of D_{16} and D_{26} in the governing differential equation and the boundary conditions renders a closed form solution impossible. That is, in analogy to both bending and buckling of a symmetric angle-ply (or anisotropic) plate, the variation in lateral displacement, δw cannot be separated into a function of x alone times a function of y alone. Again, however, the Rayleigh-Ritz approach is quite useful. The expression

$$\delta w = \sum_{m=1}^{\infty} \sum_{n=1}^{\infty} A_{mn} \sin\frac{m\pi x}{a} \sin\frac{n\pi y}{b} \quad (5.101)$$

satisfies the geometric boundary conditions ($w = 0$ on all edges), but not the natural boundary conditions ($M_n = 0$ on all edges) or the governing differential equation. Therefore, the use of Eq. (5.101) in the appropriate energy expression may result in rather slow convergence toward the exact solution. No numerical results are presently available for this laminate class.

5.5.3 Antisymmetric Cross-Ply Laminates

Antisymmetric cross-ply laminates were found in Sec. 4.3.3 to have extensional stiffnesses A_{11} A_{12}, $A_{22} = A_{11}$, and A_{66}, bending-extension coupling stiffnesses B_{11} and $B_{22} = -B_{11}$, and bending stiffnesses $D_{11}, D_{12}, D_{22} = D_{11}$, and D_{66}. The new terms here in comparison to a specially orthotropic laminate are B_{11} and B_{22}. Because of this bending-extension coupling, the vibration differential equations are coupled:

$$A_{11}\delta u_{,xx} + A_{66}\delta u_{,yy} + (A_{12} + A_{66})\delta v_{,xy} - B_{11}\delta w_{,xxx} = 0 \tag{5.102}$$

$$(A_{12} + A_{66})\delta u_{,xy} + A_{66}\delta v_{,xx} + A_{11}\delta v_{,yy} + B_{11}\delta w_{,yyy} = 0 \tag{5.103}$$

$$D_{11}(\delta w_{,xxxx} + \delta w_{,yyyy}) + 2(D_{12} + 2D_{66})\delta w_{,xxyy}$$
$$- B_{11}(\delta u_{,xxx} - \delta v_{,yyy}) + \rho\delta w_{,tt} = 0 \tag{5.104}$$

Whitney (Refs. 5-8 and 5-9) observed that the variations in displacement

$$\delta u(x,y,t) = \bar{u} \cos\frac{m\pi x}{a} \sin\frac{n\pi y}{b} e^{i\omega t}$$

$$\delta v(x,y,t) = \bar{v} \sin\frac{m\pi x}{a} \cos\frac{n\pi y}{b} e^{i\omega t} \tag{5.105}$$

$$\delta w(x,y,t) = \bar{w} \sin\frac{m\pi x}{a} \sin\frac{n\pi y}{b} e^{i\omega t}$$

satisfy simply supported edge boundary condition $S2$ at all time

$$x = 0, a \begin{cases} \delta w = 0 \quad \delta M_x = B_{11}\delta u_{,x} - D_{11}\delta w_{,xx} - D_{12}\delta w_{,yy} = 0 \\ \delta v = 0 \quad \delta N_x = A_{11}\delta u_{,x} + A_{12}\delta v_{,y} - B_{11}\delta w_{,xx} = 0 \end{cases} \tag{5.106}$$

$$y = 0, b \begin{cases} \delta w = 0 \quad \delta M_y = -B_{11}\delta v_{,y} - D_{12}\delta w_{,xx} - D_{11}\delta w_{,yy} = 0 \\ \delta u = 0 \quad \delta N_y = A_{12}\delta u_{,x} + A_{11}\delta v_{,y} + B_{11}\delta w_{,yy} = 0 \end{cases} \tag{5.107}$$

and the governing differential equations if

$$\omega^2 = \frac{\pi^4}{\rho}\left(T_{33} + \frac{2T_{12}T_{23}T_{13} - T_{22}T_{13}^2 - T_{11}T_{23}^2}{T_{11}T_{22} - T_{12}^2}\right) \tag{5.108}$$

where

$$T_{11} = A_{11}\left(\frac{m}{a}\right)^2 + A_{66}\left(\frac{n}{b}\right)^2$$

$$T_{12} = (A_{12} + A_{66})\left(\frac{m}{a}\right)\left(\frac{n}{b}\right)$$

$$T_{13} = -B_{11}\left(\frac{m}{a}\right)^3 \tag{5.109}$$

$$T_{22} = A_{11}\left(\frac{n}{b}\right)^2 + A_{66}\left(\frac{m}{a}\right)^2$$

$$T_{23} = B_{11}\left(\frac{n}{b}\right)^3$$

$$T_{33} = D_{11}\left[\left(\frac{m}{a}\right)^4 + \left(\frac{n}{b}\right)^4\right] + 2(D_{12} + 2D_{66})\left(\frac{m}{a}\right)^2\left(\frac{n}{b}\right)^2$$

(5.109)
(cont'd.)

Note that if $B_{11} = 0$, then T_{13} and T_{23} are also zero so Eq. (5.108) reduces to the specially orthotropic solution, Eq. (5.97), if $D_{11} = D_{22}$. Because T_{11}, T_{12}, and T_{22} are functions of both m and n and appear in the denominator of Eq. (5.108), no simple conclusion can be drawn about the values of m and n for the lowest frequency. Instead, Eq. (5.108) must be treated as a function of the discrete variables m and n and minimized accordingly. As a matter of fact, for the results presented in the numerical example, the fundamental frequency does correspond to both m and n equal one. Caution is urged against a general conclusion.

Numerical results from Eq. (5.108) are presented in Fig. 5-20 for graphite/epoxy composites with $E_1/E_2 = 40$, $G_{12}/E_2 = .5$, and $\nu_{12} = .25$. The effect of coupling between bending and extension is to lower the vibration frequencies. For example, the fundamental frequency of a square plate is reduced by about 40 percent from the specially orthotropic solution to the exact solution for a two-layered plate. More importantly, the specially orthotropic approximation is too high by 60 percent! As the number of layers increases, the effect of coupling

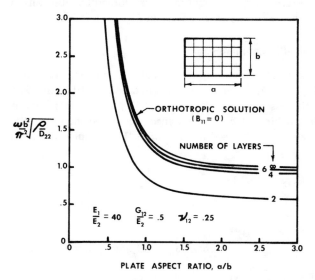

FIG. 5-20. Fundamental vibration frequencies for rectangular antisymmetric cross-ply laminated plates. (After Jones, Ref. 5-15.)

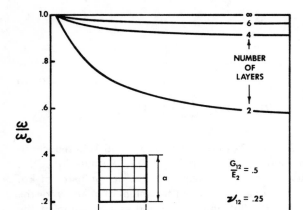

FIG. 5-21. Relative fundamental natural frequencies of square antisymmetric cross-ply laminated plates. (*After Jones, Ref. 5-15.*)

decreases. For example, a six-layered plate has a fundamental frequency only 5 percent smaller than the specially orthotropic approximation. Nevertheless, it is readily apparent that coupling between bending and extension must be considered in laminated plate vibrations. This conclusion is reinforced by observation of the vibration results for other composite materials in Fig. 5-21.

5.5.4 Antisymmetric Angle-Ply Laminates

Antisymmetric angle-ply laminates were found in Sec. 4.3.3 to have extensional stiffnesses A_{11}, A_{12}, A_{22}, and A_{66}, bending-extension coupling stiffnesses B_{16} and B_{26}, and bending stiffnesses D_{11}, D_{12}, D_{22}, and D_{66}. Thus, this type of laminate exhibits a different kind of bending-extension coupling than does the antisymmetric cross-ply laminate discussed in Sec. 5.5.3. The coupled vibration differential equations are

$$A_{11}\delta u_{,xx} + A_{66}\delta u_{,yy} + (A_{12} + A_{66})\delta v_{,xy}$$
$$- 3B_{16}\delta w_{,xxy} - B_{26}\delta w_{,yyy} = 0 \quad (5.110)$$

$$(A_{12} + A_{66})\delta u_{,xy} + A_{66}\delta v_{,xx} + A_{22}\delta v_{,yy}$$
$$- B_{16}\delta w_{,xxx} - 3B_{26}\delta w_{,xyy} = 0 \quad (5.111)$$

$$D_{11}\delta w_{,xxxx} + 2(D_{12} + 2D_{66})\delta w_{,xxyy} + D_{22}\delta w_{,yyyy}$$
$$- B_{16}(3\delta u_{,xxy} + \delta v_{,xxx}) - B_{26}(\delta u_{,yyy} + 3\delta v_{,xyy}) + \rho\delta w_{,tt} = 0$$
$$(5.112)$$

Whitney (Refs., 5-8 and 5-9) observed that the variations in displacement

$$\delta u(x, y, t) = \bar{u} \sin \frac{m\pi x}{a} \cos \frac{n\pi y}{b} e^{i\omega t}$$

$$\delta v(x, y, t) = \bar{v} \cos \frac{m\pi x}{a} \sin \frac{n\pi y}{b} e^{i\omega t} \qquad (5.113)$$

$$\delta w(x, y, t) = \bar{w} \sin \frac{m\pi x}{a} \sin \frac{n\pi y}{b} e^{i\omega t}$$

satisfy the simply supported edge boundary condition $S3$ at all time

$$x = 0, a \begin{cases} \delta w = 0 & \delta M_x = B_{16}(\delta v_{,x} + \delta u_{,y}) - D_{11}\delta w_{,xx} - D_{12}\delta w_{,yy} = 0 \\ \delta u = 0 & \delta N_{xy} = A_{66}(\delta v_{,x} + \delta u_{,y}) - B_{16}\delta w_{,xx} - B_{26}\delta w_{,yy} = 0 \end{cases}$$

$$(5.114)$$

$$y = 0, b \begin{cases} \delta w = 0 & \delta M_y = B_{26}(\delta v_{,x} + \delta u_{,y}) - D_{12}\delta w_{,xx} - D_{22}\delta w_{,yy} = 0 \\ \delta v = 0 & \delta N_{xy} = A_{66}(\delta v_{,x} + \delta u_{,y}) - B_{16}\delta w_{,xx} - B_{26}\delta w_{,yy} = 0 \end{cases}$$

$$(5.115)$$

and the governing differential equations if

$$\omega^2 = \frac{\pi^4}{\rho} \left(T_{33} + \frac{2T_{12}T_{23}T_{13} - T_{22}T_{13}^2 - T_{11}T_{23}^2}{T_{11}T_{22} - T_{12}^2} \right) \qquad (5.116)$$

where

$$T_{11} = A_{11}\left(\frac{m}{a}\right)^2 + A_{66}\left(\frac{n}{b}\right)^2$$

$$T_{12} = (A_{12} + A_{66})\left(\frac{m}{a}\right)\left(\frac{n}{b}\right)$$

$$T_{13} = -\left[3B_{16}\left(\frac{m}{a}\right)^2 + B_{26}\left(\frac{n}{b}\right)^2\right]\left(\frac{n}{b}\right)$$

$$T_{22} = A_{22}\left(\frac{n}{b}\right)^2 + A_{66}\left(\frac{m}{a}\right)^2 \qquad (5.117)$$

$$T_{23} = -\left[B_{16}\left(\frac{m}{a}\right)^2 + 3B_{26}\left(\frac{n}{b}\right)^2\right]\left(\frac{m}{a}\right)$$

$$T_{33} = D_{11}\left(\frac{m}{a}\right)^4 + 2(D_{12} + 2D_{66})\left(\frac{m}{a}\right)^2\left(\frac{n}{b}\right)^2 + D_{22}\left(\frac{n}{a}\right)^4$$

Note that if B_{16} and B_{26} are zero, then T_{13} and T_{23} are also zero so Eq. (5.116) reduces to the specially orthotropic solution, Eq. (5.65). The character of Eq. (5.116) is the same as that of Eqs. (5.81), (5.91), and (5.108) so the remarks in Sec. 5.4.3 are equally valid here.

Numerical results for graphite/epoxy composites with $E_1/E_2 = 40$, G_{12}/E_2 = .5, and $\nu_{12} = .25$ are given as a function of lamination angle in Fig. 5-22. As with B_{11} in Sec. 5.5.3, the effect of the coupling stiffnesses B_{16} and B_{26} is to lower the fundamental vibration frequencies. For example, the fundamental frequency of a square plate with $\theta = 45°$ and two layers is about 40 percent less than the specially orthotropic solution which is valid when the number of layers is infinite. Put another way, the specially orthotropic solution is too high by about 80 percent! Again, as the number of layers increases, the coupling between bending and extension decreases. For a six-layered plate the difference between the specially orthotropic solution and the exact solution is about 4 percent. Obviously, coupling between bending and extension can be quite important for unsymmetrically laminated plates. This conclusion is unchanged when other composite materials are considered as in Fig. 5-23.

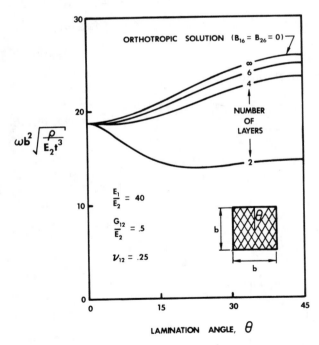

FIG. 5-22. Fundamental vibration frequencies for square antisymmetric angle-ply laminated plates. (*After Jones, Morgan, and Whitney, Ref. 5-17.*)

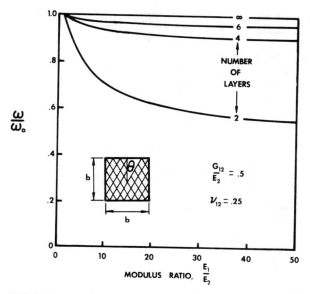

FIG. 5-23. Relative fundamental natural frequencies of square antisymmetric angle-ply laminated plates.

5.6 SUMMARY REMARKS ON EFFECTS OF STIFFNESSES

The presence of coupling between bending and extension in a laminate generally increases deflections. Hence, coupling decreases the effective stiffnesses of a laminate. At the same time, this coupling reduces buckling loads and vibration frequencies significantly.

Similarly, for laminates with twist-curvature coupling, the deflections are increased, the buckling loads decreased, and the vibration frequencies decreased. In both cases of bending-extension coupling and twist-curvature coupling, the effect on deflections, buckling loads, and vibration frequencies for a fixed-thickness antisymmetric or symmetric laminate, respectively, dies out rapidly as the number of layers increases. For more general laminates, specific investigation is required.

A more general class of laminates, unsymmetric cross-ply laminates, was discussed by Jones (Ref. 5-15). All geometric and material property symmetry requirements of the preceding sections are relaxed. Still, the restriction to cross-ply laminates (of arbitrary layer thickness and $0°$ and $90°$ stacking sequence) enables a simple exact solution to be obtained. However, because of the infinite complexity of this class of laminates, general results are impossible. Instead, consider the cross sections of the contrived, but representative unsymmetric laminate example in Fig. 5-24. There, the fibers in the second layer from the bottom are oriented at $90°$ and the fibers in all other layers are oriented at

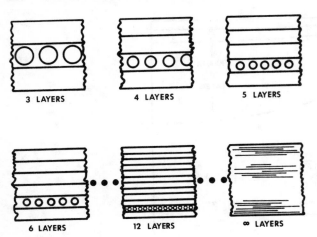

3 LAYERS   **4 LAYERS**   **5 LAYERS**

6 LAYERS   **12 LAYERS**   **∞ LAYERS**

FIG. 5-24. Unsymmetric laminate example. (*After Jones, Ref. 5-15.*)

$0°$ to the plate x axis. Thus, for a constant thickness laminate, the $90°$ layer gets thinner and moves toward the bottom of the laminate as the number of layers increases. This example is probably never encountered in engineering practice, but it is a simple, straightforward example of unsymmetric laminates that is amenable to comprehensive parametric study.

Normalized buckling loads for graphite/epoxy unsymmetric cross-ply laminated rectangular plates are shown in Fig. 5-25. The plate aspect ratio is 2, a value for which the results are the most strikingly different from baseline results.

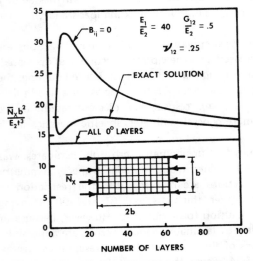

FIG. 5-25. Uniaxial buckling loads of graphite/epoxy unsymmetric cross-ply laminated rectangular plates. (*After Jones, Ref. 5-15.*)

One of the baseline comparison values is the buckling load for a laminate with all $0°$ layers. The other comparison case is an unsymmetric laminate for which coupling between bending and extension is ignored ($B_{ij} = 0$). Results for the actual unsymmetric laminate, for which coupling is considered, are labeled exact solution. The actual laminate has, not surprisingly, less buckling resistance than a laminate with $B_{ij} = 0$. On the other hand, the actual laminate has more buckling resistance than a laminate with all $0°$ layers. The reason for this curious result is related to the relative values of the stiffnesses in the x- and y-directions as well as to the plate aspect ratio. As will be seen subsequently, D_{11} is decreased somewhat by the presence of a $90°$ layer, but D_{22} is increased by factors of up to an order of magnitude. The influence of D_{22} is most easily examined for a three-layered plate which is, of course, symmetric. Thus, the exact solution corresponds to the orthotropic solution ($B_{ij} = 0$); for example, see the horizontal hash mark at three layers in Fig. 5-25 where the two curves coincide. At that point, the orthotropic solution for a plate with an aspect ratio of 2 which buckles into the $m = 1, n = 1$ mode is

$$\bar{N}_x L^2 = \pi^2 \left[\frac{D_{11}}{4} + 2(D_{12} + 2D_{66}) + 4D_{22} \right]$$

(5.118)

which is readily obtained from Eq. (5.81). The term in Eq. (5.118) involving D_{11} decreases by less than 4 percent when the actual laminate is considered as opposed to the all $0°$ layer laminate. The net result is a normalized buckling load that is 33 percent bigger for the actual laminate than for the all $0°$ layer laminate.

The differences between the exact, orthotropic, and all $0°$ layer predictions in Fig. 5-25 range from 48 percent less than the orthotropic solution at 6 layers to 18 percent less at 40 layers to 6 percent less at 100 layers. In addition, the exact results range from about 30 percent more than the all $0°$ layer laminate results at 30 layers to about 18 percent more at 100 layers. Such differences are well within the consideration of usual engineering design practice. However, the differences exist for many more layers than coupling between bending and extension was believed to be important. That belief was established on the basis of extrapolating antisymmetric cross-ply results. Obviously, such extrapolation is invalid.

As an aid to understanding the buckling behavior in Fig. 5-25, the extensional, coupling, and bending stiffnesses are plotted versus the number of layers in Fig. 5-26. The stiffnesses in the x-direction (with which most fibers are aligned), A_{11} and D_{11}, are nearly independent of the number of layers. However, the $90°$ layer causes the stiffnesses in the y-direction, A_{22} and D_{22}, to deviate by up to an order of magnitude from the values for all $0°$ layers (which are the same as for an infinite number of layers). These discrepancies die out very slowly as the number of layers increases. Moreover, the normalized stiffness for coupling between bending and extension, which can be shown to be

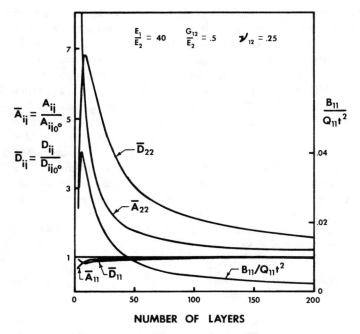

FIG. 5-26. Normalized stiffnesses of example unsymmetric cross-ply graphite/epoxy laminate. (*After Jones, Ref. 5-15.*)

$$\frac{B_{11}}{Q_{11}t^2} = \left[\frac{1}{2(NL)^2}\right]\left(1 - \frac{E_2}{E_1}\right)(NL - 3) \tag{5.119}$$

(where NL is the number of layers) and which appears in Eq. (5.81) and thereby enables stiffnesses A_{22} and D_{22} to influence the buckling load, also dies out slowly. Do not attempt to compare the magnitudes of, for example, D_{22} and $B_{11}/(Q_{11}t^2)$ quantitatively. They do not have the same base for normalization (B_{11} cannot be normalized relative to its value for all $0°$ layers since there its value is zero). Thus, the effect of a single unsymmetrically placed $90°$ layer on the buckling load dies out very slowly as the number of layers increases.

The approximation of a general laminate by a specially orthotropic laminate can result in errors as big as a factor of 3. Thus, use of the approximation must be carefully proven to be justified for the case under consideration. Always remember that the specially orthotropic approximation yields unconservative results. Thus, the only general rule is that coupling should be included in every analysis of laminated plates unless coupling is proven to be insignificant.

The effects of the twist coupling stiffnesses on deflections, buckling loads, and vibration frequencies of anisotropic plates can be assessed more accurately than in this book by use of a procedure due to Whitney (Ref. 5-18). He uses a double Fourier series expansion of the unknown displacement to satisfy the

TABLE 5-4. Buckling of simply supported
square graphite/epoxy anistropic plates
under biaxial compression *†

Number of terms in series $m = n$	$\bar{N}b^2/Q_{22}\,t^3$	
	Whitney's Fourier analysis	Ashton's Rayleigh-Ritz analysis
1	6.763	21.438
3	8.115	13.013
5	8.318	11.565
7	8.418	11.060
9	8.481	—
11	8.521	—
13	8.556	—

*$Q_{11}/Q_{22} = 25$, $Q_{12}/Q_{22} = .25$, $Q_{66}/Q_{22} = .5$
principal material directions at 45° to plate sides.
†After Whitney, Ref. 5-18.

natural boundary conditions and thereby speed convergence to precise results. Note that the Rayleigh-Ritz method described here does not satisfy the natural boundary conditions and therefore convergence is very slow and sometimes not to the correct solution. As an example, consider a simply supported square graphite/epoxy plate ($Q_{11}/Q_{22} = 25$, $Q_{12}/Q_{22} = .25$, and $Q_{66}/Q_{22} = .5$) with principal material directions at 45° to the plate sides. When the plate is subjected to biaxial in-plane compression, $\bar{N}_x = \bar{N}_y$, the buckling loads are given in Table 5-4 for Whitney's approach and the Rayleigh-Ritz approach due to Ashton (Ref. 5-19). Obviously, Whitney's results converge rapidly although there is some oscillation. In contrast, the Rayleigh-Ritz results converge (?) slowly and to an incorrect value. Ashton (Ref. 5-20) shows that the rate of convergence of a Rayleigh-Ritz method for simply supported anisotropic plates is dependent on the orthotropy ratio, E_1/E_2, when the natural boundary conditions are not satisfied. Thus, his results in Table 5-4 are to be expected. The main point is that Whitney has provided a means to solve the problem accurately.

PROBLEM SET 5

Exercise 5.1 Derive Eqs. (5.6) through (5.8).
Exercise 5.2 Derive the analogy of Eqs. (5.6) to (5.8) for the buckling differential equations, Eqs. (5.13) to (5.15).
Exercise 5.3 Derive the analogy of Eqs. (5.6) to (5.8) for the vibration differential equations, Eqs. (5.22) to (5.24).
Exercise 5.4 Derive Eq. (5.30).
Exercise 5.5 Verify Eq. (5.31).
Exercise 5.6 Obtain the coefficients A_{mn}, B_{mn}, and C_{mn} in Eq. (5.49).
Exercise 5.7 Obtain the coefficients A_{mn}, B_{mn}, and C_{mn} in Eq. (5.58).
Exercise 5.8 Derive Eq. (5.65).

Exercise 5.9 Derive Eq. (5.81). Hint: for a set of homogeneous equations to have a nontrivial solution, the determinant of the coefficients must be zero.

Exercise 5.10 Derive Eq. (5.91). See hint in Ex. 5.9.

Exercise 5.11 Derive Eq. (5.97).

Exercise 5.12 Derive Eq. (5.108). See hint in Ex. 5.9.

Exercise 5.13 Derive Eq. (5.116). See hint in Ex. 5.9.

REFERENCES

5-1   Ashton, J. E., and J. M. Whitney: "Theory of Laminated Plates," Technomic Publishing Co., Westport, Conn., 1970.

5-2   Timoshenko, S. P., and S. Woinowsky-Krieger: "Theory of Plates and Shells," McGraw-Hill Book Company, New York, 1959.

5-3   Almroth, B. O.: Influence of Edge Conditions on the Stability of Axially Compressed Cylindrical Shells, *AIAA Journal*, January, 1966, pp. 134-140.

5-4   Wang, James Ting-shun: On the Solution of Plates of Composite Materials, *J. Composite Materials*, July, 1969, pp. 590-592.

5-5   Ashton, J. E.: Anisotropic Plate Analysis, *General Dynamics Research and Engineering Report*, FZM-4899, 12 October 1967.

5-6   Langhaar, Henry L.: "Energy Methods in Applied Mechanics," John Wiley, New York, 1962.

5-7   Ashton, J. E.: An Analogy for Certain Anisotropic Plates, *J. Composite Materials*, April, 1969, pp. 355-358.

5-8   Whitney, J. M., and A. W. Leissa: Analysis of Heterogeneous Anisotropic Plates, *J. Appl. Mech.*, June, 1969, pp. 261-266.

5-9   Whitney, James Martin: "A Study of the Effects of Coupling Between Bending and Stretching on the Mechanical Behavior of Layered Anisotropic Composite Materials," Ph.D. thesis, Department of Engineering Mechanics, The Ohio State University, Columbus, Ohio, 1968. (Available from University Microfilms, Inc., Ann Arbor, Michigan, as no. 69-5000.)

5-10   Whitney, J. M.: Bending-Extension Coupling in Laminated Plates Under Transverse Loading, *J. Composite Materials*, January, 1969, pp. 20-28.

5-11   Timoshenko, Stephen P., and James M. Gere: "Theory of Elastic Stability," McGraw-Hill Book Company, New York, 1961.

5-12   Ashton, J. E., and M. E. Waddoups: Analysis of Anisotropic Plates, *J. Composite Materials*, January, 1969, pp. 148-165.

5-13   Chamis, C. C.: Thermostructural Response, Structural and Material Optimization of Particulate Composite Plates, *Case Western Reserve Univ. Rep.* No. SMSMDD 21, 1968.

5-14   Mandell, J. F.: Experimental Investigation of the Buckling of Anisotropic Fiber Reinforced Plastic Plates, *Air Force Materials Laboratory Tech. Rep.* AFML-TR-68-281, October, 1968.

5-15   Jones, Robert M.: Buckling and Vibration of Rectangular Unsymmetrically Laminated Cross-Ply Plates, *AIAA Journal*, December, 1973, pp. 1626-1632.

5-16   Jones, Robert M.: Plastic Buckling of Eccentrically Stiffened Multilayered Circular Cylindrical Shells, *AIAA Journal*, February, 1970, pp. 262-270.

5-17   Jones, Robert M., Harold S. Morgan, and James M. Whitney: Buckling and Vibration of Antisymmetrically Laminated Angle-Ply Rectangular Plates, *J. Appl. Mech.*, December, 1973, pp. 1143-1144.

5-18   Whitney, J. M.: On the Analysis of Anisotropic Rectangular Plates, *Air Force Materials Laboratory Tech. Rep.* AFML-TR-72-76, August, 1972.

5-19 Ashton, J. E.: Clamped Skew Plates of Orthotropic Material Under Transverse Load, in "Developments in Mechanics," vol. 5, The Iowa State University Press, Ames, Iowa, 1969, pp. 297-306.

5-20 Ashton, J. E.: Anisotropic Plate Analysis-Boundary Conditions, *J. Composite Materials,* April, 1970, pp. 162–171.

Chapter 6
SUMMARY AND
OTHER TOPICS

6.1 REVIEW OF CHAPTERS 1 THROUGH 5

The basic nature of composite materials was introduced in Chap. 1. An overall classification scheme was presented, and the mechanical behavior aspects of composite materials that differ from those of conventional materials were described in a qualitative fashion. The book was then restricted to laminated fiber-reinforced composite materials. The basic definitions and how such materials are made were then treated. Finally, the current and potential advantages of composite materials were discussed.

The macromechanical behavior of a lamina was quantitatively described in Chap. 2. The basic three-dimensional stress-strain relations for elastic anisotropic and orthotropic materials were examined. Subsequently, those relations were specialized for the plane stress state normally found in a lamina. The plane stress relations were then transformed in the plane of the lamina to enable treatment of composite laminates with laminae at various angles. The various strengths of a lamina were discussed and subsequently used in biaxial strength theories to predict the off-axis strength of a lamina.

The micromechanical behavior of a lamina was treated in Chap. 3. Both a mechanics of materials and an elasticity approach were used to predict the stiffnesses of a lamina. Mechanics of materials approaches were used to predict some of the strengths of a lamina.

The basic building block, a lamina, was put together to form a laminate in Chap. 4. The behavioral restrictions were covered in the section on classical lamination theory. Special cases of laminates were discussed to learn about laminate characteristics and behavior. Predicted and experimental laminate stiffnesses were favorably compared. Next, the strength of laminates was discussed and found to be reasonably predictable. Interlaminar stresses were analyzed because of their apparent strong influence on laminate strength. Finally, in laminate design, the topics of invariant stiffnesses and joints were briefly addressed.

The use of composite laminate characteristics in analysis of bending, buckling, and vibration of plates was examined in Chap. 5. The governing differential equations were introduced. Then, each of the basic structural problems was analyzed for orthotropic, anisotropic, antisymmetric cross-ply, and antisymmetric angle-ply laminated, simply supported, rectangular plates. Thus, the effects of anisotropy or orthotropy and coupling between bending and extension were evaluated.

Obviously, the foregoing description of problems in the mechanics of composite materials is incomplete. Some topics do not fit well within the logical framework just described. Other topics are too advanced for an introductory book, even at the graduate level. The rest of this chapter is devoted to a brief discussion of some basic lamina and laminate characteristics not included in preceding chapters for one reason or another.

6.2 FATIGUE

Fatigue of a structural element may be a significant design parameter. The aircraft crashes due to fatigue failures are well known. Thus, the obvious question is how do composite materials fatigue characteristics stack up against those of conventional metals? The answer is, in brief, much better! The material or internal damping in composite materials is high, yet the fatigue characteristics are good. One of the main reasons for this fortunate circumstance is schematically depicted in Fig. 6-1 from Ref. 6-1. There, the initial imperfections in composite materials such as broken fibers, delamination, matrix cracking, fiber debonding, voids, etc., can be much larger than corresponding imperfections in conventional metals such as cracks. However, the growth of damage in a metal is typically much more abrupt as evidenced by Fig. 6-1 and hence potentially more dangerous than in a composite material. Accordingly, the typical *S-N* (stress versus number of cycles) curves shown in Fig. 6-2 from Ref. 6-2 result. The boron/epoxy composite curve is much flatter than the aluminum curve as well as

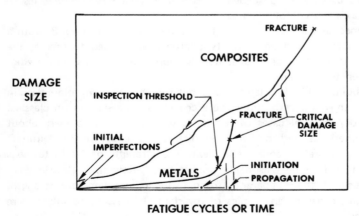

FIG. 6-1. Fatigue damage behavior of composites and metals.
(*After Salkind, Ref. 6-1.*)

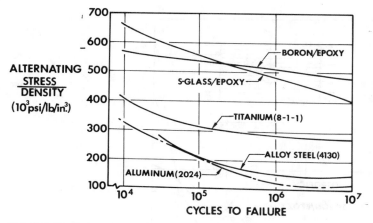

FIG. 6-2. Typical tension-tension fatigue data. (*After Pinckney, Ref. 6-2.*)

being flatter than the curves for any of the metals shown. The susceptibility of composite materials to effects of stress concentrations such as those caused by notches, holes, etc., is much less than for metals. Thus, the initial advantage of higher strength of boron/epoxy over aluminum in Fig. 6-2 actually *increases* under fatigue conditions. This advantage of increased life as well as increased specific strength and stiffness over conventional metals is one of the principal reasons for the rapidly expanding use of composite materials. Salkind (Ref. 6-3) reviewed fatigue characterisitcs of composite materials. Eisenmann, Kaminski, Reed, and Wilkins (Ref. 6-4) recently used fatigue characteristics as a base for a composite materials reliability procedure.

6.3 HOLES IN LAMINATES

Laminates, as any structure, have holes to serve various purposes. An obvious purpose is to accommodate a bolt. Another reason is to provide access from one side of the laminate to the other. The solution for the stresses around holes is quite difficult.

One of the first solutions to the problem of stresses around an elliptical hole in an infinite anisotropic plate was given by Lekhnitskii (Ref. 6-5). A more recent and comprehensive summary of the problem and many others is Savin's monograph (Ref. 6-6). Numerous results by Lekhnitskii are shown in his books (Refs. 6-7 and 6-8). Two special cases are of particular interest.

First, an orthotropic plate has stress applied in the principal material direction as in Fig. 6-3. There, Greszczuk (Ref. 6-9) plotted the circumferential stress around the hole for an isotropic material and several unidirectional composite materials. Observe that the usual isotropic material stress concentration factor is 3; that is, $\sigma_{\theta 1}/\sigma_1 = 3$ at $\theta = 90°$. For composite materials, the

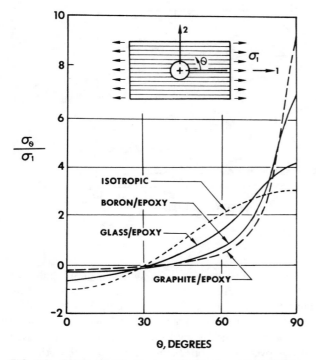

FIG. 6-3. Effect of material properties on circumferential stress σ_θ at edge of a circular hole in an orthotropic plate subjected to σ_1. (*After Greszczuk, Ref. 6-9.*)

stress concentration factor is much higher (4 for glass/epoxy, about 6 for boron/epoxy, and about 9 for graphite/epoxy). Moreover, the circumferential stress at $\theta = 0°$ is reduced for composites relative to isotropic materials. Because of isotropy of material properties, the key factor in failure of isotropic plates with holes is the magnitude of the stress concentration factor from which the maximum (failure) stress is obtained. However, for orthotropic materials, a combined stress failure criterion instead of a maximum stress failure criterion must be used as noted in Sec. 2.9. Thus, stress concentration factors alone, as in Fig. 6-3, are insufficient for failure prediction of orthotropic (and anisotropic) plates. Moreover, if the plates are laminated, the comparison of stress states with failure stress states must be done on a layer-by-layer basis.

The second special case is an orthotropic lamina loaded at angle α to the fiber direction. Effectively, such a situation is an anisotropic lamina under load. Stress concentration factors for boron/epoxy were obtained by Greszczuk (Ref. 6-9) in Fig. 6-4. There, the circumferential stress around the edge of the circular hole is plotted versus angular position around the hole. The circumferential stress is normalized by σ_α, the applied stress. The results for $\alpha = 0°$ are identical to those in Fig. 6-3. As α approaches $90°$, the peak stress concentration factor

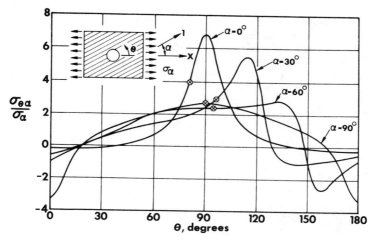

FIG. 6-4. Stress concentration at edge of a circular hole in an anisotropic plate subjected to stress at angle α to principal material direction. (*After Greszczuk, Ref. 6-9.*)

decreases and shifts location around the hole. However, as shown, the combined stress state at failure, upon application of a failure criterion, always occurs near $\theta = 90°$. Implicitly, the analysis of failure due to stress concentrations around holes in a lamina is quite involved.

The next obvious step is to extend the analysis to a laminate. Greszczuk (Ref. 6-9) considered a symmetric cross-ply laminate which is subjected to tension in a fiber direction, tension at 45° to a fiber direction, or in-plane shear. The resulting stress concentrations are shown in Fig. 6-5. As before, the circumferential stress is normalized by the applied stress. However, the circumferential stress is not a maximum lamina stress but a gross stress on the entire laminate; that is, it is actually N_θ/t where N_θ is a circumferential force per unit width and t is the laminate thickness. The stresses in each lamina are then found by use of the concepts in Sec. 4.2, classical lamination theory. Failure is determined by application of a biaxial strength criterion to *each* layer. Thus, the effect of holes on laminate behavior is much more complex than on lamina or plate behavior. The interlaminar stresses in the manner of Sec. 4.6 are ignored. Accordingly, the predicted stresses are not accurate within about one laminate thickness from the edge.

6.4 FRACTURE MECHANICS

The strength of any material is inherently related to flaws which are always present. Specifically, the strengths of composites are governed by their flaw-initiated characteristics. Thus, the mechanics of fracture including crack propagation or extension are of extreme importance in the design analysis of

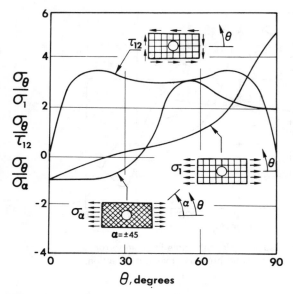

FIG. 6-5. Stress concentrations at the edge of a circular hole in a cross-ply laminate. (*After Greszczuk, Ref. 6-9.*)

composite structures. Fracture mechanics criteria are now a part of every metal airplane design. This step was made by the Air Force as a result of recent fracture and fatigue problems on F-111, C5-A, Electra, etc. The prospectus for composite materials applications in the near future is that they, too, will have fracture mechanics design criteria imposed.

The fracture process generally takes place in three stages. First, a microcrack is initiated (or an initial flaw or imperfection can be present). Second, the microcrack grows in a stable fashion and may link with other microcracks to attain macrocrack size. Third, the macrocrack propagates in an unstable fashion at a critical stress level. These three stages are found and clearly defined only in ductile materials. Some of the stages, for example, stage two, are not found in brittle materials. A prominent characteristic of composite materials is their high resistance to crack propagation because of the matrix ductility.

Fracture is caused by higher stresses around flaws or cracks than in the surrounding material. However, fracture mechanics is much more than the study of stress concentration factors. Such factors are useful in determining the influence of relatively large holes in bodies (see Sec. 6.3, "Holes in Laminates"), but are not particularly helpful when the body has sharp notches or cracklike flaws. For composite materials, fracture has a new dimension as opposed to homogeneous isotropic materials because of the presence of two or more constituents. Fracture can be a fracture of the individual constituents or a separation of the interface between the constituents.

The discussion of fracture mechanics will be divided in two parts. First, basic principles of fracture mechanics will be described. Second, the application of fracture mechanics concepts to composite materials will be discussed. In both parts, the basic approach is that of Wu (Ref. 6-10).

6.4.1 Basic Principles of Fracture Mechanics

The acknowledged father of fracture mechanics is A.A. Griffith (Ref. 6-11). His principal contribution is an analysis of crack stability based on energy equilibrium. If a crack is in equilibrium, the decrease of strain energy U must be equal to the increase of surface energy S due to crack extension, that is,

$$\frac{\partial U}{\partial a} = \frac{\partial S}{\partial a} \qquad (6.1)$$

where a is the crack length. The strain energy release rate, $\partial U/\partial a$, is actually the crack extension force. Prior to Griffith's approach, the application of classical elasticity concepts led to infinite stresses at the crack tip.

Irwin (Ref. 6-12) extended Griffith's theory to elastic-plastic materials and pointed out the three kinematically admissible crack extension modes shown in Fig. 6-6. These modes, opening, forward shear, and parallel shear, can be summed to obtain any crack.

Attention will be restricted to the strain energy release rate for the opening mode. This mode occurs for the plate with a centrally located crack of length $2a$ under load P in Fig. 6-7. The strain energy in the plate is

$$U = \tfrac{1}{2}Pe \qquad (6.2)$$

where e is the elongation between loading points separated by distance L. The spring constant of the plate is

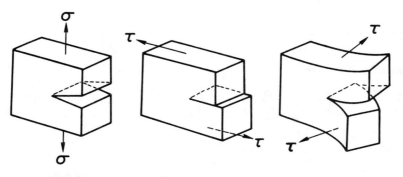

OPENING MODE   **FORWARD SHEAR MODE**   **PARALLEL SHEAR MODE**

FIG. 6-6. Crack extension modes.

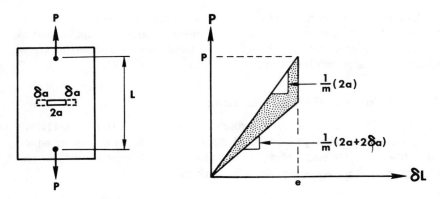

FIG. 6-7. Cracked plate and load-deformation diagram. (*After Wu, Ref. 6-10.*)

$$\frac{1}{m} = \frac{P}{e} \qquad (6.3)$$

The strain energy release rate due to crack extension $2\delta a$ is the shaded area in Fig. 6-7 if the loading head does not move during crack extension, that is,

$$\frac{\partial e}{\partial a} = 0 \qquad (6.4)$$

From Eq. (6.2),

$$\frac{\partial U}{\partial a} = \frac{1}{2} e \frac{\partial P}{\partial a} + \frac{1}{2} P \frac{\partial e}{\partial a} \qquad (6.5)$$

but, because of Eq. (6.4),

$$\frac{\partial U}{\partial a} = \frac{1}{2} e \frac{\partial P}{\partial a} \qquad (6.6)$$

Then, from Eq. (6.3)

$$\frac{\partial P}{\partial a} = \frac{1}{m} \frac{\partial e}{\partial a} + e \frac{\partial}{\partial a}\left(\frac{1}{m}\right) = -\frac{P}{m} \frac{\partial m}{\partial a} \qquad (6.7)$$

so that

$$\frac{\partial U}{\partial a} = -\frac{P^2}{m} \frac{\partial m}{\partial a} \qquad (6.8)$$

Irwin (Ref. 6-12) calls the strain energy release rate $-\,\mathcal{G}$ so

$$\mathcal{G} = \frac{P^2}{m} \frac{\partial m}{\partial a} \tag{6.9}$$

which can be measured since P and m can be measured. The same value of $\mathcal{G}$ results if the load is held constant during crack extension.

The strain energy release rate was expressed in terms of stresses around a crack tip by Irwin. He considered a crack under a plane stress loading of σ^∞, a symmetric stress relative to the crack, and τ^∞ a skew-symmetric stress relative to the crack in Fig. 6-8. The stresses have a superscript $^\infty$ because they are applied an infinite distance from the crack. The stress distribution very near the crack can be shown by use of classical elasticity theory to be, for example,

$$\sigma_x = \frac{\sigma^\infty \sqrt{a}}{\sqrt{2r}} \cos \frac{\theta}{2} \left(1 - \sin \frac{\theta}{2} \sin \frac{3\theta}{2}\right) - \frac{\tau^\infty \sqrt{a}}{\sqrt{2r}} \sin \frac{\theta}{2} \left(2 + \cos \frac{\theta}{2} \cos \frac{3\theta}{2}\right)$$

$$(6.10)$$

The stress singularity for this and the other stress components σ_y and τ_{xy} is of order $1/\sqrt{r}$ Moreover, the terms $\sigma^\infty \sqrt{a}$ and $\tau^\infty \sqrt{a}$ are called symmetric and skew-symmetric stress intensity factors:

$$k_1 = \sigma^\infty \sqrt{a}$$
$$k_2 = \tau^\infty \sqrt{a} \tag{6.11}$$

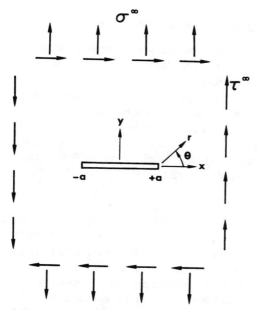

FIG. 6-8. Cracked plate with symmetric and skew-symmetric stresses at infinity.

The symmetric stress intensity factor k_1 is associated with the opening mode of crack extension in Fig. 6-6. The skew-symmetric stress intensity factor k_2 is associated with the forward shear mode. These plane stress intensity factors must be supplemented by another stress intensity factor to describe the parallel shear mode. The stress intensity factors depend on the applied loads, body geometry, and crack geometry. For plane loads, the stress distribution around the crack tip can always be separated into symmetric and skew-symmetric distributions.

The stress intensity factors are quite different from stress concentration factors. For the same circular hole, the stress concentration factor is 3 under uniaxial tension, 2 under biaxial tension, and 4 under pure shear. Thus, the stress concentration factor, which is a single scalar parameter, cannot characterize the stress state, a second-order tensor. However, the stress intensity factor exists in all stress components so is a useful concept in stress-type fracture processes. For example,

$$\mathcal{G} = \frac{\pi k_1^2}{E} \tag{6.12}$$

for a opening mode crack that extends parallel to itself. Other such relations can be obtained for plane strain and the other crack modes. The point is that the stress intensity factors appear in the strain energy release rate.

6.4.2 Application of Fracture Mechanics to Composite Materials

Composite materials have many distinctive characteristics relative to isotropic materials that render application of linear elastic fracture mechanics difficult. The anisotropy and heterogeneity, both from the standpoint of the fibers versus the matrix, and from the standpoint of multiple laminae of different orientations, are the principal problems. The extension of homogeneous anisotropic materials should be straightforward because none of the basic principles used in fracture mechanics is then changed. Thus, the approximation of composite materials by homogeneous anisotropic materials is often made. Then, stress intensity factors for anisotropic materials are calculated by use of complex variable mapping techniques.

Wu (Ref. 6-10) derives the stress distribution around a crack tip in an anisotropic material. He finds the intensities of the stresses σ_x, σ_y, and τ_{xy} are controlled not only by the parameters $\sigma^\infty \sqrt{a}$ and $\tau^\infty \sqrt{a}$ but also by functions of the anisotropic material properties and the orientation of the crack relative to the principal material directions. A simplification occurs when the crack maintains a constant orientation relative to the principal material directions (a likely circumstance if the flaws or possible paths of crack extension are, for example, all parallel to the fibers in a unidirectional lamina). However, unless the material is orthotropic and the crack is parallel to a principal material direction, the opening mode has both symmetric and skew-symmetric stresses in the stress distribution around the crack tip.

Wu (Ref. 6-10) performed a series of experiments to determine the applicability of linear elastic fracture mechanics to composite materials. He subjected unidirectionally reinforced fiberglass/epoxy plates with centrally located cracks in the fiber direction to tension, pure shear, and combined tension and shear as in Fig. 6-9. He recorded the critical load and crack length at incipient rapid crack extension and noted that the cracks propagated colinear with the original crack. Moreover, the symmetric loads led to the crack opening mode and the skew-symmetric loads led to the forward shear or sliding mode. This distinction is clearer than for isotropic materials! For load path 1, the stress intensity factors are

$$k_1 = \sigma^\infty \sqrt{a}$$

$$k_2 = 0$$

(6.13)

and the critical stress intensity factors are

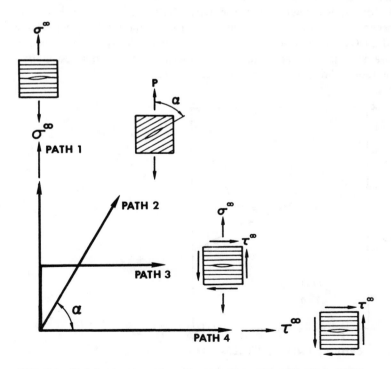

FIG. 6-9. Wu's load paths and crack orientations. (*After Wu, Ref. 6-10.*)

$$k_{1c} = \sigma_c^\infty \sqrt{a_c}$$

$$k_{2c} = 0 \qquad\qquad (6.14)$$

where σ_c^∞ is the critical stress and a_c the critical crack length at incipient rapid crack propagation. If k_{1c} is truly a material constant, as we would hope it is, then the experimental data on a plot of $\log \sigma_c^\infty$ versus $\log a_c$ should be a straight line with slope $-1/2$ since Eq. (6.14) can be written

$$\log k_{1c} = \log \sigma_c^\infty + \tfrac{1}{2} \log a_c \qquad\qquad (6.15)$$

Indeed, the slope of Fig. 6-10 is actually $-.49$ so the theory is apparently applicable to an orthotropic lamina with cracks in the fiber direction. This contention is further substantiated by tests for the other loading paths shown in Fig. 6-9.

Other researchers have substantially advanced the state-of-the-art of fracture mechanics applied to composite materials. Tetelman (Ref. 6-13) and Corten (Ref. 6-14) discuss fracture mechanics from the point of view of micro-mechanics. Sih and Chen (Ref. 6-15) treat the mixed mode fracture problem for noncollinear crack propagation. Waddoups, Eisenmann, and Kaminski (Ref. 6-16) and Konish, Swedlow, and Cruse (Ref. 6-17) extend the concepts of fracture mechanics to laminates. Impact resistance of unidirectional composites is discussed by Chamis, Hansen, and Serafini (Ref. 6-18). They use strain energy and fracture strength concepts along with micromechanics to assess impact resistance in longitudinal, transverse, and shear modes.

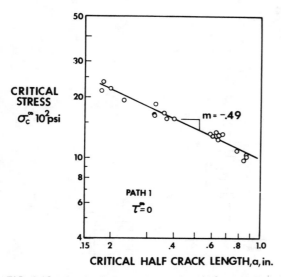

FIG. 6-10.   Cracked plate under tension. (*After Wu, Ref. 6-10.*)

6.5 TRANSVERSE SHEAR EFFECTS

The effect of transverse shearing stresses, τ_{xz} and τ_{yz}, can be more important for laminated composite plates and shells than for isotropic plates and shells. Composite materials typically have a low matrix Young's modulus in comparison to the fiber modulus and even in comparison to the overall laminae moduli. Since the matrix material is the bonding agent between laminae, the shearing effect on the entire laminate is built up by summation of the contributions of each interlaminar zone of matrix material. This summation effect cannot be ignored because laminates can have 100 or more layers!

Study of transverse shearing stress effects is divided in two parts. First, exact elasticity solutions for composite laminates in cylindrical bending are examined. These solutions are limited in applicability to practical problems but are extremely useful as checkpoints for more broadly applicable approximate theories. Second, various approximations for treatment of transverse shearing stresses in plate theory are discussed.

6.5.1 Exact Solutions for Cylindrical Bending

Pagano (Ref. 6-19) studied cylindrical bending of laminated composite plates. Each layer is orthotropic and has principal material directions aligned with the plate axes. The plate is infinitely long in the y-direction (see Fig. 6-11). When subjected to a transverse load $p = p(x)$, that is, p is independent of y, the plate deforms into a cylinder:

$$u = u(x) \quad v = 0 \quad w = w(x) \tag{6.16}$$

Thus, the plate is in a state of plane strain in the xz plane.

Pagano's exact solution for the stresses and displacements is too complex to present here. The corresponding classical lamination theory result stems from the equilibrium equations, Eqs. (5.6) to (5.8), which simplify to

$$A_{11}u_{,xx} - B_{11}w_{,xxx} = 0$$
$$D_{11}w_{,xxxx} - B_{11}u_{,xxx} = p \tag{6.17}$$

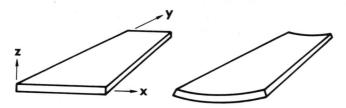

FIG. 6-11. Cylindrical bending of an infinitely long strip.

when the orthotropy and Eq. (6.16) are accounted for. These equilibrium equations can be uncoupled by differentiating the first equation to give

$$u_{,xxx} = \frac{B_{11}}{A_{11}} w_{,xxxx} \tag{6.18}$$

and substituting it in the second equation to yield

$$w_{,xxxx} = \frac{A_{11}}{D} p \tag{6.19}$$

where

$$D = A_{11}D_{11} - B_{11}^2 \tag{6.20}$$

When $p = p_0 \sin nx$, the solutions to Eqs. (6.18) and (6.19) are

$$u = \frac{B_{11}p_0}{Dn^3} \cos nx$$

$$\tag{6.21}$$

$$w = \frac{A_{11}p_0}{Dn^4} \sin nx$$

whereupon the only strain is

$$\epsilon_x = u_{,x} - zw_{,xx} = \left(\frac{A_{11}z - B_{11}}{Dn^2}\right) p_0 \sin nx \tag{6.22}$$

The stresses in each layer are

$$\sigma_{x_k} = \frac{p_0 Q_{11}^k (A_{11}z - B_{11})}{Dn^2} \sin nx$$

$$\tag{6.23}$$

$$\sigma_{y_k} = \frac{p_0 Q_{12}^k (A_{11}z - B_{11})}{Dn^2} \sin nx$$

Even though in classical lamination theory by virtue of the Kirchhoff hypothesis we assume the stresses τ_{xz} and σ_z are zero, we can still obtain these stresses approximately by integration of the stress equilibrium equations:

$$\tau_{xz,x} = -\sigma_{x,x} \tag{6.24}$$

$$\sigma_{z,z} = -\tau_{xz,x}$$

to obtain

$$\tau_{xz}^k = -\frac{p_0 Q_{11}^k}{Dn}\left(\frac{A_{11}}{2}z^2 - B_{11}z + H_k\right)\cos nx$$

$$\sigma_z^k = -\frac{p_0 Q_{11}^k}{D}\left(\frac{A_{11}}{6}z^3 - \frac{B_{11}}{2}z^2 + H_k z + L_k\right)\sin nx$$

(6.25)

where the constants H_k and L_k are determined from the surface and interlaminar boundary conditions on the stresses.

Pagano (Ref. 6-19) presented numerical results for several laminates made of a high modulus graphite/epoxy composite with

$$E_1 = 25 \times 10^6 \text{ psi} \qquad E_2 = 1 \times 10^6 \text{ psi}$$

$$G_{12} = .5 \times 10^6 \text{ psi} \qquad G_{23} = .2 \times 10^6 \text{ psi}$$

$$\nu_{12} = \nu_{23} = .25$$

His loading was $p = p_0 \sin(\pi x/L)$, that is, $n = \pi/L$, on a symmetric three-layer laminate. First, the normalized transverse deflection w is plotted versus the span-to-depth ratio, L/t, in Fig. 6-12. The deviation of the actual solution from the approximate classical lamination theory solution is quite substantial at low span-to-thickness ratios, $S = L/t$. Even at $S = 20$ where ordinarily classical plate theory would be accurate for isotropic materials, the deviation is about 20 percent.

The normal stress σ_x is plotted through the thickness for span-to-thickness ratios of 4 and 10 in Fig. 6-13a and b, respectively. The deviation of the classical lamination theory solution from the exact solution is quite drastic for $S = 4$, but not particularly large for $S = 10$. The transverse shearing stress τ_{xz} is plotted through the thickness for $S = 4$ and $S = 10$ in Figs. 6-14a and b, respectively. The differences between classical lamination theory and the exact solution are not large for $S = 4$ and are fairly small for $S = 10$. The in-plane displacement u is

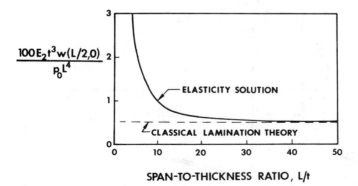

SPAN-TO-THICKNESS RATIO, L/t

FIG. 6-12. Normalized deflection versus span-to-thickness ratio. (After Pagano, Ref. 6-19.)

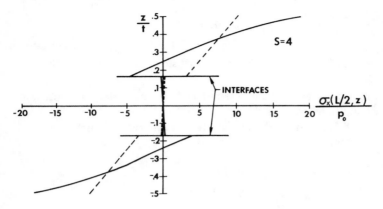

FIG. 6-13a. Variation of σ_x through the thickness for $S = 4$. (*After Pagano, Ref. 6-19.*)

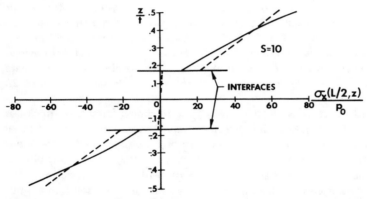

FIG. 6-13b. Variation of σ_x through the thickness for $S = 10$. (*After Pagano, Ref. 6-19.*)

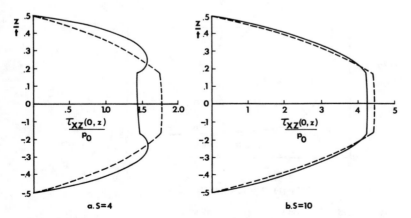

FIG. 6-14. Variation of τ_{xz} through the thickness. (*After Pagano, Ref. 6-19.*)

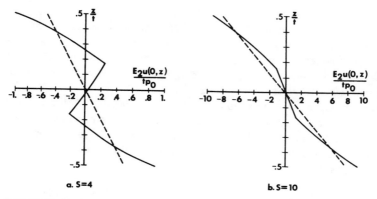

FIG. 6-15. Variation of u through the thickness. (*After Pagano, Ref. 6-19.*)

plotted through the thickness for $S = 4$ and $S = 10$ in Fig. 6-15a and b, respectively. Obviously, the displacement varies almost linearly in each layer, but certainly not through the laminate when $S = 4$. When $S = 10$, the deviation from linearity through the laminate is not great. Thus, the Kirchhoff hypothesis of nondeformable normals is not appropriate for low values of S.

Obviously, the classical lamination theory stresses in Pagano's example converge to the exact solution much more rapidly than do the displacements as the span-to-thickness ratio increases. The stress errors are on the order of 10 percent or less for S as low as 20. The displacements are severely underestimated for S between 4 and 30, common values for laboratory characterization specimens. Thus, a practical means of accounting for transverse shearing deformations is required. That objective is attacked in the next section.

First, other work on the exact solutions to special problems will be reviewed. Pagano extended his theory to plates in Ref. 6-20; that is, his strip was of finite length. Then, in Ref. 6-21, he included the effect of in-plane shear coupling in order to treat angle-ply laminates. Pagano and Wang (Ref. 6-22) extended the orthotropic laminate solution to more general loadings. Finally, Pagano and Hatfield (Ref. 6-23) examined laminated plates with many layers.

6.5.2 Approximate Treatment of Transverse Shear Effects

The preceding subsection was devoted to a comparison of a special exact elasticity solution with classical lamination theory results. The importance of transverse shear effects was clearly demonstrated. However, that demonstration was for a special problem of rather narrow interest. The objective of this subsection is to display approaches and results for the approximate consideration of transverse shear effects for general laminated plates.

The treatment of transverse shear stress effects in plates made of isotropic materials stems from the classical papers by Reissner (Ref. 6-24) and Mindlin

(Ref. 6-25). Extension of Reissner's theory to plates made of orthotropic materials is due to Girkmann and Beer (Ref. 6-26). Ambartsumyan (Ref. 6-27) treated symmetrically laminated plates with orthotropic laminae having their principal material directions aligned with the plate axes. Whitney (Ref. 6-28) extended Ambartsumyan's analysis to symmetrically laminated plates with orthotropic laminae of arbitrary orientation.

The basic approaches as summarized by Ashton and Whitney (Ref. 6-29) will now be discussed. First, a symmetric laminate with orthotropic laminae having principal material directions aligned with the plate axes will be treated. The transverse normal strain can be found from the orthotropic stress-strain relations, Eq. (2.15), as

$$\epsilon_z = \frac{1}{C_{33}} (\sigma_z - C_{13}\epsilon_x - C_{23}\epsilon_y) \tag{6.26}$$

which can be used to eliminate ϵ_z from the stress-strain relations for the kth layer leaving

$$
\begin{Bmatrix}
\sigma_x \\
\sigma_y \\
\tau_{yz} \\
\tau_{xz} \\
\tau_{xy}
\end{Bmatrix}_k
=
\begin{bmatrix}
Q_{11} & Q_{12} & 0 & 0 & 0 \\
Q_{12} & Q_{22} & 0 & 0 & 0 \\
0 & 0 & Q_{44} & 0 & 0 \\
0 & 0 & 0 & Q_{55} & 0 \\
0 & 0 & 0 & 0 & Q_{66}
\end{bmatrix}_k
\begin{Bmatrix}
\epsilon_x \\
\epsilon_y \\
\gamma_{yz} \\
\gamma_{xz} \\
\gamma_{xy}
\end{Bmatrix}_k
\tag{6.27}
$$

where, if σ_z is neglected as in classical lamination theory,

$$
Q_{ij} =
\begin{cases}
C_{ij} - \dfrac{C_{i3}C_{j3}}{C_{33}} \, , & \text{if } i,j = 1, 2 \\
\\
C_{ij}, & \text{if } i, j = 4, 5, 6
\end{cases}
\tag{6.28}
$$

The transverse shearing stress distribution is then approximated by

$$\tau_{xz}^k = [Q_{55}^k f(z) + a_{55}^k] \Phi_x (x, y) \tag{6.29}$$
$$\tau_{yz}^k = [Q_{44}^k f(z) + a_{44}^k] \Phi_y (x, y)$$

where $f(z) = f(-z)$ because of laminate symmetry. Also, a_{44}^k and a_{55}^k are determined from the equilibrium conditions that the shearing stresses vanish at the top and bottom surfaces of the plate $[f(t/2) = f(-t/2) = 0]$ and are continuous at layer interfaces. The shearing strains are obtained from the stress-strain relations as

$$
\gamma_{xz}^k = \left[f(z) + \frac{a_{55}^k}{Q_{55}^k} \right] \Phi_x
$$

$$
\gamma_{yz}^k = \left[f(z) + \frac{a_{44}^k}{Q_{44}^k} \right] \Phi_y
$$

(6.30)

Then, integration of the strain-displacement relation, Eq. (2.2), with respect to z (with w assumed to be independent of z) yields

$$
u^k = -zw_{,x} + [J(z) + g_1^k(z)] \Phi_x
$$

$$
v^k = -zw_{,y} + [J(z) + g_2^k(z)] \Phi_y
$$

(6.31)

where

$$
J(z) = \int f(z)\,dz \qquad g_1^k(z) = \frac{a_{55}^k}{Q_{55}^k} z + b_1^k \qquad g_2^k(z) = \frac{a_{44}^k}{Q_{44}^k} z + b_2^k
$$

(6.32)

The constants b_1^k and b_2^k are determined from continuity conditions for u and v at layer interfaces and the symmetry condition that u and v vanish at the laminate middle surface. Obviously, because of the presence of Φ_x and Φ_y, u and v are not linear functions of z as in classical lamination theory.

The moment relations are obtained from integration of the stress-strain relations, Eq. (6.27), after the strain-displacement relations, Eq. (6.22), and the displacement relations, Eq. (6.31), are substituted:

$$
M_x = -D_{11}w_{,xx} - D_{12}w_{,yy} + (F_{11} + H_{111})\Phi_{x,x} + (F_{12} + H_{122})\Phi_{y,y}
$$

$$
M_y = -D_{12}w_{,xx} - D_{22}w_{,yy} + (F_{12} + H_{121})\Phi_{x,x} + (F_{22} + H_{222})\Phi_{y,y}
$$

$$
M_{xy} = -2D_{66}w_{,xy} + (F_{66} + H_{661})\Phi_{x,y} + (F_{66} + H_{662})\Phi_{y,x}
$$

(6.33)

where the D_{ij} are the usual bending stiffnesses and

$$
F_{ij} = \int_{-t/2}^{t/2} Q_{ij}^k z J(z)\,dz \qquad\qquad i, j = 1, 2, 6
$$

$$
\qquad\qquad\qquad\qquad\qquad\qquad \ell = 1, 2
$$

$$
H_{ij\ell} = \int_{-t/2}^{t/2} Q_{ij}^k z g_\ell^k(z)\,dz
$$

(6.34)

The shear resultants are

$$Q_x = \int_{-t/2}^{t/2} \tau_{xz}^k \, dz = K_{55} \Phi_x$$

$$Q_y = \int_{-t/2}^{t/2} \tau_{yz}^k \, dz = K_{44} \Phi_y$$

(6.35)

where

$$K_{ii} = \int_{-t/2}^{t/2} [Q_{ii}^k f(z) + a_{ii}^k] \, dz \quad i = 4, 5$$

(6.36)

The equilibrium equations are

$$M_{x,x} + M_{xy,y} - Q_x = 0$$
$$M_{xy,x} + M_{y,y} - Q_y = 0$$
$$Q_{x,x} + Q_{y,y} + p + N_x w_{,xx} + 2N_{xy} w_{,xy} + N_y w_{,yy} = 0$$

(6.37)

or, in terms of the present variables,

$$D_{11} w_{,xxx} + (D_{12} + 2D_{66}) w_{,xyy} - (F_{11} + H_{111}) \Phi_{x,xx} - (F_{66} + H_{661}) \Phi_{x,yy}$$
$$- (F_{12} + F_{66} + H_{122} + H_{662}) \Phi_{y,xy} + K_{55} \Phi_x = 0$$
$$(D_{12} + 2D_{66}) w_{,xxy} + D_{22} w_{,yyy} + (F_{12} + F_{66} + H_{121} + H_{661}) \Phi_{x,xy} \quad (6.38)$$
$$+ (F_{66} + H_{662}) \Phi_{y,xx} + (F_{22} + H_{222}) \Phi_{y,yy} + K_{44} \Phi_y = 0$$
$$K_{55} \Phi_{x,x} + K_{44} \Phi_{y,y} + p + N_x w_{,xx} + 2N_{xy} w_{,xy} + N_y w_{,yy} = 0$$

The boundary conditions for these equilibrium equations are more complicated than for classical lamination theory. However, they are more logical since the Kirchhoff free edge condition, in which a combination of shearing force and twisting moment appears, is replaced by those conditions themselves. In summary, the new boundary conditions along each edge are

$$Q_n = 0 \quad \text{or} \quad w = 0 \quad M_n = 0 \quad \text{or} \quad w_{,n} = 0 \quad M_{nt} = 0 \quad \text{or} \quad u_{t,z}\big|_{z=0} = 0$$

(6.39)

where n and t denote directions normal and transverse to the edge, respectively.

For a simply supported laminated rectangular plate subjected to the distributed load

$$p = p_0 \sin \frac{m\pi x}{a} \sin \frac{n\pi y}{b} \qquad (6.40)$$

the displacement and rotations

$$w = A \sin \frac{m\pi x}{a} \sin \frac{n\pi y}{b}$$

$$\Phi_x = B \cos \frac{m\pi x}{a} \sin \frac{n\pi y}{b} \qquad (6.41)$$

$$\Phi_y = C \sin \frac{m\pi x}{a} \cos \frac{n\pi y}{b}$$

exactly satisfy the boundary conditions

$$M_x = v,z|_{z=0} = w = 0 \quad \text{on} \quad x = 0, a$$
$$M_y = u,z|_{z=0} = w = 0 \quad \text{on} \quad y = 0, b \qquad (6.42)$$

The shearing stresses are assumed on the basis of elasticity results (Ref. 6-19) to vary approximately as a segment of a parabola in each layer, that is,

$$f(z) = 1 - 4\left(\frac{z}{t}\right)^2 \qquad (6.43)$$

Then, the overall problem is determinate and reduces to the solution of the following set of simultaneous algebraic equations for A, B, and C:

$$[D_{11}m^2 + (D_{12} + 2D_{66})n^2R^2]\left(\frac{m\pi}{a}\right)A - \Big[(F_{11} + H_{111})m^2 + (F_{66}$$

$$+ H_{661})n^2R^2 + \frac{K_{55}R^2S^2}{\pi^2}\Big]B - (F_{12} + F_{66} + H_{122} + H_{662})mnRC = 0$$

$$[(D_{12} + 2D_{66})m^2 + D_{22}n^2R^2]\left(\frac{n\pi R}{a}\right)A - (F_{12} + F_{66} + H_{121} + H_{661})mnRB$$

$$- \Big[(F_{66} + H_{662})m^2 + (F_{22} + H_{222})n^2R^2 + \frac{K_{44}R^2S^2}{\pi^2}\Big]C = 0$$

$$K_{55}mB + K_{44}nRC = \frac{p_0 a}{\pi} \qquad (6.44)$$

in which $R = a/b$ and $S = a/t$.

Whitney (Ref. 6-28) obtained results from Eq. (6.44) for deflection of a square four-layered symmetric cross-ply $[0°/90°/90°/0°]$ laminated graphite/

epoxy plate under the load $p = p_0 \sin (\pi x/a) \sin (\pi y/a)$. The material properties are typical of a high modulus graphite/epoxy:

$$\frac{E_1}{E_2} = 40 \qquad \frac{G_{12}}{E_2} = .6 \qquad \frac{G_{13}}{E_2} = .5 \qquad \nu_{12} = .25$$

The results shown in Fig. 6-16 for the present shear deformation approach versus classical lamination theory are quite similar qualitatively to the comparison between the exact cylindrical bending solution and classical lamination theory in Fig. 6-12.

A more direct comparison of Whitney's shear deformation solution (Ref. 6-28) for deflection of an antisymmetric cross-ply infinite strip with the elasticity solution and the classical lamination theory solution is shown in Fig. 6-17. Obviously, Whitney's shear deformation theory solution is quite good for prediction of deflections. However, Whitney's shearing stress distribution through the thickness at the edge of the infinite strip in Fig. 6-18 does not agree well with the elasticity solution. If, instead of an equation analogous to Eq. (6.43), the shearing stresses are calculated from the stresses σ_x, σ_y, and τ_{xy} by the elasticity equations, then the better agreement in Fig. 6-18 between the modified shear deformation theory and the elasticity solution is obtained.

In other work, Whitney and Pagano (Ref. 6-30) extended Yang, Norris, and Stavsky's work (Ref. 6-31) to the treatment of coupling between bending and extension. Whitney (Ref. 6-32) uses a higher order stress theory to obtain improved predictions of σ_x, σ_y, and τ_{xy} and displacements at low width-to-

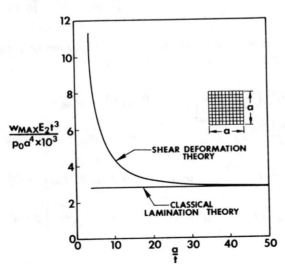

FIG. 6-16. Deflection of a square four-layer symmetric cross-ply graphite/epoxy plate under $p_0 \sin \pi x/a \sin \pi y/a$. (After Whitney, Ref. 6-28.)

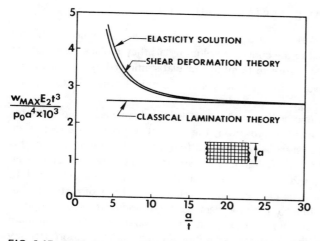

FIG. 6-17. Deflection of an infinite two-layer cross-ply graphite/epoxy strip under $p_0 \sin \pi x/a$. (*After Whitney, Ref. 6-28.*)

thickness ratios. Reissner (Ref. 6-33) used his variational theorem to derive a consistent set of equations for inclusion of transverse shearing deformation effects in symmetrically laminated plates. Finally, Ambartsumyan (Ref. 6-34) extended his treatment of transverse shearing deformation effects from plates to shells.

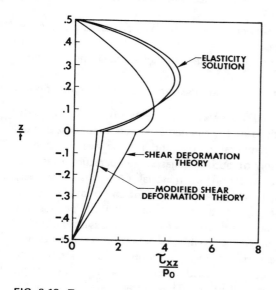

FIG. 6-18. Transverse shear stress distribution along edges of an infinite two-layer graphite/epoxy strip under $p_0 \sin \pi x/a$ with $a/t = 4$. (*After Ashton and Whitney, Ref. 6-29.*)

6.6 ENVIRONMENTAL EFFECTS

Composite materials must survive in the environment to which they are subjected at least as well as the conventional materials they replace. Some of the environments include exposure to humidity or water immersion, salt spray, jet fuel, hydraulic fluid, stack gas (includes sulfur dioxide), fire, lightning, and gunfire.

Humidity or water immersion can lead to degraded stiffnesses and strengths (Ref. 6-35). However, after dehydration, the original properties are recovered. Some of the same but irreversible effects are found for salt spray (although it is somewhat corrosive), jet fuel, hydraulic fluid, and stack gas. Fire is, of course, an extreme environment, and its damage is obvious. Aircraft are subjected to lightening strikes so must be protected and certainly not built of materials that are particularly susceptible to ligthning damage. Lightning tests were conducted on aluminum, fiberglass, boron/epoxy, and graphite/epoxy. The aluminum sustained little damage; the fiberglass had minor surface marks; the boron/epoxy had some cracks and bubble delaminations; and the graphite/epoxy had the resin stripped and a hole burned in another place. Surface coatings and/or lightning rods could reduce these effects substantially.

Tsai (Ref. 6-36) listed and quantitatively discussed specific degradation mechanisms for environmental exposure:

"A. Loss of strength of the reinforcing fibers by a stress-corrosion mechanism.

B. Degradation of the fiber-matrix interface resulting in loss of adhesion and interfacial bond strength.

C. Permeability of the matrix material to corrosive agents such as water vapor which affects both A and B above.

D. Normal viscoelastic dependence of matrix modulus and strength on time and temperature.

E. Combined action of temperature and moisture accelerated degradation."

Whitney and Ashton (Ref. 6-37) studied the effects on laminated plate deflections, buckling, and vibrations of environmental effects that cause expansional strains. Such effects are temperature rise and matrix swelling due to water vapor or sudden expansion of absorbed gases. Temperature rises and matrix expansion can cause degradation of material properties. However, Whitney and Ashton held the material properties constant and treated only the effects of temperature and swelling on structural response. Both phenomena cause buckling; thus, buckling because of thermal expansion was assessed directly. They also analyzed the effect of matrix swelling on vibration frequencies and bending deflections. A limiting case of the vibration problem is buckling because of matrix swelling. Swelling caused reduced buckling loads and vibration frequencies and increased bending deflections. These effects are all significant enough to warrant attention in composite structures design analysis.

6.7 SHELLS

Work on analysis of the common structural shell element made of composite materials is very extensive. Contributions will be mentioned that parallel the developments in Chap. 5 on plates. Some of the first analyses of laminated shells are by Dong, Pister, and Taylor (Ref. 6-38) and the monograph by Ambartsumyan (Ref. 6-34). Further efforts include the buckling work on laminated shells by Cheng and Ho (Ref. 6-39) and on eccentrically stiffened laminated shells by Jones (Ref. 6-40).

Classical solutions to laminated shell buckling and vibration problems in the manner of Chap. 5 were obtained by Jones and Morgan (Ref. 6-41). Their results are presented as normalized buckling loads or fundamental natural frequency versus the Batdorf shell curvature parameter. They showed that, for antisymmetrically laminated cross-ply shells as for plates, the effect of coupling between bending and extension on buckling loads and vibration frequencies dies out rapidly as the number of layers increases. However, for unsymmetrically laminated cross-ply shells, the effect of coupling dies out very, very slowly. Thus, analyses of all unsymmetrically laminated plates and shells should include the effects of coupling between bending and extension. Otherwise, serious overestimates of buckling loads and vibration frequencies can be obtained. Similarly, serious underestimates of plate and shell deflections, and hence stresses, can occur if coupling is ignored.

6.8 MISCELLANEOUS TOPICS

Some basic lamina and laminate behavioral characteristics were deliberately overlooked in the preceding. Among them are plastic or nonlinear deformations, viscoelastic behavior, and wave propagation.

Shear stress–shear strain curves typical of fiber-reinforced epoxy resins are quite nonlinear. All other stress-strain curves are essentially linear. Hahn and Tsai (Ref. 6-42) analyzed lamina behavior with this nonlinear deformation behavior. Hahn (Ref. 6-43) extended the analysis to laminate behavior. Inelastic effects in micromechanics analyses were examined by Adams (Ref. 6-44).

Viscoelastic characteristics of composite materials are usually due to a viscoelastic matrix material such as epoxy resin. General stress analysis of viscoelastic composites was discussed by Schapery (Ref. 6-45). An important application to laminated plates was made by Sims (Ref. 6-46).

Wave propogation in an inhomogeneous anisotropic material such as a fiber-reinforced composite material is a very complex subject. However, its study is motivated by many important applications such as the use of fiber-reinforced composites in reentry vehicle nosetips, heatshields, and other protective systems. Chou (Ref. 6-47) gives an introduction to analysis of wave propagation in composite materials. Others have applied wave propagation theory to shell stress problems.

Design of composite structures was given only a brief overview. The analysis of a few basic structural elements such as tension and compression members and plates was discussed. However, torsion members, shear panels, shells, and stiffened plates and shells were neglected. Moreover, the integration of these structural elements in a structural component such as an aircraft wing or fuselage section was ignored. Such topics are far beyond the scope of this book. In fact, design is such a complex and yet inspecific subject that few references are available.

REFERENCES

6-1   Salkind, M. J.: VTOL Aircraft, in "Applications of Composite Materials," ASTM STP 524, American Society for Testing and Materials, 1973, pp. 76-107.

6-2   Pinckney, R. L.: Helicopter Rotor Blades, in "Applications of Composite Materials," ASTM STP 524, American Society for Testing and Materials, 1973, pp. 108-133.

6-3   Salkind, M. J.: Fatigue of Composites, in "Composite Materials: Testing and Design (Second Conference)," ASTM STP 497, American Society for Testing and Materials, 1972, pp. 143-169.

6-4   Eisenmann, J. R., B. E. Kaminski, D. L. Reed, and D. J. Wilkins: Toward Reliable Composites: An Examination of Design Methodology, *J. Composite Materials*, July, 1973, pp. 298-308.

6-5   Lekhnitskii, S. G.: Stresses in Infinite Anisotropic Plate Weakened by Elliptical Hole, *DAN SSSR*, vol. 4, no. 3, 1936.

6-6   Savin, G. N.: "Stress Distribution Around Holes," Naukova Dumka Press, Kiev, 1968. Also *NASA* TT F-601, November, 1970.

6-7   Lekhnitskii, S. G.: "Theory of Elasticity of an Anisotropic Elastic Body," Government Publishing House for Technical-Theoretical Works, Moscow and Leningrad, 1950. Also P. Fern (trans.), Holden-Day, San Francisco, 1963.

6-8   Lekhnitskii, S. G.: "Anisotropic Plates," S. W. Tsai and T. Cheron (trans.), Gordon and Breach, New York, 1968.

6-9   Greszczuk, L. B.: Stress Concentrations and Failure Criteria for Orthotropic and Anisotropic Plates with Circular Openings, in "Composite Materials: Testing and Design (Second Conference)," ASTM STP 497, American Society for Testing and Materials, 1972, pp. 363-381.

6-10   Wu, Edward M.: Fracture Mechanics of Anisotropic Plates, in S. W. Tsai, J. C. Halpin, and Nicholas J. Pagano (eds.), "Composite Materials Workshop," Technomic Publishing Co., Westport Conn., 1968, pp. 20-43.

6-11   Griffith, A. A.: The Phenomena of Rupture and Flow in Solids, *Philosophical Trans. Roy. Soc.*, vol. 221A, October, 1920, pp. 163-198.

6-12   Irwin, G. R.: Fracture, in "Handbuch der Physik," vol. V, Springer, New York, 1958.

6-13   Tetelman, A. S.: Fracture Processes in Fiber Composite Materials, in "Composite Materials: Testing and Design," ASTM STP 460, American Society for Testing and Materials, 1969, pp. 473-502.

6-14   Corten, Herbert T.: Micromechanics and Fracture Behavior of Composites, in Lawrence J. Broutman and Richard H. Krock (eds.), "Modern Composite Materials," Addison-Wesley, New York, 1967, pp. 27-105.

6-15   Sih, G. C., and E. P. Chen: Fracture Analysis of Unidirectional Composites, *J. Composite Materials*, April, 1973, pp. 230-244.

6-16   Waddoups, M. E., J. R. Eisenmann, and B. E. Kaminski: Macroscopic Fracture Mechanics of Advanced Composite Materials, *J. Composite Materials*, October, 1971, pp. 446-454.

6-17 Konish, H. J., Jr., J. L. Swedlow, and T. A. Cruse: Experimental Investigation of Fracture in an Advanced Fiber Composite, *J. Composite Materials*, January, 1972, pp. 114-124.

6-18 Chamis, C. C., M. P. Hanson, and T. T. Serafini: Impact Resistance of Unidirectional Fiber Composites, in "Composite Materials: Testing and Design (Second Conference)," ASTM STP 497, American Society for Testing and Materials, 1972, pp. 324-349.

6-19 Pagano, N. J.: Exact Solutions for Composite Laminates in Cylindrical Bending, *J. Composite Materials*, July, 1969, pp. 398-411.

6-20 Pagano, N. J.: Exact Solutions for Rectangular Bidirectional Composites and Sandwich Plates, *J. Composite Materials*, January, 1970, pp. 20-34.

6-21 Pagano, N. J.: Influence of Shear Coupling in Cylindrical Bending of Anisotropic Laminates, *J. Composite Materials*, July, 1970, pp. 330-343.

6-22 Pagano, N. J., and A. S. D. Wang: Further Study of Composite Laminates Under Cylindrical Bending, *J. Composite Materials*, October, 1971, pp. 521-528.

6-23 Pagano, N. J., and Sharon J. Hatfield: Elastic Behavior of Multilayered Bidirectional Composites, *AIAA Journal*, July, 1972, pp. 931-933.

6-24 Reissner, Eric: The Effect of Transverse Shear Deformation on the Bending of Elastic Plates, *J. Appl. Mech.*, June, 1945, pp. A-69—A-77.

6-25 Mindlin, R. D.: Influence of Rotatory Inertia and Shear on Flexural Motions of Isotropic, Elastic Plates, *J. Appl. Mech.*, March, 1951, pp. 31-38.

6-26 Girkmann, K., and R. Beer: Application of Eric Reissner's Refined Plate Theory to Orthotropic Plates, *Osterr. Ingenieur-Archiv.*, vol. 12, 1958, pp. 101-110. Robert M. Jones (trans.), Dept. of Theoretical and Applied Mechanics, University of Illinois, Urbana, 1962.

6-27 Ambartsumyan, S. A.: "Theory of Anisotropic Plates," J. E. Ashton (ed.), T. Cheron (trans.), Technomic Publishing Co., Stamford, Conn., 1970. (Russian publication date unspecified.)

6-28 Whitney, J. M.: The Effect of Transverse Shear Deformation on the Bending of Laminated Plates, *J. Composite Materials*, July, 1969, pp. 534-547.

6-29 Ashton, J. E., and J. M. Whitney: "Theory of Laminated Plates," Technomic Publishing Co., Westport, Conn., 1970, chapter VII.

6-30 Whitney, J. M., and N. J. Pagano: Shear Deformation in Heterogeneous Anisotropic Plates, *J. Appl. Mech.*, December, 1970, pp. 1031-1036.

6-31 Yang, P. Constance, Charles H. Norris, and Yehuda Stavsky: Elastic Wave Propagation in Heterogeneous Plates, *Int. J. Solids and Structures*, October, 1966, pp. 665-684.

6-32 Whitney, James M.: Stress Analysis of Thick Laminated Composite and Sandwich Plates, *J. Composite Materials*, October, 1972, pp. 426-440.

6-33 Reissner, E.: A Consistent Treatment of Transverse Shear Deformations in Laminated Anisotropic Plates, *AIAA Journal*, May, 1972, pp. 716-718.

6-34 Ambartsumyan, S. A.: "Theory of Anisotropic Shells," State Publishing House for Physical and Mathematical Literature, Moscow, 1961. Also *NASA* TT F-118, May, 1964.

6-35 Fried, N.: Degradation of Composite Materials: The Effect of Water on Glass-Reinforced Plastic, in W. Wendt, H. Liebowitz, and N. Perrone (eds.), "Mechanics of Composite Materials," *Proc. 5th Symp. Naval Structural Mechanics, 1967*, Pergamon, New York, 1970, pp. 813-837.

6-36 Tsai, Stephen W.: Environmental Factors in the Design of Composite Materials, in F. W. Wendt, H. Liebowitz, and N. Perrone (eds.), "Mechanics of Composite Materials," *Proc. 5th Symp. Naval Structural Mechanics, 1967*, Pergamon, New York, 1970, pp. 749-767.

6-37 Whitney, J. M., and J. E. Ashton: Effect of Environment on the Elastic Response of Layered Composite Plates, *AIAA Journal,* September, 1971, pp. 1708-1713.

6-38 Dong, S. B., K. S. Pister, and R. L. Taylor: On the Theory of Laminated Anisotropic Shells and Plates, *J. Aerospace Sciences*, August, 1962, pp. 969-975.

6-39 Cheng, S., and B. P. C. Ho: Stability of Heterogeneous Aeolotropic Cylindrical Shells Under Combined Loading, *AIAA Journal*, April, 1963, pp. 892-898.

6-40 Jones, Robert M.: Buckling of Circular Cylindrical Shells with Multiple Orthotropic Layers and Eccentric Stiffeners, *AIAA Journal*, December, 1968, pp. 2301-2305. Errata, October, 1969, p. 2048.

6-41 Jones, Robert M., and Harold S. Morgan: Buckling and Vibration of Cross-Ply Laminated Circular Cylindrical Shells, *AIAA Paper No.* 74-33, 12th Aerospace Sciences Meeting, Washington, D.C., 30 January–1 February 1974. *AIAA Journal*, to be published.

6-42 Hahn, Hong T., and Stephen W. Tsai: Nonlinear Elastic Behavior of Unidirectional Composite Laminae, *J. Composite Materials*, January, 1973, pp. 102-118.

6-43 Hahn, Hong T.: Nonlinear Behavior of Laminated Composites, *J. Composite Materials*, April, 1973, pp. 257-271.

6-44 Adams, Donald F.: Inelastic Analysis of a Unidirectional Composite Subjected to Transverse Normal Loading, *J. Composite Materials*, July, 1970, pp. 310-328.

6-45 Schapery, R. A.: Stress Analysis of Viscoelastic Composite Materials, in S. W. Tsai, J. C. Halpin, and Nicholas J. Pagano (eds.), "Composite Materials Workshop," Technomic Publishing Co., Westport, Conn., 1968, pp. 153-192. Also *J. Composite Materials*, July, 1967, pp. 228-267.

6-46 Sims, David Ford: "Viscoelastic Creep and Relaxation Behavior of Laminated Composite Plates," Ph.D. dissertation, Dept. of Mechanical Engineering and Solid Mechanics Center, SMU Institute of Technology, Dallas, Texas, 1972. (Also available from Xerox University Microfilms as Order 72-27, 298.)

6-47 Chou, Pei Chi: Introduction to Wave Propagation in Composite Materials, in S. W. Tsai, J. C. Halpin, and Nicholas H. Pagano (eds.), "Composite Materials Workshop," Technomic Publishing Co., Westport, Conn., 1968, pp. 193-216.

Appendix A
MATRICES
AND TENSORS

Matrix and tensor notation is useful when dealing with systems of equations. Matrix theory is a straightforward set of operations for linear algebra and is covered in Sec. A.1. Tensor notation, treated in Sec. A.2, is a classification scheme in which the complexity ranges upward from scalars (zero order tensors) and vectors (first order tensors) through second order tensors and beyond.

The mathematical operations in the study of composite materials are strongly dependent on use of matrix theory. Tensor theory is a convenient tool, although such formal notation can be avoided without great loss. However, some of the properties of composite materials are more readily apparent and appreciated if the reader is conversant with tensor theory.

A.1 MATRIX ALGEBRA

A.1.1 Matrix Definitions

A *matrix* is a rectangular array of elements. The array has m rows and n columns and is called a rectangular matrix of order (m,n). If $m = n$, the array is called a *square matrix* of order n. The elements of an array $[A]$ are called A_{ij}, that is, the element in the ith row and j^{th} column of $[A]$. Thus, a matrix is an array:

$$[A] = \begin{bmatrix} A_{11} & A_{12} & \ldots & A_{1n} \\ A_{21} & A_{22} & \ldots & A_{2n} \\ \cdot & \cdot & & \cdot \\ \cdot & \cdot & & \cdot \\ \cdot & \cdot & & \cdot \\ A_{m1} & A_{m2} & \ldots & A_{mn} \end{bmatrix} \tag{A.1}$$

Two arrays $[A]$ and $[B]$ are equal only if they have the same number of rows and columns and all their corresponding elements are equal, that is,

$$A_{ij} = B_{ij} \quad i = 1, m \quad j = 1, n \tag{A.2}$$

A *row matrix* consists of a single row and has order $(1, n)$:

$$[A] = [A_1 \, A_2 \ldots A_n] \tag{A.3}$$

A *column matrix* has a single column and has order $(n, 1)$:

$$[A] = \{A\} = \begin{Bmatrix} A_1 \\ A_2 \\ \cdot \\ \cdot \\ \cdot \\ A_n \end{Bmatrix} \tag{A.4}$$

where the braces are ordinarily used to distinguish a column matrix from a general matrix.

The *transpose* of a matrix is denoted by a superscript T:

$$[A]^T = \begin{bmatrix} A_{11} & A_{21} & \ldots & A_{m1} \\ A_{12} & A_{22} & \ldots & A_{m2} \\ \cdot & \cdot & & \cdot \\ \cdot & \cdot & & \cdot \\ \cdot & \cdot & & \cdot \\ A_{1n} & A_{2n} & \ldots & A_{mn} \end{bmatrix} \tag{A.5}$$

and is obtained by interchanging rows and columns of Eq. (A.1).

For a square matrix, the *principal or main diagonal* goes from the upper left to the lower righthand corner of the matrix. Thus, the principal diagonal has elements A_{ii}. A *symmetric (square) matrix* has elements that are symmetric about the principal diagonal, that is,

$$A_{ij} = A_{ji} \tag{A.6}$$

Another way of saying the same thing is $[A] = [A]^T$.

A *diagonal matrix* is a square matrix with zero elements everywhere except on the principal diagonal (that is, all off-diagonal elements are zero):

$$
\begin{bmatrix}
A_{11} & 0 & \cdots & 0 \\
0 & A_{22} & \cdots & 0 \\
\vdots & \vdots & & \vdots \\
0 & 0 & & A_{nn}
\end{bmatrix}
\tag{A.7}
$$

If all the elements along the principal diagonal of a diagonal matrix are equal, the matrix is called a *scalar matrix*. One important scalar matrix has all ones on the principal diagonal and is called the *identity or unit matrix*:

$$
[I] =
\begin{bmatrix}
1 & 0 & \cdots & 0 \\
0 & 1 & \cdots & 0 \\
\vdots & \vdots & & \vdots \\
0 & 0 & \cdots & 1
\end{bmatrix}
\tag{A.8}
$$

The *determinant* of a square matrix of order two is called a determinant of order two and is defined as

$$
D = \begin{vmatrix} A_{11} & A_{12} \\ A_{21} & A_{22} \end{vmatrix} = A_{11}A_{22} - A_{12}A_{21}
\tag{A.9}
$$

whereas for a determinant of order three,

$$
D = \begin{vmatrix} A_{11} & A_{12} & A_{13} \\ A_{21} & A_{22} & A_{23} \\ A_{31} & A_{32} & A_{33} \end{vmatrix} = A_{11}\begin{vmatrix} A_{22} & A_{23} \\ A_{32} & A_{33} \end{vmatrix} - A_{12}\begin{vmatrix} A_{21} & A_{23} \\ A_{31} & A_{33} \end{vmatrix} + A_{13}\begin{vmatrix} A_{21} & A_{22} \\ A_{31} & A_{32} \end{vmatrix}
\tag{A.10}
$$

and, by mathematical induction, for a determinant of order n, if M_{1i} is the determinant of order $n - 1$ formed by striking out the first row and i^{th} column of D,

$$D = \begin{vmatrix} A_{11} & A_{12} & \dots & A_{1n} \\ A_{21} & A_{22} & \dots & A_{2n} \\ \vdots & \vdots & & \vdots \\ A_{n1} & A_{n2} & \dots & A_{nn} \end{vmatrix} = A_{11}M_{11} - A_{12}M_{12} + \dots + (-1)^{1+n}A_{1n}M_{1n}$$

$$\text{(A.11)}$$

That is, a determinant of order n is obviously defined in terms of determinants of order $n - 1$. In Eq. (A.11), the determinant M_{1i} is called the *minor* of element A_{1i} and the quantity $(-1)^{1+i}M_{1i}$ is called the *cofactor* of A_{1i}, that is,

$$a_{ij} = (-1)^{i+j}M_{ij} \qquad \text{(A.12)}$$

A determinant can be evaluated by expansion along any row or column, that is,

$$D = \sum_{i=1}^{n} A_{ij}a_{ij} = \sum_{j=1}^{n} A_{ij}a_{ij} \qquad \text{(A.13)}$$

where the free index is not summed.

Some elementary properties of determinants include:

(a) If each element in a row or column is multiplied by k, the determinant is multiplied by k.

(b) If two rows or two columns are proportional, the determinant is zero.

(c) If two rows or two columns are interchanged, the determinant changes sign.

(d) If rows and columns are interchanged, the determinant is not changed.

The *cofactor matrix* of a square matrix is the matrix of cofactors of each element, that is,

$$[a] = \begin{bmatrix} a_{11} & a_{12} & \dots & a_{1n} \\ a_{21} & a_{22} & \dots & a_{2n} \\ \cdot & \cdot & & \cdot \\ \cdot & \cdot & & \cdot \\ \cdot & \cdot & & \cdot \\ a_{n1} & a_{n2} & \dots & ann \end{bmatrix} \qquad \text{(A.14)}$$

A.1.2 Matrix Operations

Addition

Two matrices, $[A]$ and $[B]$, can be added only if they have the same number of rows and columns, respectively. Then, the sum $[C]$ is obtained by adding the corresponding elements of $[A]$ and $[B]$:

$$C_{ij} = A_{ij} + B_{ij} \tag{A.15}$$

For example,

$$\begin{bmatrix} C_{11} & C_{12} \\ C_{21} & C_{22} \end{bmatrix} = \begin{bmatrix} A_{11} & A_{12} \\ A_{21} & A_{22} \end{bmatrix} + \begin{bmatrix} B_{11} & B_{12} \\ B_{21} & B_{22} \end{bmatrix} = \begin{bmatrix} A_{11} + B_{11} & A_{12} + B_{12} \\ A_{21} + B_{21} & A_{22} + B_{22} \end{bmatrix} \tag{A.16}$$

Obviously, addition is both commutative, that is,

$$[A] + [B] = [B] + [A] \tag{A.17}$$

and associative, that is,

$$\big([A] + [B]\big) + [C] = [A] + \big([B] + [C]\big) \tag{A.18}$$

Subtraction

The difference of two matrices is obtained by subtraction of the corresponding elements of $[A]$ and $[B]$:

$$C_{ij} = A_{ij} - B_{ij} \tag{A.19}$$

and is subject to the requirement that the number of rows and columns be the same for $[A]$ and $[B]$. Also, subtraction is both commutative and associative.

Multiplication

The simplest form of matrix multiplication is the product of a scalar, s, and a matrix, $[A]$, wherein all elements of $[A]$ are multiplied by s:

$$s[A] = s \begin{bmatrix} A_{11} & A_{12} \\ A_{21} & A_{22} \end{bmatrix} = \begin{bmatrix} sA_{11} & sA_{12} \\ sA_{21} & sA_{22} \end{bmatrix} \tag{A.20}$$

The product, $[A][B]$, of two matrices is defined only when the number of rows in $[B]$ equals the number of columns in $[A]$. Here, $[B]$ is said to be premultiplied by $[A]$ or, alternatively, $[A]$ is said to be postmultiplied by $[B]$. The product, $[A][B]$, is obtained by first multiplying each element of the ith row of $[A]$ by the corresponding element of the jth column of $[B]$ and then adding those results:

$$[C] = [A] [B] = [A_{ik} B_{kj}] \tag{A.21}$$

where the summation on k goes from 1 to the number of columns in $[A]$. For example,

$$\begin{bmatrix} A_{11} & A_{12} & A_{13} \\ A_{21} & A_{22} & A_{23} \\ A_{31} & A_{32} & A_{33} \end{bmatrix} \begin{Bmatrix} B_{11} \\ B_{21} \\ B_{31} \end{Bmatrix} = \begin{bmatrix} A_{11}B_{11} + A_{12}B_{21} + A_{13}B_{31} \\ A_{21}B_{11} + A_{22}B_{21} + A_{23}B_{31} \\ A_{31}B_{11} + A_{32}B_{21} + A_{33}B_{31} \end{bmatrix} = \begin{Bmatrix} C_{11} \\ C_{21} \\ C_{31} \end{Bmatrix} \tag{A.22}$$

For a more complicated $[B]$ matrix, which has, say, n columns whereas $[A]$ has m rows (remember $[A]$ must have p columns and $[B]$ must have p rows), the $[C]$ matrix will have m rows and n columns. That is, the multiplication in Eqs. (A.21) and (A.22) is repeated as many times as there are columns in $[B]$. Note that, although the product $[A] [B]$ can be found as in Eq. (A.21), the product $[B] [A]$ is not simultaneously defined unless $[B]$ and $[A]$ have the same number of rows and columns. Thus, $[A]$ cannot be premultiplied by $[B]$ if $[A] [B]$ is defined unless $[B]$ and $[A]$ are square. Moreover, even if both $[A] [B]$ and $[B] [A]$ are defined, there is no guarantee that $[A] [B] = [B] [A]$. That is, matrix multiplication is not necessarily commutative.

Inversion
The *inverse* of a square matrix is denoted by a superscript -1 and is defined as

$$[A]^{-1} = \frac{[a]^T}{|A|} \tag{A.23}$$

(the transpose of the cofactor matrix is called the *adjoint matrix*) and has the property

$$[A] [A]^{-1} = [A]^{-1} [A] = [I] \tag{A.24}$$

The determinant of $[A]$ in Eq. (A.23) cannot vanish; otherwise Eq. (A.23) is not defined and $[A]$ is said to be a *singular matrix*.

Solution of Linear Equations
The principal use of the inverse matrix is in solution of linear equations or the application of transformations. If

$$\{Y\} = [A]\{X\} \tag{A.25}$$

where $\{Y\}$ and $\{X\}$ are column matrices, then

$$[A]^{-1}\{Y\} = [A]^{-1} [A]\{X\} = \{X\} \tag{A.26}$$

The foregoing result along with Eq. (A.23) is known as Cramer's rule. If $\{Y\}$ in Eq. (A.25) is zero, then the system of equations

$$[A]\{X\} = 0 \tag{A.27}$$

is said to be *homogeneous*. If matrix $[A]$ is nonsingular (so its inverse $[A]^{-1}$ exists) then

$$\{X\} = [A]^{-1}\{0\} = \{0\} \tag{A.28}$$

This solution for $\{X\}$ in which all the unknowns are zero is called the *trivial solution*. A nontrivial solution to Eq. (A.27) exists, therefore, only when matrix $[A]$ is singular, that is, when $|A| = 0$.

Miscellaneous

Some other matrix operations of interest include

$$\left([A]^{-1}\right)^T = \left([A]^T\right)^{-1} \tag{A.29}$$

that is, the transpose of the inverse of a matrix is equal to the inverse of the transpose. Also,

$$\left([A][B][C]\right)^T = [C]^T[B]^T[A]^T \tag{A.30}$$

$$\left([A][B][C]\right)^{-1} = [C]^{-1}[B]^{-1}[A]^{-1} \tag{A.31}$$

which are known as the *reversal laws* of transposition and inversion, respectively.

A.2 TENSORS

Vectors are commonly used for description of many physical quantities such as force, displacement, velocity, etc. However, vectors alone are not sufficient to represent all physical quantities of interest. For example, stress, strain, and the stress-strain laws cannot be represented by vectors. Tensors are an especially useful generalization of vectors. The key feature of tensors is that they *transform,* on rotation of coordinates, in special manners. Tsai (Ref. A-1) gives a complete treatment of the tensor theory useful in composite materials analysis. What follows are the essential fundamentals.

Cartesian tensors, that is, tensors in a cartesian coordinate system, will be discussed. Three independent quantities are required to describe the position of a point in cartesian coordinates. This set of quantities is x_i where x_i is (x_1, x_2, x_3). The index i in x_i has values 1, 2, and 3 because of the three coordinates in three-dimensional space. The indices i and j in a_{ij} mean, therefore, that a_{ij} has nine components. Similarly, b_{ijk} has 27 components, and c_{ijkl} has 81 components, etc. The indices are part of what is called *index notation*. The number of subscripts on the symbol denotes the *order* of the tensor. For example, a is a zero order tensor (a scalar), a_i a first order tensor (vector), a_{ij} a second order tensor, a_{ijkl} a fourth order tensor, etc. The number of components, N, necessary for description of a tensor of order k in n-dimensional space is

$$N = n^k \tag{A.32}$$

The range convention and the summation convention will be used. The *range convention* is any subscript that appears only once on one side of an expression takes on values 1, 2, and 3. The *summation convention* is any subscript that appears twice on one side of an expression is summed from 1 to 3. The repeated index is called the *dummy index*.

A.2.1 Transformation of Coordinates

Consider the behavior of various tensors under the transformation of coordinates in Fig. A-1 where a rotation about the z axis is made. That is, the x, y, z coordinates are transformed to the x', y', z' coordinates where the z-direction coincides with the z'-direction. The *direction cosines* for this transformation are

$$[\alpha_{ij}] = [T] = \begin{bmatrix} \cos\theta & \sin\theta & 0 \\ -\sin\theta & \cos\theta & 0 \\ 0 & 0 & 1 \end{bmatrix} \tag{A.33}$$

where α_{ij} is the cosine of the angle between the ith direction in the x', y', z' systems and the jth direction in the x, y, z system, that is, for all transformations (not just the foregoing special rotation),

$$\alpha_{ij} = \cos(x'_i, x_j) \tag{A.34}$$

Thus, the transformation of coordinates can be written in index notation as

$$x'_i = \alpha_{ij}x_j \tag{A.35}$$

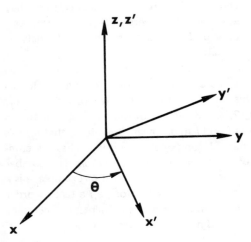

FIG. A-1. Rotation of coordinates in *xy* plane.

or

$$x_1' = \alpha_{11}x_1 + \alpha_{12}x_2 + \alpha_{13}x_3$$

$$x_2' = \alpha_{21}x_1 + \alpha_{22}x_2 + \alpha_{23}x_3 \qquad \text{(A.36)}$$

$$x_3' = \alpha_{31}x_1 + \alpha_{32}x_2 + \alpha_{33}x_3$$

or in matrix form as

$$\begin{Bmatrix} x_1' \\ x_2' \\ x_3' \end{Bmatrix} = \begin{bmatrix} \alpha_{11} & \alpha_{12} & \alpha_{13} \\ \alpha_{21} & \alpha_{22} & \alpha_{23} \\ \alpha_{31} & \alpha_{32} & \alpha_{33} \end{bmatrix} \begin{Bmatrix} x_1 \\ x_2 \\ x_3 \end{Bmatrix} \qquad \text{(A.37)}$$

This type of transformation will be used to assist in the definition of various orders of tensors. Each tensor will be defined on the basis of the type of transformation it satisfies. Tensors transform according to the relations

$$x_{ij\ldots kl}' = \alpha_{im}\alpha_{jn} \ldots \alpha_{ko}\alpha_{lp}x_{mn\ldots op} \qquad \text{(A.38)}$$

A.2.2 Definition of Various Tensor Orders

A *scalar* is a tensor of order zero and has $3^0 = 1$ component. Since it has only magnitude and not direction, no transformation relations are needed. Examples of scalars include speed (but not velocity), work, and energy.

A *vector* is a tensor of first order and has $3^1 = 3$ components. Vectors transform according to

$$A_i' = \alpha_{ij}A_j \qquad \text{(A.39)}$$

where A_i' is the transformed vector, α_{ij} the direction cosines of the transformation, and A_j the original vector. Examples of vectors include displacements, coordinates, velocity, forces, and moments.

A *tensor of second order* has $3^2 = 9$ components and transforms according to

$$A_{ij}' = \alpha_{ik}\alpha_{jl}A_{kl} \qquad \text{(A.40)}$$

Stress and strain are both second order tensors.

A *tensor of fourth order* has $3^4 = 81$ components and transforms according to

$$A_{ijkl}' = \alpha_{im}\alpha_{jn}\alpha_{ko}\alpha_{lp}A_{mnop} \qquad \text{(A.41)}$$

The stiffnesses and compliances in stress-strain relations are fourth order tensors because they relate two second order tensors:

$$\sigma_{ij} = C_{ijkl}\epsilon_{kl} \qquad \text{(A.42)}$$

$$\epsilon_{ij} = S_{ijkl}\sigma_{kl} \tag{A.43}$$

A.2.3 Contracted Notation

Contracted notation is a rearrangement of terms such that the number of indices is reduced although their range increases. For second order tensors, the number of indices is reduced from 2 to 1 and the range increased from 3 to 9. The stresses and strains, for example, are contracted as in Table A-1. Similarly, the fourth order tensors for stiffnesses and compliances in Eqs. (A.42) and (A.43) have 2 instead of 4 free indices with a new range of 9. The number of components remains 81 ($3^4 = 9^2$).

In contracted notation, the stress-strain relations, Eqs. (A.42) and (A.43) are written as

$$\sigma_i = C_{ij}\epsilon_j \tag{A.44}$$

$$\epsilon_j = S_{ij}\sigma_j \tag{A.45}$$

Obviously, the number of free indices no longer denotes the order of the tensor. Also, the range on the indices no longer denotes the number of spatial dimensions. If the stress and strain tensors are symmetric (they are if no body couples act on an element), then

$$\sigma_{ij} = \sigma_{ji} \qquad \epsilon_{ij} = \epsilon_{ji} \tag{A.46}$$

and, therefore, the number of independent stresses and strains is reduced to six each as in Table 2-1. This type of symmetry leads to a reduction of the number of independent components of C_{ij} and S_{ij} from 81 to 36 in 3-space. The C_{ij} and S_{ij} can further be shown to be symmetric (see Sec. 2.2), that is,

$$C_{ij} = C_{ji} \qquad S_{ij} = S_{ji} \tag{A.47}$$

whereupon the number of independent components of C_{ij} and S_{ij} is further reduced from 36 to 21 in 3-space. The stiffness matrix is then (the compliance matrix is similar):

TABLE A-1. Contraction of stresses and strains

Stresses		Strains	
Tensor notation	Contracted notation	Tensor notation	Contracted notation
σ_{11}	σ_1	ϵ_{11}	ϵ_1
σ_{22}	σ_2	ϵ_{22}	ϵ_2
σ_{33}	σ_3	ϵ_{33}	ϵ_3
$\sigma_{23} = \tau_{23}$	σ_4	$2\epsilon_{23} = \gamma_{23}$	ϵ_4
$\sigma_{31} = \tau_{31}$	σ_5	$2\epsilon_{31} = \gamma_{31}$	ϵ_5
$\sigma_{12} = \tau_{12}$	σ_6	$2\epsilon_{12} = \gamma_{12}$	ϵ_6
$\sigma_{32} = \tau_{32}$	σ_7	$2\epsilon_{32} = \gamma_{32}$	ϵ_7
$\sigma_{13} = \tau_{13}$	σ_8	$2\epsilon_{13} = \gamma_{13}$	ϵ_8
$\sigma_{21} = \tau_{21}$	σ_9	$2\epsilon_{21} = \gamma_{21}$	ϵ_9

$$C_{ij} = [C] = \begin{bmatrix} C_{11} & C_{12} & C_{13} & C_{14} & C_{15} & C_{16} \\ C_{12} & C_{22} & C_{23} & C_{24} & C_{25} & C_{26} \\ C_{13} & C_{23} & C_{33} & C_{34} & C_{35} & C_{36} \\ C_{14} & C_{24} & C_{34} & C_{44} & C_{45} & C_{46} \\ C_{15} & C_{25} & C_{35} & C_{45} & C_{55} & C_{56} \\ C_{16} & C_{26} & C_{36} & C_{46} & C_{56} & C_{66} \end{bmatrix} \qquad \text{(A.48)}$$

wherein the relation of a component of C_{ij} to that of C_{ijkl} is rather complex.

A.2.4 Matrix Form of Tensor Transformations

Tensors can easily be written in matrix form. For example, a vector a_i can be represented by a column matrix:

$$a_i = [A] = \begin{Bmatrix} A_1 \\ A_2 \\ A_3 \end{Bmatrix} \qquad \text{(A.49)}$$

or a row matrix

$$a_i = [A] = [A_1 \quad A_2 \quad A_3] \qquad \text{(A.50)}$$

Also, a second order tensor can be written

$$a_{ij} = [A] = \begin{bmatrix} A_{11} & A_{12} & A_{13} \\ A_{21} & A_{22} & A_{23} \\ A_{31} & A_{32} & A_{33} \end{bmatrix} \qquad \text{(A.51)}$$

or in contracted notation as a column (or row) matrix:

$$a_i = [A] = \begin{Bmatrix} A_1 \\ A_2 \\ A_3 \\ A_4 \\ A_5 \\ A_6 \\ A_7 \\ A_8 \\ A_9 \end{Bmatrix} \qquad \text{(A.52)}$$

A fourth order tensor can be written as a 9 × 9 array in analogy to Eq. (A.51) but, by use of contracted notation, is sometimes drastically simplified to a 6 × 6 symmetric array.

The stress-strain relations in this book are typically expressed in matrix form by use of contracted notation. Both the stresses and strains as well as the stress-strain relations must be transformed. First, the stresses transform for a rotation about the z axis as in Fig. A-1 according to

$$\{\sigma'\} = [T] \ \{\sigma\} \tag{A.53}$$

or

$$
\begin{Bmatrix} \sigma'_1 \\ \sigma'_2 \\ \sigma'_3 \\ \sigma'_4 \\ \sigma'_5 \\ \sigma'_6 \end{Bmatrix}
=
\begin{bmatrix}
\cos^2\theta & \sin^2\theta & 0 & 0 & 0 & 2\cos\theta\sin\theta \\
\sin^2\theta & \cos^2\theta & 0 & 0 & 0 & -2\cos\theta\sin\theta \\
0 & 0 & 1 & 0 & 0 & 0 \\
0 & 0 & 0 & \cos\theta & -\sin\theta & 0 \\
0 & 0 & 0 & \sin\theta & \cos\theta & 0 \\
-\cos\theta\sin\theta & \cos\theta\sin\theta & 0 & 0 & 0 & \cos^2\theta - \sin^2\theta
\end{bmatrix}
\begin{Bmatrix} \sigma_1 \\ \sigma_2 \\ \sigma_3 \\ \sigma_4 \\ \sigma_5 \\ \sigma_6 \end{Bmatrix}
$$

$$\tag{A.54}$$

In two dimensions, this rotation simplifies to

$$
\begin{Bmatrix} \sigma'_1 \\ \sigma'_2 \\ \tau'_{12} \end{Bmatrix}
=
\begin{bmatrix}
\cos^2\theta & \sin^2\theta & 2\cos\theta\sin\theta \\
\sin^2\theta & \cos^2\theta & -2\cos\theta\sin\theta \\
-\cos\theta\sin\theta & \cos\theta\sin\theta & \cos^2\theta - \sin^2\theta
\end{bmatrix}
\begin{Bmatrix} \sigma_1 \\ \sigma_2 \\ \tau_{12} \end{Bmatrix}
\tag{A.55}
$$

which in graphical form is the well-known Mohr's Circle. The strains transform in a similar manner as is shown in Sec. 2.6 for a case of plane stress. The stiffness and compliance transformations are very complex even for a simple rotation about an axis as in Eq. (A.33). The complete expressions are given by Tsai (Ref. A-1). For plane stress states, the transformations of the reduced stiffnesses are given in Sec. 2.6.

REFERENCE

A-1 Tsai, Stephen W.: Mechanics of Composite Materials, part II, Theoretical Aspects, *Air Force Materials Laboratory Tech. Rept.* AFML-TR-66-149, November, 1966.

MAXIMA AND MINIMA
OF FUNCTIONS OF A
SINGLE VARIABLE

Most engineering students are well aware that the first derivative of a function is zero at a maximum or minimum of the function. Fewer recall that the sign of the second derivative signifies whether the stationary value determined by a zero first derivative is a maximum or a minimum. Even fewer are aware of what to do if the second derivative happens to be zero. Thus, this appendix is presented to put the determination of relative maxima and minima of a function on a firm foundation.

Consider a function $V(x)$ for which a stationary value occurs at $x = x_1$, that is,

$$\left.\frac{dV}{dx}\right|_{x=x_1} = 0 \tag{B.1}$$

Such a stationary value of V can be a relative maximum, relative minimum, neutral point, or an inflection point as shown in Fig. B-1. There, Eq. (B-1) is satisfied at points 1, 2, 3, 4, and 5. By inspection, the function $V(x)$ has a relative minimum at points 1 and 4, a relative maximum at point 3, and an inflection point at point 2. Also shown in Fig. B-1 is a succession of neutral points for which all derivatives of $V(x)$ vanish. A simple physical example of

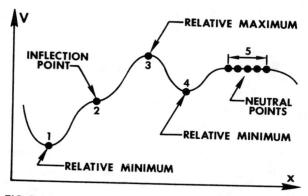

FIG. B-1. Stationary values of $V(x)$.

stationary values is a bead on a wire shaped as in Fig. B-1. That is, a minimum of $V(x)$ (the total potential energy of the bead) corresponds to stable equilibrium, a maximum or inflection point to unstable equilibrium, and a neutral point to neutral equilibrium.

To determine the specific character of a stationary value, first expand $V(x)$ in a Taylor series about the stationary point $x = x_1$:

$$V(x_1 + h) = V(x_1) + h \left.\frac{dV}{dx}\right|_{x=x_1} + \frac{h^2}{2!} \left.\frac{d^2 V}{dx^2}\right|_{x=x_1} + \frac{h^3}{3!} \left.\frac{d^3 V}{dx^3}\right|_{x=x_1} + \cdots$$

(B.2)

where h is the expansion parameter about x_1 as shown in Fig. B-2. Then, the change in V, ΔV, is

$$\Delta V = h \left.\frac{dV}{dx}\right|_{x=x_1} + \frac{h^2}{2!} \left.\frac{d^2 V}{dx^2}\right|_{x=x_1} + \frac{h^3}{3!} \left.\frac{d^3 V}{dx^3}\right|_{x=x_1} + \cdots \qquad \text{(B.3)}$$

However, in accordance with the definition of a stationary point, Eq. (B-1), the first term in Eq. (B-3) for ΔV vanishes irrespective of the value or size of h so that

$$\Delta V = \frac{h^2}{2!} \left.\frac{d^2 V}{dx^2}\right|_{x=x_1} + \frac{h^3}{3!} \left.\frac{d^3 V}{dx^3}\right|_{x=x_1} + \frac{h^4}{4!} \left.\frac{d^4 V}{dx^4}\right|_{x=x_1} + \cdots \qquad \text{(B.4)}$$

The character of ΔV will determine the type of stationary value at $x = x_1$. Specifically, the dominant term in the Taylor series for ΔV must be examined in order to determine whether ΔV is always positive (a relative minimum), always negative (a relative maximum), sometimes negative and sometimes positive (an inflection point), or always zero (a neutral point). For ΔV to be positive, the leading term in the Taylor series, Eq. (B-4), which is by inspection the largest term because h is a very small number, must be positive, that is,

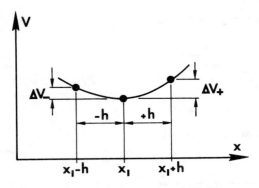

FIG. B-2. Taylor series expansion of $V(x)$ about $x = x_1$.

$$\frac{h^2}{2!} \frac{d^2 V}{dx^2}\bigg|_{x=x_1} > 0 \tag{B.5}$$

But even though h can have positive or negative values, since h is squared,

$$\frac{d^2 V}{dx^2}\bigg|_{x=x_1} > 0 \tag{B.6}$$

is sufficient for a minimum of $V(x)$ at $x = x_1$. Similar reasoning leads to

$$\frac{d^2 V}{dx^2}\bigg|_{x=x_1} < 0 \tag{B.7}$$

as sufficient for a maximum of $V(x)$ at $x = x_1$. However, if

$$\frac{d^2 V}{dx^2}\bigg|_{x=x_1} = 0 \tag{B.8}$$

then the second derivative term in the Taylor series is no longer the dominant term.

Accordingly, the next term in the Taylor series

$$\frac{h^3}{3!} \frac{d^3 V}{dx^3}\bigg|_{x=x_1} \tag{B.9}$$

must be examined. Obviously, since h can have positive or negative values and is cubed, the term in Eq. (B-9) can be positive or negative irrespective of the (nonzero) value of the third derivative. Thus, a nonzero third derivative of $V(x)$ at $x = x_1$ corresponds to an inflection point of $V(x)$ since ΔV can be either positive or negative.

If the term involving the third derivative, Eq. (B-9), in the Taylor series is zero, then the next higher term

$$\frac{h^4}{4!} \frac{d^4 V}{dx^4}\bigg|_{x=x_1} \tag{B.10}$$

is the dominant term. Obviously, the conclusions reached for the second derivative term are also valid for the fourth derivative term, Eq. (B-10).

Thus, by mathematical induction, the rules for determining the character of a stationary value of $V(x)$ at $x = x_1$ are

1. If the first nonzero derivative evaluated at $x = x_1$ is even and greater than zero, then $V(x_1)$ is a relative minimum.
2. If the first nonzero derivative evaluated at $x = x_1$ is even and less than zero, then $V(x_1)$ is a relative maximum.

3. If the first nonzero derivative evaluated at $x = x_1$ is odd, then $V(x_1)$ is an inflection point.

4. If all derivatives are zero, then $V(x_1)$ is a neutral point.

The following simple examples are useful aids to understanding the foregoing rules.

(1) $V = x^2$ (See Fig. B-3)

  $V' = 2x$

  $V' = 0$ at $x = 0$

  $V''|_{x=0} = 2$

  $\therefore V(0)$ is a relative minimum

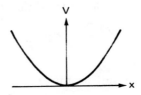

FIG. B-3. $V = x^2$.

(2) $V = x^3$ (See Fig. B-4)

  $V' = 3x^2$

  $V' = 0$ at $x = 0$

  $V''|_{x=0} = 6x|_{x=0} = 0$

  $V'''|_{x=0} = 6$

  $\therefore V(0)$ is an inflection point

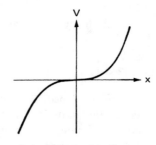

FIG. B-4. $V = x^3$.

(3) $V = x^4$ (See Fig. B-5)

  $V' = 4x^3$

  $V' = 0$ at $x = 0$

  $V''|_{x=0} = 12x^2|_{x=0} = 0$

  $V'''|_{x=0} = 24x|_{x=0} = 0$

  $V''''|_{x=0} = 24$

  $\therefore V(0)$ is a minimum

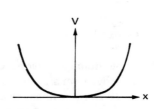

FIG. B-5. $V = x^4$.

(4) $V = x^{10}$ (See Fig. B-6)

$V' = 10x^9$

$V' = 0$ at $x = 0$

By mathematical induction,

$V''\big|_{x=0} = 0$

$\vdots$

$V^{IX}\big|_{x=0} = 0$

$V^{X}\big|_{x=0} = 10!$

$\therefore V(0)$ is a minimum

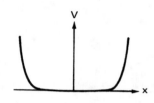

FIG. B-6. $V = x^{10}$.

If V is a function of more than one variable, then more complex criteria for determining maxima and minima will be obtained. Generally, but not always, the second partial derivatives of the function with respect to all its variables are sufficient to determine the character of a stationary value of V. For such functions, the theory of quadratic forms as described by Langhaar (Ref. B-1) should be examined.

REFERENCE

B-1 Langhaar, Henry L.: "Energy Methods in Applied Mechanics," Wiley, New York, 1962, pp. 308-328.

Appendix C
TYPICAL
STRESS-STRAIN CURVES

Typical stress-strain curves are shown for the commonly used fiber-reinforced materials fiberglass/epoxy, boron/epoxy, and graphite/epoxy.

C.1 FIBERGLASS/EPOXY STRESS–STRAIN CURVES

The curves for 3M XP251S fiberglass/epoxy are shown in Figs. C-1 through C-5 (pp. 334 through 336). Curves are given for both tensile and compressive behavior of the direct stresses. Note that the behavior in the fiber direction is essentially linear in both tension and compression. Transverse to the fiber direction, the behavior is nearly linear in tension, but very nonlinear in compression. The shear stress-strain curve is highly nonlinear. The Poisson's ratios (not shown) are essentially constant with values $\nu_{12} = .25$ and $\nu_{21} = .09$.

C.2 BORON/EPOXY STRESS–STRAIN CURVES

The curves for boron/epoxy are shown in Figs. C-6 through C-11 (pp. 336 through 339). As with fiberglass/epoxy, the behavior in the fiber direction is essentially linear in both tension and compression; in the direction transverse to the fibers the behavior is nearly linear in tension and fairly nonlinear in compression; and highly nonlinear in shear. The Poisson's ratio, ν_{12}, decreases in tension and increases in compression.

C.3 GRAPHITE/EPOXY STRESS–STRAIN CURVES

The curves for NARMCO 5605 graphite/epoxy shown in Figs. C-12 through C-17 (pp. 339 through 342) are analogous in form to the boron/epoxy curves.

REFERENCES

C-1 Plastics for Aerospace Vehicles, part 1, Reinforced Plastics, "Military Handbook," MIL-HDBK-17A, January, 1971.
C-2 Ashton, J. E., J. C. Halpin, and P. H. Petit: "Primer on Composite Materials: Analysis," Technomic Publishing Co., Westport, Conn., 1969.

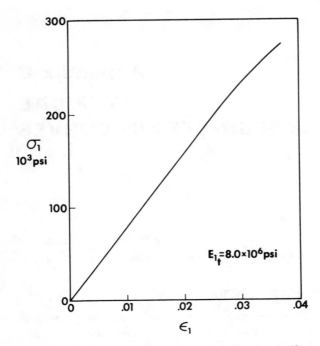

FIG. C-1. Tensile $\sigma_1 - \epsilon_1$ curve for 3M XP251S fiberglass/epoxy. (*Adapted from Ref. C-1.*)

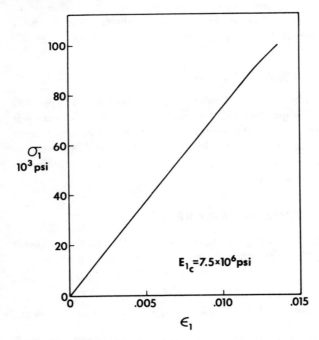

FIG. C-2. Compressive $\sigma_1 - \epsilon_1$ curve for 3M XP251S fiberglass/epoxy. (*Adapted from Ref. C-1.*)

334

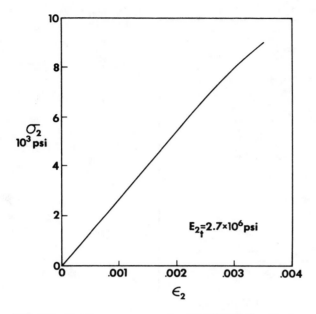

FIG. C-3. Tensile $\sigma_2 - \epsilon_2$ curve for 3M XP251S fiberglass/epoxy.
(*Adapted from Ref. C-1.*)

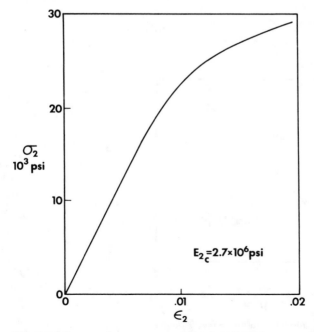

FIG. C-4. Compressive $\sigma_2 - \epsilon_2$ curve for 3M XP251S fiberglass/epoxy.
(*Adapted from Ref. C-1.*)

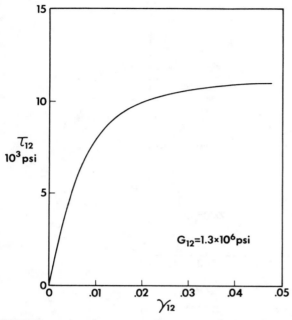

FIG. C-5. Shear stress-strain curve for 3M XP251S fiberglass/epoxy. (*Adapted from Ref. C-1.*)

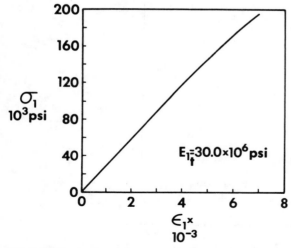

FIG. C-6. Tensile $\sigma_1 - \epsilon_1$ curve for boron/epoxy. (*Adapted from Ref. C-2.*)

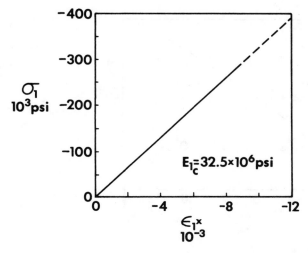

FIG. C.7. Compressive $\sigma_1 - \epsilon_1$ curve for boron/epoxy. (*Adapted from Ref. C-2.*)

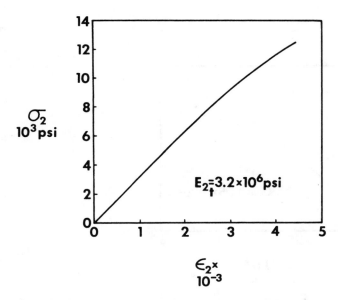

FIG. C-8. Tensile $\sigma_2 - \epsilon_2$ curve for boron/epoxy. (*Adapted from Ref. C-2.*)

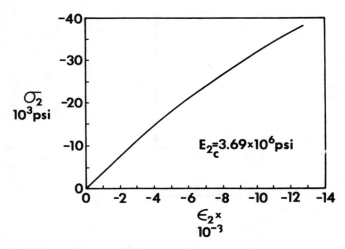

FIG. C-9. Compressive $\sigma_2 - \epsilon_2$ curve for boron/epoxy. (*Adapted from Ref. C-2.*)

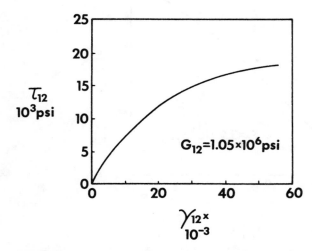

FIG. C-10. Shear stress-strain curve for boron/epoxy. (*Adapted from Ref. C-2.*)

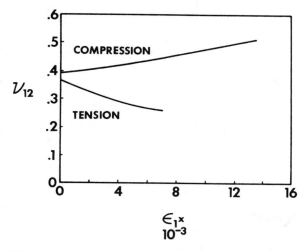

FIG. C-11. Poisson's ratio curves for boron/epoxy. (*Adapted from Ref. C-2.*)

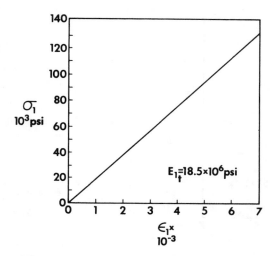

FIG. C-12. Tensile $\sigma_1 - \epsilon_1$ curve for Narmco 5605 graphite/epoxy. (*Adapted from Ref. C-2.*)

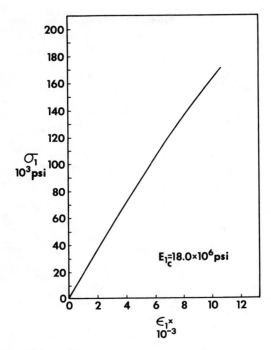

FIG. C-13. Compressive $\sigma_1 - \epsilon_1$ curve for Narmco 5605 graphite/epoxy. (*Adapted from Ref. C-2.*)

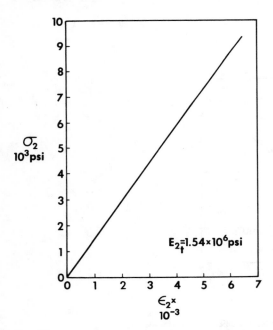

FIG. C-14. Tensile $\sigma_2 - \epsilon_2$ curve for Narmco 5605 graphite/epoxy. (*Adapted from Ref. C-2.*)

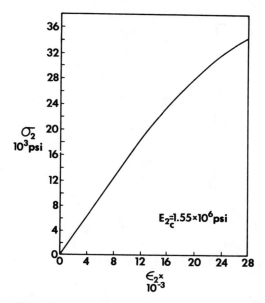

FIG. C-15. Compressive $\sigma_2 - \epsilon_2$ curve for Narmco 5605 graphite/epoxy.
(*Adapted from Ref. C-2.*)

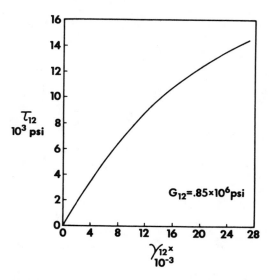

FIG. C-16. Shear stress-strain curve for Narmco 5605 graphite/epoxy.
(*Adapted from Ref. C-2.*)

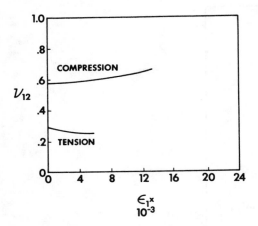

FIG. C-17. Poisson's ratio curves for Narmco 5605 graphite/epoxy. (*Adapted from Ref. C-2.*)

BIBLIOGRAPHY

Following is a list of most of the books published on mechanics of composite materials. Since new books are appearing each year, the list is necessarily incomplete. No attempt has been made to assemble a bibliography of papers and reports. The open literature on composites is rapidly expanding. The most common sources of papers are the usual journals, *Journal of Composite Materials, Journal of Applied Mechanics, AIAA Journal*, etc., as well as the ASTM conference proceedings listed below.

1. Ambartsumyan, S. A.: "Theory of Anisotropic Shells," State Publishing House for Physical and Mathematical Literature, Moscow, 1961. Also *NASA* TT F-118, May, 1964.
2. Ambartsumyan, S. A.: "Theory of Anisotropic Plates," Technomic Publishing Co., Westport, Conn., 1970.
3. Ashton, J. E., J. C. Halpin, and P. H. Petit: "Primer on Composite Materials: Analysis," Technomic Publishing Co., Westport, Conn., 1969.
4. Ashton, J. E. and J. M. Whitney: "Theory of Laminated Plates," Technomic Publishing Co., Westport, Conn., 1970.
5. Benjamin, B. S.: "Structural Design with Plastics," Van Nostrand Reinhold, New York, 1969.
6. Broutman, Lawrence J., and Richard H. Krock (eds.): "Composite Materials," a multivolume treatise including Arthur G. Metcalfe (ed.), "Interfaces in Metal Matrix Composites," vol. 1; G. P. Sendeckyj (ed.), "Mechanics of Composite Materials," vol. 2; Bryan R. Noton (ed.), "Engineering Applications of Composites," vol. 3; Kenneth G. Kreider (ed.), "Metallic Matrix Composites," vol. 4; Lawrence J. Broutman (ed.), "Fracture and Fatigue," vol. 5; Edwin P. Pleuddemann (ed.), "Interfaces in Polymer Matrix Composites," vol. 6; C. C. Chamis (ed.), "Structural Design and Analysis—Part I," vol. 7; C. C. Chamis (ed.), "Structural Design and Analysis—Part II," vol. 8. Academic Press, New York, 1974.
7. Broutman, L. J., and R. H. Krock (eds.): "Modern Composite Materials," Addison-Wesley, Reading, Mass., 1967.
8. Bush, Spencer H. (symposium chairman): "Fiber Composite Materials," American Society for Metals, Cleveland, Ohio, 1965.
9. Calcote, Lee R.: "The Analysis of Laminated Composite Structures," Van Nostrand Reinhold, New York, 1969.
10. Corten, H. T. (conference chairman): "Composite Materials: Testing and Design (Second Conference)," ASTM STP 497, American Society for Testing and Materials, 1972.
11. Dietz, A. G. H. (ed.): "Engineering Laminates," John Wiley, New York, 1949.
12. Dietz, A. G. H.: "Composite Materials," 1965 Edgar Marburg Lecture, American Society for Testing and Materials, 1965.

13. Dietz, A. G. H. (ed.): "Composite Engineering Laminates," M.I.T. Press, Cambridge, Mass., 1969.
14. Hearmon R. F. S.: "Applied Anisotropic Elasticity," Oxford University Press, Oxford, 1961.
15. Holliday, Leslie (ed.): "Composite Materials," Elsevier, Amsterdam, 1966.
16. Lekhnitskii, S. G.: "Theory of Elasticity of an Anisotropic Elastic Body," Holden-Day, San Francisco, 1963.
17. Lekhnitskii, S. G.: "Anisotropic Plates," Gordon and Breach, New York, 1968.
18. McCullough, R. L.: "Concepts of Fiber-Resin Composites," Marcel Dekker, New York, 1971.
19. Rosato, D. V., and C. S. Grove, Jr.: "Filament Winding: Its Development, Manufacture, Applications, and Design," Wiley Interscience, 1964.
20. Schwartz, R. T., and H. S. Schwartz (eds.): "Fundamental Aspects of Reinforced Plastic Composites," Wiley Interscience, 1968.
21. Scala, E.: "Composite Materials for Combined Functions," Hayden Book Co., Rochelle Park, N. J., 1973.
22. Spencer, A. J. M.: "Deformations of Fibre-Reinforced Materials," Oxford University Press, Oxford, 1972.
23. Tsai, S. W., J. C. Halpin, and N. J. Pagano (eds.): "Composite Materials Workshop," Technomic Publishing Co., Westport, Conn., 1968.
24. Wendt, F. W., H. Liebowitz, and N. Perrone (eds.): "Mechanics of Composite Materials," *Proc. 5th Symp. Naval Structural Mechanics 1967*, Pergamon, New York, 1970.
25. Whitney, J. M. (symposium chairman): "Analysis of the Test Methods for High Modulus Fibers and Composites," ASTM STP 521, American Society for Testing and Materials, 1973.
26. Yurenka, Steven (symposium chairman): "Composite Materials: Testing and Design," ASTM STP 460, American Society for Testing and Materials, 1969.

INDEX

INDEX